DIGITAL SIMULATION OF DYNAMIC SYSTEMS
A CONTROL THEORY APPROACH

DIGITAL SIMULATION
OF DYNAMIC SYSTEMS
A CONTROL THEORY APPROACH

TOM T. HARTLEY
GUY O. BEALE
STEPHEN P. CHICATELLI

PTR Prentice Hall
Englewood Cliffs, New Jersey 07632

Library of Congress Cataloging–in–Publication Data

Hartley, T. T. (Tom T.), 1957–
    Digital simulation of dynamic systems : a control theory approach
  / Tom T. Hartley, Guy O. Beale, Stephen P. Chicatelli.
      p.   cm.
    Includes index.
    ISBN 0–13–219957–2
    1. Digital computer simulation. 2. Control theory. I. Beale,
  Guy O. II. Chicatelli, S. P. (Stephen P.), 1964– . III. Title.
  QA76.9.C65H36   1994
  003′ .85′0113—dc20   94–2660
                                   CIP

Acquisitions editor: Karen Gettman
Cover design: Jeannette Jacobs Design
Cover design director: Eloise Starkweather-Muller
Manufacturing manager: Alexis R. Heydt
Compositor/Production Services: Pine Tree Composition, Inc.

©1994 PTR Prentice Hall
Prentice-Hall, Inc.
A Paramount Communications Company
Englewood Cliffs, New Jersey 07632

The publisher offers discounts on this book when ordered in bulk
quantities. For more information, contact Corporate Sales Department,
PTR Prentice Hall, 113 Sylvan Avenue, Englewood Cliffs, NJ 07632.
Phone: 201–592–2863; FAX: 201–592–2249.

Printed in the United States of America

10 9 8 7 6 5 4 3 2 1

ISBN   0-13-219957-2

Prentice-Hall International (UK) Limited, *London*
Prentice-Hall of Australia Pty. Limited, *Sydney*
Prentice-Hall Canada Inc., *Toronto*
Prentice-Hall Hispanoamericana, S.A., *Mexico*
Prentice-Hall of India Private Limited, *New Delhi*
Prentice-Hall of Japan, Inc., *Tokyo*
Simon & Schuster Asia Pte. Ltd., *Singapore*
Editora Prentice-Hall do Brasil, Ltda., *Rio de Janeiro*

We dedicate this book to our families, our teachers,
and our students, from whom we have learned much.

Table of Contents

Preface   **xi**

1. Introduction   **1**

    1.1 Objectives of the book   1
    1.2 What is digital simulation?   2
    1.3 Why simulate?   3
    1.4 Simulation characteristics   4
    1.5 Historical perspective   5
    1.6 Organization of the text   6

2. Z-Transform Methods and Simulation   **8**

    2.1 Introduction to z-transform methods   8
    2.2 Review of z-transforms   8
    2.3 Difference equations and transfer functions   12
    2.4 Transfer function·response—Output sampling   14
    2.5 Input sampling   15
    2.6 Reconstruction   16
    2.7 The data reconstruction simulation method   22
    2.8 Conclusions   23
    Problems   23

3. Operational Substitution Methods   24

3.1   Introduction   24
3.2   Forward-looking rectangles (Euler's method)   25
3.3   General procedure   30
3.4   Other associative methods   32
3.5   Derivation of nonassociative methods   37
3.6   Matrix methods for operational substitution algebra   39
3.7   The error in operational substitution   43
3.8   Application to multielement systems   45
3.9   Simulation of closed-loop systems   47
3.10 Time-varying systems   53
3.11 Operational methods for time-varying systems   54
3.12 Operational methods for nonlinear systems   56
3.13 Other approaches for transfer function representations   57
3.14 Conclusions   59
       Appendix: Families of Substitution Operator Matrices   59
       Problems   64

4. Linear Multistep Methods   66

4.1   Introduction   66
4.2   Initial-value problems   67
4.3   The integration problem   67
4.4   Introduction to linear multistep methods   68
4.5   Derivation of linear multistep methods with Taylor series   73
4.6   Control theory insight   76
4.7   Introduction to the Adams family   78
4.8   Introduction to the backward difference family   80
4.9   Properties of Linear Multistep Methods   81
4.10 Closed-loop integration   86
4.11 Control theory revisited   92
4.12 Weak stability theory (closed-loop integration)   94
4.13 Introduction to stability regions   98
4.14 The $\hat{s}$-plane   104
4.15 A design example   104
4.16 Characteristics of implicit methods   111
4.17 Introduction to predictor-corrector methods   114
4.18 Weak stability of predictor-corrector methods   116
4.19 An example predictor-corrector stability study   118
4.20 Modifiers in predictor-corrector methods   120
4.21 Weak stability for predictor-corrector methods with modifiers   124
4.22 Variable timestep and variable order methods   124
4.23 Conclusions   126
       Problems   126

5. Stability Regions Revisited 129

5.1 Introduction 129
5.2 More about frequency responses 129
5.3 Stability regions and the inverse Nyquist plot 136
5.4 Inside the λ-T-plane 141
5.5 A robustness analysis of the simulation process 148
5.6 Discussion of specific stability regions 153
5.7 Predictor-corrector stability regions 153
5.8 Predictor-corrector-modifier stability regions 153
5.9 Hardware-in-the-loop stability 154
5.10 Conclusions 160
Problems 160

6. Runge-Kutta Methods 161

6.1 Introduction 161
6.2 Graphical derivation of Runge-Kutta methods 162
6.3 The general Runge-Kutta method 165
6.4 Linear system with piecewise constant input 166
6.5 Nonlinear system with piecewise constant input 169
6.6 Linear system with multirate input sampling 174
6.7 Transfer functions and weak stability 177
6.8 Stability regions for Runge-Kutta methods 182
6.9 Variable timestep Runge-Kutta methods 184
6.10 Conclusions 188
Problems 188

7. Stiff Systems 190

7.1 Introduction 190
7.2 Definition and examples 191
7.3 More stability definitions 194
7.4 Stability region placement methods 202
7.5 Matrix stability region placement methods 209
7.6 Runge-Kutta methods 216
7.7 Inverse Nyquist array 224
7.8 Conclusions 238
Problems 239

8. Nonlinear Systems 241

8.1 Introduction 241
8.2 First order continuous-time systems 242
8.3 Linearization 243
8.4 Second-order systems 244
8.5 Third-order systems 247

8.6   Higher-order systems   251
8.7   Discrete-time systems   252
8.8   Simulation of nonlimit cycling systems   258
8.9   Simulation of systems with limit cycles   266
8.10 Simulation of chaotic systems   269
8.11 Simulation of systems with discontinuous nonlinearities   270
8.12 Simulation of time-delay systems   272
8.13 Divergence preserving methods   273
8.14 Conclusions   275
      Problems   275

9. **Multiple Integration**   **276**

9.1   Introduction   276
9.2   Motivating problems for double integration   276
9.3   General linear multistep double integrator   278
9.4   Weak stability for the double integrator   280
9.5   Stability regions for double integrators   280
9.6   Predictor-corrector methods for double integration   282
9.7   Triple integration   282
9.8   Higher-order integration   284
9.9   Substitution methods revisited   285
9.10 Conclusions   286
      Problems   286

10. **Concluding Discussion**   **290**

10.1 Competing criteria   290
10.2 Review of basic approaches   292
10.3 System characteristics   293
10.4 Choosing a method and T 295
10.5 Available software   298
      Problems   299

Appendices

A. **References**   **300**

B. **Stability Region Plots**   **303**

C. **Listing of Provided Software**   **373**

D. **Collection of Dynamic Systems**   **379**

Index   **385**

Preface

As everyone knows, the use of digital computers has grown enormously in the past half century. From their initial development in the 1940s, when digital computers were large, slow, and unreliable, we now have personal computers and workstations that are compact, very fast, and highly reliable. Almost everyone has access to the computing power of digital computers.

This same time frame has seen the development of the space program, highly sophisticated military and civilian aircraft, and automated factories. The parallel development of digital computers and very complex engineered systems has created both the need for and the resources for a new discipline, namely, digital simulation. Digital simulation refers to the numerical solution of the differential equations representing some dynamic system on a digital computer. The purpose of the simulation is to analyze the behavior of the system under certain operating conditions without actually exposing the system to those conditions, or perhaps even without building the system. The aircraft and space industries are natural examples of where digital simulation plays a crucial role in the successful design of systems, but there are many other examples as well. Among these are chemical processing plants, nuclear generating stations, and control system evaluation.

The major characteristics of digital simulation are accuracy, stability, and computation time. These three characteristics are highly interrelated. The required accuracy depends on the application, as does the computation time. However, the simulation must be stable to obtain any meaningful results. The accuracy, stability, and computation time all depend on the numerical algorithm used to solve the particular set of differential equations and on the value of the simulation timestep. To obtain usable simulation results for complex systems, the engineer must have a thorough understanding of the effects that the choice of numerical algorithm and timestep have on the simulation results.

Many powerful software packages allow the user to describe mathematically the system to be simulated. Once the problem is put into the proper format, which may be a graphical interface, the software package does the rest. The numerical integration method to be used is selected from one of the several preprogrammed methods by the software package. In many situations, this provides acceptable results. However, in the simulation of complex, high-order nonlinear systems, in the simulation of systems with peculiar characteristics such as stiffness, and in real-time simulation, these software packages may require a more knowledgeable user in order to obtain acceptable results. In any event, the engineer needs the background to be able to determine whether or not the simulation results are correct.

The purpose of this book is to provide the knowledge necessary to design and perform simulations of dynamics systems and to analyze the results obtained from simulations from the perspective of accuracy, stability, and computation time. A strong connection between the mathematics of digital simulation and automatic control theory is made throughout the book. This is meant to make the numerical methods used in digital simulation more accessible to the engineer.

This book is intended to be used in two primary ways. The first is as a text in a one-semester course on digital simulation at the senior undergraduate or first-year graduate level in an engineering curriculum. It is hoped that the student would have had an introductory course in control systems design. The book can also be used as a reference in more advanced graduate courses, such as optimal control, where simulation is used to model system behavior numerically. The other main audience of the book is practicing engineers who are involved in simulating dynamic systems or in analyzing the results of simulations. These engineers can obtain a solid foundation in the mathematics and techniques used in digital simulation.

A chart expressing the relationships among the chapters follows. Chapter 1 introduces simulation and provides some interesting historical background. Part of Chapter 2 may be review for some students, but the sections on reconstruction methods should be covered since they are shown to be related to simulation. Chapter 3 deals with operational substitution methods, where discrete-time transfer functions are obtained from continuous-time transfer functions. Many of the characteristics of simulation are introduced here. Chapter 4 is the main chapter for numerical integration. Various algorithms are introduced, and quantitative methods of representing simulation accuracy and stability are developed. Chapter 5 goes into the concept of stability regions in more detail and can be omitted if time does not allow its inclusion. Chapter 6 presents Runge-Kutta methods, deriving the accuracy constraints and stability regions for these methods, which are among the most popular integration methods. Chapter 7 deals with the special problem of stiff systems, which are very difficult to simulate without particular care in the choice of simulation method and timestep. Chapter 8 discusses nonlinear systems. The different types of behavior, including chaos, that can be obtained from nonlinear systems are described, and the special care that is necessary in simulating them is discussed. Simulation of these nonlinear systems can use the methods described in the other chapters, as shown in the chart. Chapter 9 discusses the development of multiple integration methods, primarily drawing on the material from chapters 3 and 4. Chapter 10 presents a discussion on the trade-offs required in designing a simulation and on the system characteristics that affect this trade-off. Subjective advice is given on how to select a simulation method and the value for the simulation timestep.

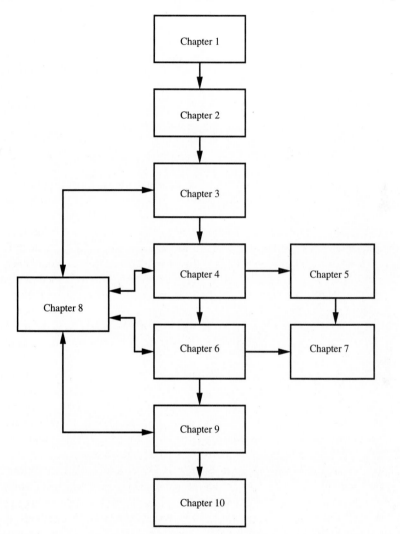

We would like to thank Karen Gettman, senior editor, for her support and encouragement during this project. Thanks also go to the reviewers for the time they spent in providing very constructive comments on our manuscript, as well as to the many students who endured earlier presentations of this material. Finally, we would like to thank our families, Karen and John, Susie and Michael, and Amy for the many hours they allowed us to work on this book.

DIGITAL SIMULATION
OF DYNAMIC SYSTEMS
A CONTROL THEORY APPROACH

1

Introduction

1.1 OBJECTIVES OF THE BOOK

The overall objective of the book is to provide the practical knowledge, as well as the theoretical basis when necessary, required to design and evaluate digital simulations of continuous-time dynamic systems. A byproduct will be the ability to understand the capabilities and limitations of available simulation methods.

The text is intended both as a classroom teaching tool for engineering curricula and as a self-study guide for practicing engineers interested in learning how to get accurate, meaningful simulations of real physical systems. The systems to be considered in this text are those described by linear or nonlinear ordinary differential equations. It is assumed that these equations are familiar to engineers of almost every discipline. Attention is given to systems described by stiff nonlinear differential equations, which are used to model many real systems of interest to engineers. The problems of real-time simulation are also discussed.

The development of this material is a mixture of classical numerical analysis and the theory of automatic control. We feel that this approach gives the engineer a much clearer insight into the principles of digital simulation than that obtained from a more traditional course in numerical methods. This text is not offered in competition to the theoretical derivations and proofs fundamental to the mathematical foundations of simulation. Rather, it is aimed at giving engineers (present and future) the definitions and techniques necessary to apply these theoretical concepts to simulate a system of interest.

1.2 WHAT IS DIGITAL SIMULATION?

Most systems of concern to engineers are represented by differential equations; the solution to these equations describes the time response of the system to the given initial conditions and input history. Electrical, mechanical, thermal, and other systems that contain energy storage elements can all be represented by differential equations. Often the assumption is made that the system is lumped-parameter so that it can be described by ordinary differential equations (ODEs) rather than partial differential equations (PDEs). In this text, discussions are limited to systems represented by ordinary differential equations.

Digital simulation refers to the solution on a digital computer of the equations that describe a particular system. This solution is obtained either from the numerical integration of the system's differential equations or from the solution of an equivalent difference equation. It is often assumed that the solution values are being generated at equally spaced points in time. However, it will be shown that in certain applications, variable timestep methods are useful. The time interval between points is referred to as the simulation timestep or step size, and it is always given the symbol T. In general, this numerical solution is only approximate. The digital computer can produce solutions only at discrete points in time, and the solution at a particular point is limited to one of a finite number of values. The real system represented by the ODE, however, has a solution at every instant of time, and the solution value can come from a continuum of possible values. Thus, there will be an error between the simulation of the system produced by the computer and the response of the actual system.

As will be presented in the following chapters, this error is influenced by a number of factors. The particular method chosen to integrate the ODE numerically or to generate the equivalent difference equation and the time interval between solution points are the most critical factors in establishing the accuracy of the simulation. Each of these factors will be covered in detail in the remainder of the text. The way in which numbers are represented in the computer is also a factor in determining accuracy through its influence on round-off error. However, this topic will not be discussed in this text. It is generally much less of a problem for simulation stability and accuracy unless extremely small values of T are used.

The field of digital simulation can be broken into different areas depending on the requirements of the problem. For some problems, the equations must be solved in the same actual time in which the system itself responds. This is generally referred to as real-time simulation. If the actual elapsed time that the simulation takes is not critical, the simulation would be non–real-time, batch, or off-line simulation. If a human operator has to interface with the simulation while it is running, the simulation is said to have "a person in the loop." Generally, person-in-the-loop simulation is also done in real time, for example, the training of an airline pilot seated in a model of the cockpit. Based on the pilot's controls, a computer would solve the equations of motion for the aircraft and provide updated displays based on the solution. The equations would have to be solved in real time to give the pilot an accurate feeling of actually flying the airplane. An analogous situation is "hardware-in-the-loop" simulation, in which the simulation must interact with a piece of equipment, for example, a control system under evaluation. Some advanced control techniques, such as model reference adaptive control, require that a mathematical model of the controlled system be solved in real time as part of the on-line control procedure. This is equivalent to real-time simulation.

1.3 WHY SIMULATE?

The simulation of physical systems is an integral part of the modern design process. Consider the design of a new model aircraft, spacecraft, or some other complex system. The design has been based on physical principles, experimental data from similar models, and the various specifications imposed on the system. According to the design, the system should perform as intended. However, before a pilot is allowed to fly a newly designed aircraft, it would be prudent to verify its operation without risking the pilot's life and the craft itself. Indeed, before the aircraft is even built, the manufacturer would want to verify the design before spending a large amount of time and money in its fabrication.

This verification can be done through digital simulation. The differential equations that represent the system can be solved by the computer for a large number of initial conditions, input trajectories, and parameter values. Thus, the simulation not only can be used to provide information about a particular design but can also be used in the design process itself by allowing different values for various parameters. This design assistance and design verification result in reduced cost and time, and the possible saving of human life is obvious. Digital simulation is widespread throughout the aircraft and space industries for design verification as well as for the training of pilots and astronauts.

Another example of the use of simulation is in the design of control systems for industrial processes. Consider an existing chemical process and its controller. An engineer has designed a new model controller for the process, perhaps to improve its steady-state accuracy or transient performance. One way of testing the new controller is to substitute it in place of the existing controller. However, since most designs turn out to be an iterative process, the new controller probably would not immediately produce the desired performance from the process. There would be some period of time during which the process would not be producing the product within specifications. This tends to make plant managers very upset. Trying to debug the controller, knowing that the product line was being held up, can cause a lot of anxiety and wasted time. In an extreme case, the new controller could produce a situation that would be dangerous to the plant employees.

A better way of initially testing the new controller is to interact with a real-time simulation of the process it is to control. Necessary corrections to the design can be based on the results of the simulation, and savings in time and product can be realized. Also, the plant manager will probably be more amenable to new designs in the future if the changeover from old to new can be done efficiently.

Many other examples can be given to illustrate the utility, and sometimes necessity, of simulating a given system. One possible example, which would be of more interest to the mathematician than to the engineer, is a nonlinear differential equation that has no closed-form solution. In this case, the only way to generate a solution is numerically. This equation may not represent a real physical system and, therefore, would lie outside the scope of this text. Since this text is intended for engineers, we concentrate on the simulation of physical systems.

In summary, digital simulation can be used beneficially whenever it is necessary or desirable to save time, money, materials, and human life. Any field of application involving the design of a system, the verification of the design, or the training of its operator can benefit from the educated use of digital simulation.

1.4 SIMULATION CHARACTERISTICS

The three major characteristics of digital simulation are (1) stability, (2) accuracy, and (3) computation time. Stability of the simulation is concerned with the question of whether the numerical solution computed during the simulation would converge to the exact solution, in the absence of round-off error, or whether the difference between the approximate and exact solutions would become unbounded. Just as with closed-loop control systems, instability can occur in digital simulation even when the system being simulated is inherently stable and the algorithm used to perform the numerical integration is also stable. Therefore, stability is an important topic in the study of simulation since an unstable simulation of a stable system will produce worthless results. It will be seen that stability of the simulation depends on the relationship among the numerical integration method, the characteristics of the system being simulated, and the timestep chosen for the integration.

Accuracy is a measure of how closely the numerical results of the simulation match the values that would be obtained if the system equations could be solved exactly. The major sources of inaccuracy in a stable simulation are truncation error, due to approximating an infinite series with a finite number of terms, and round-off error, due to the fact that computations are being performed with finite precision arithmetic. Accuracy is somewhat subjective in nature since the intended use of the simulation results will influence how much error in the simulation can be tolerated. When the system being simulated is inherently unstable, the user would be interested in how long the system could be simulated before the error became unacceptably large. Quantitative measures of accuracy can be developed in terms of the simulation timestep and integration algorithm.

Generally, the same number and type of calculations are performed at each timestep of the simulation. Therefore, the time required to complete a simulation is directly related to the computation time required by the numerical integration algorithm and the function evaluations. The function evaluations are specified by the system being simulated, so other than certain programming efficiency techniques that can be used, the time required by one set of function evaluations is fixed for a particular system. Therefore, the numerical integration algorithm represents the variable quantity in determining the computation time at each timestep in the simulation. Various integration algorithms cover a wide range in the number of multiplications and additions required per integration timestep, as well as in the number of times the function evaluations have to be computed. Unfortunately, higher-accuracy algorithms generally require longer computation times, and often they impose more stringent constraints on the selection of the simulation timestep.

Often, trade-offs must be made because of the conflicting requirements of accuracy, stability, and computation time. If the simulation must be done in real time, where all computations must be completed within a prescribed time limit, accuracy may have to be sacrificed to satisfy the time constraints. Higher-accuracy algorithms also generally require a smaller simulation step size to maintain stability. This also conflicts with computation time needs. Although "real time" is not a precisely defined range of numbers, it is generally true that the dynamics of the system equations impose an upper limit on the integration timestep to obtain an accurate and stable simulation. Even when the simulation is not to be done in real time, long computation times may not be desirable, just for convenience.

1.5 HISTORICAL PERSPECTIVE

A brief outline of the important events in the evolution of simulation techniques follows. It is interesting to observe that the need to simulate dynamic systems was one of the major driving forces in the development of computing equipment and computing. An excellent review of the evolution of methods for solving differential equations appears in Paynter (1989).

Numerical methods for solving ordinary differential equations were developed during the 1700s and 1800s. Astronomers used these methods to predict Halley's comet in 1913. The computers in those days were people, performing the computations manually.

Bush's differential analyzer was developed in the late 1920s. This was an analog computer that used mechanical components to perform the calculations necessary for solving differential equations to three-digit accuracy. The equipment consisted of a series of six mechanical integrators connected in series to produce the necessary integrals of the highest derivative and the appropriate mechanical function generators to represent the differential equation itself. This device was an extension of work done 50 years earlier by Lord Kelvin. One of Bush's major contributions to the mechanical analog computer was the use of power amplifiers between integrators, isolating them from one another to reduce loading effects. Several other researchers around the world constructed similar differential analyzers, with the biggest one containing 18 integrators. One such device was made from a children's building kit similar to an Erector set (Paynter, 1989).

Shortly after Bush introduced his mechanical differential analyzer, Hazen introduced an electrical circuit that was the equivalent of the mechanical system. This was the onset of replacing mechanical parts with electronic ones, which were smaller, lighter, and had no moving parts. The invention of the electronic operational amplifier by Philbrick as the basic building block for analog computers completed the demise of the mechanical differential analyzer. Electronic analog computers were used extensively during World War II for navigation and fire control, and they still find use in the simulation of continuous-time systems.

Digital computers were introduced in 1946 with the Electronic Numerical Integrator and Computer (ENIAC) at the University of Pennsylvania. High accuracy could be obtained, and heating effects, which plague analog computers, were reduced. The penalty paid was in longer computation times because the calculations are generally performed sequentially in the digital computer. Since then, computational power has increased dramatically, so that even complex simulations can be performed on personal computers in a reasonable length of time. Parallel processing capabilities of some digital computers continue to extend the applications amenable to simulation.

Operational methods for simulation were introduced in the 1950s and 1960s. These methods use techniques that are familiar to engineers, particularly electrical engineers and other engineers familiar with control system design. The methods are easy to use but are generally limited to the simulation of linear systems.

Simulation languages were introduced in the 1960s. These languages attempt to make the simulation methods transparent to the user. Some of these languages have excellent graphics capabilities and can develop the simulation equations from a block diagram of the system.

Currently, the areas of artificial intelligence and digital simulation are merging. Expert systems are being designed that have the following capabilities: (1) to configure program modules to implement a specified system block diagram; (2) to run the simulation with specified inputs; and (3) to analyze the results of the simulation in terms of stability and time-domain performance. One example is given in Beale and Kawamura (1989).

1.6 ORGANIZATION OF THE TEXT

Chapter 2 provides a review of the z-transform and its application to digital simulation. Difference equations and transfer functions for discrete-time systems are defined and discussed. Sampling of continuous-time signals to produce discrete-time signals and the reconstruction of discrete-time signals into continuous-time signals are presented next. Special attention is given to signal reconstruction through a zero-order hold device. Much of this material will be familiar to engineers who have studied discrete-time signals and systems, digital control, or digital signal processing.

In Chapter 3, integration methods for linear systems by operational calculus are developed. With operational methods, a discrete-time operator is substituted for the Laplace transform of the integration process $(1/s)$ in a transfer function representation of the system to be simulated. This produces an equivalent discrete-time system model to be used in the simulation. Various operators are defined and comparisons are given between them. The operational approach should seem particularly convenient to readers who are familiar with control system design since it uses a transfer function approach and the result of the substitution is a difference equation that can be solved recursively for the current output value.

The classical linear multistep methods of numerical integration are introduced in Chapter 4. Integration operators for approximating the mathematical integration process are derived through the Taylor series approach. Convergence of the simulated response to the true response is discussed, and the conditions required for convergence are presented. The explicit Adams-Bashforth and implicit Adams-Moulton families of multistep operators are treated in detail. A major part of the chapter deals with simulation stability. Stability is a key issue in the simulation of a system since a simulation with an unbounded error is useless. Stability of the numerical integration process is introduced from the point of view of automatic control theory; this provides a very natural approach for many engineers. Stability regions are then presented as a powerful means of analyzing both the absolute and relative stability of a simulation. The chapter concludes with the presentation of stability for predictor-corrector methods.

Chapter 5 continues with stability regions for linear multistep integration methods. The interior of these regions is treated in detail, bringing out the connection between stability regions and simulation accuracy. Predictor-corrector methods are also considered, and the chapter concludes with a discussion of simulation stability when there is a hardware system in the simulation loop.

The focus shifts in Chapter 6 from linear multistep methods to Runge-Kutta methods. Derivation of these methods through a graphical technique is presented. Their stability is discussed, and stability regions are shown for various order methods. Since the function evaluations are computed more than once per timestep with Runge-Kutta methods, the case in which the input signal can be measured each time those computations occur is discussed.

Chapter 7 deals with stiff systems. One definition of a stiff system is that the simulation timestep required for stability is much smaller than that required for accuracy of the important dynamics. The typical case involves system eigenvalues with negative real parts that cover a wide range of values. Several stability definitions that have been developed to treat stiff systems are introduced, and various methods for successfully simulating stiff systems are presented. Linear multistep methods, including stability region placement (SRP) and matrix stability region placement (MSRP) methods, Runge-Kutta methods, and other techniques are discussed.

Chapter 8 presents an in-depth coverage of the simulation of nonlinear systems. These systems are classified by their order and by the nature of their equilibrium points. Simulation of systems that exhibit chaos, a topic of much current interest, is discussed.

The chapters up to this point have discussed numerical integration from the standpoint of representing a single stage of mathematical integration by some form of discrete-time operation. Chapter 9 considers operations that represent multiple stages of integration. The double integrator is treated in detail, and higher-order integrators are also discussed.

Chapter 10 provides a discussion, sometimes subjective, of the various factors concerning simulation presented in the preceding chapters. The purpose of the chapter is to bring all the information together (integrate the data, if you will) so that a simulation designer or user can make intelligent decisions about what type of integration method and what value of timestep should be chosen. Recommendations are based on the authors' experience, and a brief discussion of simulation software is given.

Four appendixes are included in the text. Appendix A contains the list of references cited throughout the text. Appendix B provides stability region plots for a large number of numerical integration methods. These plots show the interior as well as the boundary of the regions. Appendix C contains source listings in MATLAB of a number of programs used for simulation. These include programs for plotting stability regions and programs used in examples presented in the text. Appendix D has mathematical models for several representative dynamic systems. These can be used by educators as example problems or as projects in developing and performing simulations to assign to students.

2

Z-Transform Methods
and Simulation

2.1 INTRODUCTION TO Z-TRANSFORM METHODS

The purpose of this chapter is to introduce z-transform techniques since they provide the machinery for understanding the simulation process. The fundamentals of sampled data theory are also reviewed. The concept of data sampling leads directly to the first simulation method presented in the book, namely, the use of mathematical hold devices. These methods are particularly useful for linear systems, or systems with separate linear and nonlinear blocks.

The z-transform provides the link between a signal or system representation in the discrete-time domain and its representation in the complex frequency domain. It has many of the same properties as the Laplace transform and offers the same advantages for discrete-time system analysis that the Laplace transform does for continuous-time systems. Readers who require more background in this area should refer to Jury (1964), Phillips and Nagle (1990), Ragazzini and Franklin (1958), or Rosko (1972) for a more complete discussion.

2.2 REVIEW OF Z-TRANSFORMS

This section introduces the fundamentals of z-transform techniques, beginning with a definition. The z-transform of a continuous-time function (which is zero for negative time) is defined as

$$F(z) = Z[f(t)] = \sum_{n=0}^{\infty} f(nT)z^{-n}. \qquad (2\text{--}1)$$

Here T is an appropriately chosen sampling period that will also be referred to as the timestep or stepsize. To illustrate the use of this definition, an example is given. If $f(t)$ is a ramp function, then $f(t) = t$ for positive time. Thus

$$F(z) = \sum_{n=0}^{\infty} nTz^{-n} = 0 + Tz^{-1} + 2Tz^{-2} + \ldots + nTz^{-n} + \ldots = \frac{Tz}{(z-1)^2} \qquad (2\text{--}2)$$

The final closed-form polynomial fraction is easily verified to be equivalent to the infinite series through long division. The above z-transform definition is given for completeness. Most often, any good set of transform tables should be used to facilitate the transformation process. Table 2.1 contains several of the most useful z-transform pairs and some fundamental z-transform properties. The tables given in Jury (1964) are probably the best available.

The inverse z-transform is most easily obtained from the definition above, as follows,

$$f^*(t) = \sum_{n=0}^{\infty} f(nT)\delta(t - nT) \qquad (2\text{--}3)$$

and represents a sampled version of the original function. The coefficients in this series can easily be obtained by using long division of any rational function of z. Here $\delta(t - nT)$ is the Dirac delta function of infinite height, zero width, and unity area occurring at time nT. The * indicates a sampled function. For the ramp function above,

$$f^*(t) = 0\delta(t) + T\delta(t - T) + 2T\delta(t - 2T) + \ldots \qquad (2\text{--}4)$$

The z-operator itself has some useful properties, which are stated below without proof. Some of these will be very useful as we progress to the later material.

Shift Theorem
Let $Y(z) = Z[y_n]$. Then, assuming zero initial conditions,

$$zY(z) = Z[y_{n+1}] \text{ and } z^{-1}Y(z) = Z[y_{n-1}]$$

where $y_n = y(nT) = y(t)\delta(t - nT)$.

Thus it is seen that z operates as a forward time-shift operator and that $1/z$ operates as a backward time-shift operator.

Final-Value Theorem
If $f(t)$ and its time derivative are z-transformable and it has its z-domain poles strictly inside the unit circle, then

$$\lim_{n \to \infty} f_n = \lim_{z \to 1} \left[\frac{z-1}{z} F(z) \right]$$

TABLE 2.1   BRIEF TABLE OF Z-TRANSFORM PROPERTIES AND PAIRS

$f_n, n \geq 0$		$F(z)$
$k_1 f_n + k_2 g_n$	Linearity	$k_1 F(z) + k_2 G(z)$
f_{n+m}	Forward shift	$z^m F(z) - \sum\limits_{j=0}^{m-1} f_j z^{m-j}$
f_{n-m}	Backward shift	$z^{-m} F(z)$
f_0	Initial value	$\lim\limits_{z \to \infty} F(z)$
f_∞	Final value	$\lim\limits_{z \to 1} \left[\dfrac{z-1}{z} \right] F(z)$
$\delta(n)$		1
1		$\dfrac{z}{z-1}$
nT		$\dfrac{Tz}{(z-1)^2}$
e^{-anT}		$\dfrac{z}{z - e^{-aT}}$
$e^{-anT} \sin(bnT)$		$\dfrac{e^{-aT} \sin(bT) z}{z^2 - 2e^{-aT} \cos(bT) z + e^{-2aT}}$
$e^{-anT} \cos(bnT)$		$\dfrac{z^2 - e^{-aT} \cos(bT) z}{z^2 - 2e^{-aT} \cos(bT) z + e^{-2aT}}$

Note that if the system were unstable, this formula would still provide a numerical result. However, the final value for an unstable system is unbounded, and this numerical result cannot be interpreted as the final value. Also, this result is analogous to the continuous-time final-value theorem:

$$\lim_{t \to \infty} f(t) = \lim_{s \to 0} [sF(s)]$$

Initial-Value Theorem

If $f(t)$ and its time derivative are z-transformable, its initial value is given by

$$\lim_{n \to 0} f_n = \lim_{z \to \infty} [F(z)] = \lim_{s \to \infty} [sF(s)]$$

Stability

Given a discrete-time transfer function, $G(z) = N(z)/D(z)$, this system is stable (in the bounded input–bounded output sense) if all the roots of $D(z)$ (the system poles) lie inside the z-domain unit circle. A system with isolated poles on the unit circle is considered marginally stable.

Many tests for the stability of a z-domain transfer function are available, including the Schur-Cohn test, the Jury table, and the bilinear transformed Routh-Hurwitz test (Jury, 1958, 1964; Phillips & Nagle, 1990). From a practical viewpoint, stability is most often determined through standard control system software by computing the specific poles of the z-domain transfer function.

Since digital simulation entails representing continuous-time integrators ($1/s$) in discrete time, it will be important to understand the relationship between the s-plane and the z-plane. A simple example serves to relate the two. Let

$$y(t) = x(t - T)$$

The Laplace transform of this equation can be written by using the continuous-shift operator representing the time delay as

$$Y(s) = e^{-sT}X(s)$$

Alternatively, the z-transform of this equation can also be taken as

$$Y(z) = z^{-1}X(z)$$

It then becomes clear that the following important correspondence can be made

$$z \Leftrightarrow e^{sT} \tag{2–5}$$

This correspondence is generally used to define the complex variable z in terms of the complex variable s in the definition of the z-transform. With this definition, the z-transform is seen to be the Laplace transform of the sampled signal, with the change in variable from s to z given by the preceding equation. With this correspondence, the original definition of the z-transform is obtained.

This equation also represents a conformal mapping from the s-plane into the z-plane. The entire left half of the complex s-plane, $Re(s) < 0$, maps into the interior of the unit circle in the complex z-plane, and the right half of the complex s-plane, $Re(s) > 0$ maps into the exterior of the unit circle in the complex z-plane. The imaginary axis in the s-plane, $Re(s) = 0$, maps onto the unit circle in the z-plane. In fact, each horizontal strip in the s-plane of height $2\pi/T$ maps into the entire z-plane. Thus the mapping is many-to-one. With the normal definition of $s = \sigma + j\omega$, the above expression becomes

$$z = e^{\sigma T}[\cos(\omega T) + j\sin(\omega T)] \tag{2–6}$$

It can then be seen that lines of constant σ map into circles in the z-plane, and lines of constant ωT map into radial lines emanating from the origin of the z-plane. This is clarified in Figure 2.1.

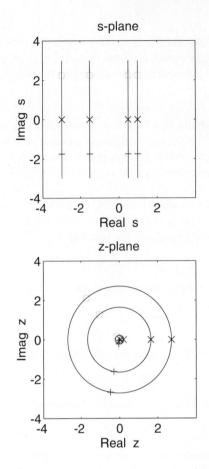

Figure 2.1a Mapping of the *s*-plane into the *z*-plane.

2.3 DIFFERENCE EQUATIONS AND TRANSFER FUNCTIONS

Z-transforms are most often used by control engineers for obtaining qualitative information about discrete-time systems to be controlled. This is usually done by *z*-transforming a given difference equation to obtain a *z*-domain transfer function. The poles, zeros, frequency response, and so on of this transfer function give useful qualitative information about the specific system.

For example, consider the following third-order difference equation,

$$y_{n+3} + a_2 y_{n+2} + a_1 y_{n+1} + a_0 y_n = b_2 u_{n+2} + b_1 u_{n+1} + b_0 u_n$$

where *u* is considered the temporal input to the system and *y* is the temporal output from the system. Z-transforming both sides of this equation and assuming that all initial conditions are zero gives the transfer function

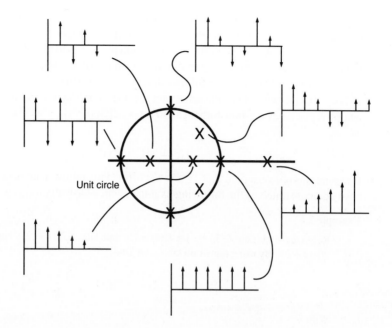

Figure 2.1b Relationship between pole locations and time responses.

$$\frac{Y(z)}{U(z)} \equiv G(z) = \frac{b_2 z^2 + b_1 z + b_0}{z^3 + a_2 z^2 + a_1 z + a_0}$$

One method of obtaining the system output for a particular input is to z-transform the input signal, multiply that $U(z)$ by $G(z)$ to obtain $Y(z)$, and then inverse transform $Y(z)$. However, with this approach the tedious algebraic inversion of the z-domain function, $Y(z)$, arising from every given input, would be required. The more appropriate method for obtaining the system output, for any possible and perhaps random inputs, is to solve the difference equation recursively. In the example above, this is done by computing y_{n+3} in terms of the previously computed outputs and previously given/measured inputs. This approach provides the output values rather than a closed-form expression for the output. Furthermore, the difference equation representing the system can be easily programmed into a digital computer, where the recursion can be done automatically in a simple loop. Thus a difference equation representation for a given continuous-time system is highly desirable. The recursive solution of difference equations is the primary solution technique used in the simulation area.

2.4 TRANSFER FUNCTION RESPONSE—OUTPUT SAMPLING

It should be remembered that the simulation problem is to represent a continuous-time system in discrete time. It will thus be important to first understand the sampling process. The discussion that follows borrows heavily from sampled data theory and is included to demonstrate that one must not carelessly use *z*-domain transfer functions. The reader should also remember that the goal of the discretization process is to obtain a recursively programmable difference equation that can operate on all possible inputs.

As seen in Figure 2.2, it follows that $Y(s) = G(s)U(s)$. When the output is sampled, $Y^*(s) = [G(s)U(s)]^*$. Since the star represents the sampling process, anything that is starred can be replaced by a *z*-transform. Thus $Y(z) = Z[G(s)U(s)]$. This is more correctly written as

$$Y(z) = Z[L^{-1}[G(s)U(s)]] \tag{2-7}$$

It should then be observed that $Y(z) \neq G(z)U(z)$. This unfortunate occurrence can be most easily demonstrated by the following example, which basically follows from Rosko (1972).

Example 2.1

Referring to Figure 2.3, we let $U(s) = 1/s$ and $G(s) = 2/(s + 2)$. Then $Y(s) = G(s)U(s) = 2/[s(s + 2)]$. By using transform tables, we find

$$Y(z) = Z[L^{-1}[2/(s(s + 2))]] = \frac{(1 - e^{-2T})z}{(z - 1)(z - e^{-2T})}$$

With $T = 1$,

$$Y(z) = \frac{0.864z}{z^2 - 1.136z + 0.136}$$

Then, inverse transforming by long division gives

$$Y(z) = 0z^0 + 0.864z^{-1} + 0.982z^{-2} + 0.998z^{-3} + \cdots$$

or

$$y(t) = 0\delta(t) + 0.864\delta(t - T) + 0.982\delta(t - 2T) + 0.998\delta(t - 3T) + \cdots$$

With transform tables, the exact solution can also be found:

$$y_n = 1 - e^{-2nT}$$

It should be noted that these solutions are equivalent and are sampled versions of the actual continuous-time response to a unit step. It should also be noted that

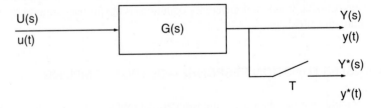

Figure 2.2

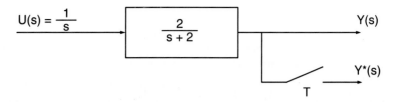

Figure 2.3

$$G(z)U(z) = 2\left(\frac{z}{z-1}\right)\left(\frac{z}{z-e^{-2T}}\right)$$

which is clearly not the same as the true $Y(z)$ given above.

2.5 INPUT SAMPLING

The above procedure for finding y_n is fairly clear and easily done by most undergraduate students. The problem that arises with this solution is that the procedure must be repeated for every input applied. That is, a discrete-time transfer function has not been obtained at all, only the $Y(z)$ that corresponds to the given $U(z)$. There is no difference equation that can be programmed into a recursive simulation that will represent the original process, $G(s)$. What can be done to solve this problem? Perhaps sampling the input would work since this will allow the multiplication of z-domain signals.

As shown in Figure 2.4, sampling the input gives $Y(s) = G(s)U^*(s)$. Then sampling the output gives $Y^*(s) = [G(s)U^*(s)]^*$, which becomes $Y^*(s) = G^*(s)U^*(s)$ since $u^*(t)$ is a sampled signal. This then directly becomes

$$Y(z) = G(z)U(z) \qquad (2–8)$$

This appears to be a nice result since the transfer function $G(z)$ is separately available. Clearly the difference equation represented by $G(z)$ can now be programmed and applied to any input to obtain the response. The following example demonstrates the utility and, unfortunately, inaccuracy of this approach.

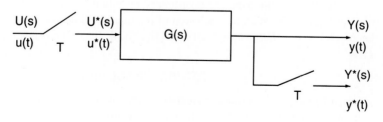

Figure 2.4

Example 2.2

Here the system of Example 2.1 is reconsidered, using the input sampling technique of Figure 2.4 for comparison. From the tables, $U(z) = z/(z - 1)$ and $G(z) = 2z/(z - e^{-2T}) = 2z/(z - 0.136)$, with $T = 1$. Then the transformed step response is

$$Y(z) = G(z)U(z) = \left(\frac{2z}{z - e^{-2T}}\right)\left(\frac{z}{z - 1}\right) = \frac{2z^2}{z^2 - 1.136z + 0.136}$$

Long division gives the response as

$$Y(z) = 2z^0 + 2.272z^{-1} + 2.309z^{-2} + 2.314z^{-3} + \ldots$$

Notice that this response is significantly different from the true response given in the previous section. Also notice that the system transfer function is directly available in the z-domain:

$$\frac{Y(z)}{U(z)} = G(z) = \frac{2z}{z - e^{-2T}}$$

Upon finding the inverse transform of the transfer function, a difference equation representing this system is obtained:

$$y_{k+1} = e^{-2T}y_k + 2u_{k+1}$$

This difference equation is then easily programmed on a digital computer. However, it should be observed that ease of solution has been obtained at a significant loss of accuracy.

2.6 RECONSTRUCTION

Example 2.2 illustrates that input sampling has reduced the accuracy of the computed output. The reason for this is that the system $G(s)$ is being driven by a series of impulses arising from the input sampling rather than from a smooth signal. Since impulses are equally exciting at all frequencies, the resulting response magnitude is expected to be much larger. It would therefore seem reasonable to add some form of low-pass filtering device to smooth the impulse train. This is usually done by adding a specialized low-pass filter called a data reconstructor. This device is added before the system and after the sampled input, as shown in the Figure 2.5 as $H(s)$. This allows a more accurate representation of the input signal while still allowing the formation of z-domain transfer functions, which act on any arbitrary input.

The effect of these data reconstructors, or hold devices, can be determined by reference to Figure 2.5 . With the inclusion of the input hold device $H(s)$, the sampled output becomes $Y^*(s) = [H(s)G(s)U^*(s)]^* = [H(s)G(s)]^*U^*(s)$. Transforming to the z-domain gives

$$Y(z) = Z[L^{-1}[H(s)G(s)]]U(z) \qquad (2\text{--}9)$$

After performing the appropriate algebra, which involves the hold device, we are left with a discrete-time transfer function, which is easily programmed as a difference equation.

The most common and useful hold device, both practically and theoretically, is prob-

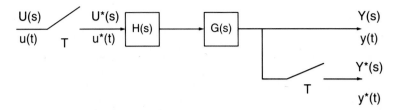

Figure 2.5

ably the zero-order hold (ZOH). It acts as a sample and hold, creating small continuous-time rectangles with a height equal to the signal at the left of the rectangle. The device is referred to as a zero-order hold because the polynomial approximation of the time function is of zero order in time, that is, a constant. Its transfer function is given as

$$H(s) = \frac{1 - e^{-sT}}{s} \qquad (2\text{–}10)$$

which is basically a system that has a finite duration rectangular impulse response. It is interesting to observe that this transfer function has no poles and an infinite number of s-plane zeros lying on the imaginary axis, which is typical of continuous-time systems with finite impulse response. The z-domain equation that results from this hold is

$$Y(z) = Z\left[L^{-1}\left[\left(\frac{1 - e^{-sT}}{s}\right)G(s)\right]\right]U(z) \qquad (2\text{–}11)$$

Removing the exponential function, since it is effectively $1/z$, gives

$$Y(z) = (1 - z^{-1})Z\left[L^{-1}\left[\frac{G(s)}{s}\right]\right]U(z)$$

or

$$(2\text{–}12)$$

$$Y(z) = \frac{z - 1}{z}Z\left[L^{-1}\left[\frac{G(s)}{s}\right]\right]U(z)$$

This is the discretization technique that most control engineers are accustomed to seeing. It should also be noted that this technique gives the exact response of any system forced by any piecewise constant input. This is easily demonstrated by the earlier example.

Example 2.3

Here the ZOH equivalent will be obtained for the system $G(s) = 2/(s + 2)$:

$$\frac{Y(z)}{U(z)} = G(z) = \frac{z - 1}{z}Z\left[L^{-1}\left[\frac{2}{s(s + 2)}\right]\right]$$

or

$$G(z) = \frac{z - 1}{z}\left(\frac{(1 - e^{-2T})z}{(z - 1)(z - e^{-2T})}\right)$$

which becomes, after cancelling terms and letting $T = 1$,

$$G(z) = \frac{1 - e^{-2T}}{z - e^{-2T}} = \frac{0.864}{z - 0.136}$$

The corresponding difference equation is

$$y_{n+1} = 0.136y_n + 0.864u_n$$

The response of this system for a step input is exactly the same as in Example 2.1, as expected, since the input is piecewise constant. Furthermore, this equation is easily programmed and applied to any arbitrary input. For inputs other than a step, there will be differences between the discrete-time and continuous-time outputs. The magnitude of these errors can be controlled by the proper choice of T. Thus we have at last obtained the desired result of a programmable difference equation for accurately simulating a continuous-time system, given arbitrary inputs.

Clearly one would expect that more accurate hold devices would yield more accurate simulations (ignoring any hardware problems). This is indeed true; however, there is usually more difficulty in obtaining the resulting $G(z)$. For example, a triangular hold device gives

$$Y(z) = \frac{(z-1)^2}{Tz} Z\left[L^{-1}\left[\frac{G(s)}{s^2} \right] \right] U(z) \qquad (2\text{–}13)$$

The added complexity here arises from an extra s-domain pole. This then requires one more term in the partial fraction expansion and in the associated algebra.

Example 2.4

Repeating Example 2.3, using the triangular hold with $T = 1$, we get

$$G(z) = \frac{(1 + e^{-2})z + (1 - e^{-2} - 2e^{-2})}{z - e^{-2}} = \frac{0.568z + 0.297}{z - 0.135}$$

The resulting difference equation is

$$y_{n+1} = 0.135y_n + 0.568u_{n+1} + 0.297u_n$$

It should be observed that the discrete-time system that results from the triangular hold requires a measurement of the current input to compute the current output. If the system represented by the preceding transfer function and difference equation is part of a closed-loop configuration, the current input value u_{n+1} might be a function of the current system output y_{n+1}. In this case, the output could not be computed without the knowledge of itself. Therefore, in block-by-block simulations of closed-loop systems, at least one of the blocks must have a transfer function that is strictly proper, that is, the degree of the numerator is less than the degree of the denominator. When at least one block is strictly proper, the simulation is known as being closed-loop realizable. If maintaining a strictly proper transfer function is important, the triangular hold is not recommended, even if it is usually more accurate. This accuracy is obtained at the expense of requiring some knowledge of the present.

Many other hold techniques are based on the idea of subinterval input reconstruction. Some of these are given in Table 2.2. It can be seen that the more complicated data reconstructors usually require more effort in the transformation process. Furthermore, the order of the transfer function is often increased along with the loss of strict properness. The reader should be reminded that each of these holds is effectively a mathematical function created to improve the approximation to the true solution provided by the difference equation. In the context of digital simulation, the hold function does not require a physical device, and many exotic holds can thus be used.

From a frequency domain viewpoint, each of these holds is an approximation to an ideal low-pass filter. These approximations share the characteristic that their DC gains are equal to the sample period T. The last two methods in Table 2.2 simply use a low-pass filter for the hold device. It is instructive to repeat our earlier example by using this method.

Example 2.5

For the plant $G(s) = \dfrac{2}{(s+2)}$, the first-order low-pass filter hold of Table 2.2 gives

$$G(z) = Z\left[L^{-1}\left[\frac{T}{1+sT}\frac{2}{s+2}\right]\right]$$

Here, it is clear that the choice of timestep is important. Since the filter has its equivalent z-plane pole at

$$z = e^{-sT} = e^{-\frac{1}{T}T} = e^{-1}$$

then T must be chosen so that the equivalent discrete-time pole of the plant is slower than this, or $|sT| < 1$. Since $s = -2$ for this plant, we will choose $T = 0.1$ so that $|sT| = 0.2 < 1$. Carrying out the required algebra, we get

$$G(z) = \frac{0.113z}{(z-0.819)(z-0.368)}$$

Although this system has an extra discrete-time pole, it would appear, from our earlier discussion, that the resulting system is a pretty good approximation.

For more information on hold devices, see Jury (1958, 1964), Kuo (1980, 1982), Phillips & Nagle (1990), Ragazzini & Franklin (1958, p. 50), and Rosko (1972).

Mathematical hold equivalents are also often performed on state space systems. The most often used zero-order hold is given here for completeness. It should be noted that any of the methods from Table 2.2 could also be used, with the necessary complications.

Let

$$\dot{x} = Ax + Bu, \text{ and } y = Cx \qquad (x(0) \text{ is given.}) \qquad (2\text{--}14)$$

where x, y, and u are continuous-vector functions of time, with x being the internal state of the system. This linear system is known to have the solution

TABLE 2.2 A COLLECTION OF MATHEMATICAL HOLDS

Zero-order hold
(Rosko, 1972)

$$H(s) = \frac{1 - e^{-sT}}{s}$$

$$G(z) = (1 - z^{-1})\, Z\left[L^{-1}\left(\frac{G(s)}{s}\right)\right]$$

First-order hold
(Rosko, 1972)

$$H(s) = \frac{1 + sT}{T}\left(\frac{1 - e^{-sT}}{s}\right)^2$$

$$G(z) = \frac{(1 - z^{-1})^2}{T}\, Z\left[L^{-1}\left(\frac{(1 + sT)\, G(s)}{s^2}\right)\right]$$

Second-order hold
(Smith, 1987)

$$H(s) = \left(\frac{2 + 3sT + 2(sT)^2}{2T^2}\right)\left(\frac{1 - e^{-sT}}{s}\right)^3$$

$$G(z) = \frac{(1 - z^{-1})^3}{2T^2}\, Z\left[L^{-1}\left(\frac{(2 + 3sT + 2(sT)^2)\, G(s)}{s^3}\right)\right]$$

Triangular hold
(Rosko, 1972)

$$H(s) = \frac{e^{sT}}{T}\left(\frac{1 - e^{-sT}}{s}\right)^2$$

$$G(z) = \frac{z(1 - z^{-1})^2}{T}\, Z\left[L^{-1}\left(\frac{G(s)}{s^2}\right)\right]$$

Parabolic hold
(Jury, 1964)

$$H(s) = \left(\frac{2e^{sT}}{T^2\,(1 + e^{-sT})}\right)\left(\frac{1 - e^{-sT}}{s}\right)^3$$

$$G(z) = \frac{2z(1 - z^{-1})^3}{T^2(1 + z^{-1})}\, Z\left[L^{-1}\left(\frac{G(s)}{s^3}\right)\right]$$

Zero-order mean
(Jury, 1958, 1964)

$$H(s) = \frac{e^{sT}\,(1 - e^{-2sT})}{2s}$$

$$G(z) = \frac{z(1 - z^{-2})}{2}\, Z\left[L^{-1}\left(\frac{G(s)}{s}\right)\right]$$

Zero-order midinterval
(Jury, 1959, 1964)

$$H(s) = \frac{e^{sT/2}\,(1 - e^{-sT})}{s}$$

$$G(z) = z^{1/2}\,(1 - z^{-1})\, Z\left[L^{-1}\left(\frac{G(s)}{s}\right)\right]$$

First-order low-pass filter
(Ragazzini & Franklin, 1958)
(should have $pT < 1$ for all poles p)

$$H(s) = T\frac{1}{1 + sT}$$

$$G(z) = Z\left[L^{-1}\left(\frac{T\, G(s)}{1 + sT}\right)\right]$$

Second-order low-pass filter
(Ragazzini & Franklin, 1958)

$$H(s) = T\,\frac{1 + \dfrac{sT}{2}}{1 + sT + \dfrac{(sT)^2}{2}}$$

$$G(z) = Z\left[L^{-1}\left(\frac{T\left(1 + \dfrac{sT}{2}\right) G(s)}{1 + sT + \dfrac{(sT)^2}{2}}\right)\right]$$

$$x(t) = e^{At}x(0) + \int_0^t e^{A(t-\tau)}Bu(\tau)d\tau \tag{2-15}$$

A discrete-time ZOH equivalent is found as follows. A sampled version of the solution is obtained as

$$x(kT) = e^{AkT}x(0) + \int_0^{kT} e^{A(kT-\tau)}Bu(\tau)d\tau \tag{2-16}$$

By sampling one step later in time, we get

$$x((k+1)T) = e^{A(k+1)T}x(0) + \int_0^{(k+1)T} e^{A((k+1)T-\tau)}Bu(\tau)d\tau \tag{2-17}$$

Multiplying the former equation by e^{AT} and subtracting it from the latter, we get

$$x((k+1)T) = e^{AT}x(kT) + \int_{kT}^{(k+1)T} e^{A((k+1)T-\tau)}Bu(\tau)d\tau \tag{2-18}$$

The corresponding output equation is $y((k+1)T) = Cx((k+1)T)$. It should be noted that the first integral was subtracted out of the second equation, thereby changing the lower limit of integration. This is now an equation giving a discrete-time state space representation. The problem is that the integral must still be evaluated for every given input. This is undesirable, as in Example 2.1, since it corresponds to output sampling. This difficulty can be resolved by using a ZOH on $u(t)$. Then the $Bu(t)$ terms are constant over the integration interval and can be taken out of the integral to give (using an easier notation)

$$x_{k+1} = e^{AT}x_k + \left[\int_0^T e^{A\tau}d\tau\right]Bu_k \equiv A_d x_k + B_d u_k \tag{2-19}$$

and

$$y_{k+1} = Cx_{k+1}$$

If the input is in fact piecewise constant over the interval, this discrete representation gives the exact output response. The system transfer function can then be found as

$$G(z) = C[zI - A_d]^{-1}B_d \tag{2-20}$$

It should be noted that the resulting discrete-time state space system is identical in input-output behavior to what would be obtained from the transfer function approach applied to the original continuous-time state space system. It is important to remember that both approaches give the same approximation to the original continuous-time system.

2.7 THE DATA RECONSTRUCTION SIMULATION METHOD

The data reconstruction approach, particularly the zero-order hold approximation, for performing digital simulation as discussed above is generally useful when dealing with linear systems. Since the zero-order hold is often used for control system design, it is discussed more completely here. Similar comments also apply to the other hold techniques.

Although the zero-order hold is known to give the exact response for a linear system being driven by a step input, there are several serious disadvantages:

1. To use this method, it is necessary to inverse transform $G(s)/s$. At the very least, it is necessary to expand this into partial fractions so that a table of transforms may be used. This can be tedious, as it often requires finding polynomial roots. It is even more tedious for higher-order holds.

2. Given a large system, it may be important to pick out several of the system variables as outputs. This being the case, the expansions discussed above must be repeated for every output.

3. The actual implementation of a multiblock process will require excessive computation times because of loss of the z-domain product.

These problems do not necessarily prohibit the use of the method but are important considerations when implementing a large simulation. Current control system software packages, such as MATLAB, can eliminate some problems, 1 in particular. They cannot, however, change the fundamental properties of the method, as in 2 and 3.

These problems can be further clarified by the example given in Figure 2.6. To get accurate simulation responses for both outputs in the figure, the following transformations are required:

$$G_1(z) = \frac{Y_1(z)}{U(z)} = Z[L^{-1}[H(s)G_1(s)]]$$

$$G_2(z) = \frac{Y_2(z)}{U(z)} = Z[L^{-1}[H(s)G_1(s)G_2(s)]]$$

Clearly this tedious process would need to be repeated for every other block that was added to the system. Furthermore, both of these equations must be programmed, which increases computation time. It should be observed that the output from one difference equation does not go into the input of the difference equation for the next block. Hence, the ZOH simulation method effectively loses the desirable block-cascading property. One might be tempted to avoid these problems by using a ZOH approximation for each block added to a system. This is certainly possible; however, the problem is that the accuracy of the later blocks continues to degrade since we are adding sample and hold devices serially into the system. In this case, an exact response is never obtained for any input, including a step. It should then be clear that the ZOH method is difficult for multiblock systems, particularly if there are any nonlinearities.

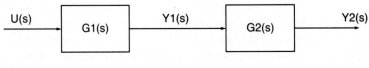

Figure 2.6

2.8 CONCLUSIONS

Z-transform techniques are powerful tools in the digital simulation of continuous-time systems, as well as in control system design. This chapter stresses the transformation of continuous-time transfer functions into discrete-time transfer functions. The goal of this process is to obtain difference equations that can be solved recursively for any input. Although this chapter presents techniques for accomplishing this, as does the remainder of the book, it should be remembered that the transformation from the s-operator to the z-operator is not a rational transformation:

$$s = \frac{1}{T} \ln(z) \qquad\qquad (2\text{--}21)$$

The consequence of this is that an exact discrete-time representation for a given continuous-time system, which is valid for all possible inputs, can never be obtained. Fortunately, the approximation can be made to be arbitrarily close if one is willing to do the associated work.

When one is specifically considering the data reconstruction approach to simulation, the following comments should be remembered. To develop a transfer function relating the input and output variables, a sampler must be placed between the input and the first block of the system. Otherwise, the input cannot be separated from the system itself, and the output transform would need to be derived for each different input applied to the system. Typically, a zero-order hold is placed at the input to the first system block and after this sampler to provide a piecewise-constant signal, although any reasonable low-pass filter is sufficient for reconstruction. Before the advent of control system software packages, the ZOH simulation method was algebraically tedious because of the need for partial fraction expansions, which implicitly requires the factoring of polynomials. Some of this difficulty is eliminated with the methods of the next chapter.

PROBLEM

1. Find a ZOH equivalent of $H(s) = 100/(s + 50)$ with $T = 1.0$.

3

Operational Substitution Methods

3.1 INTRODUCTION

Although the zero-order hold approach is mathematically accurate, it does suffer from several disadvantages, some of which are examined in the last chapter. These include the requirement to obtain a ZOH representation for every desired output signal in the system. This problem further complicates another problem, which is the necessary partial fraction expansion (which may include factoring polynomials) followed by the use of transform tables to obtain each discrete transfer function. As a solution to these problems, an alternative approach is considered in this chapter; it reduces the computational burden to the performance of some algebra. These methods, called operational substitution methods, alleviate the above problems with the ZOH method at the cost of a simulation that is not guaranteed to be stable. The wise choice of a simulation timestep however, does usually lead to a stable and accurate simulation.

Basically, the idea behind operational substitution methods is to take an ideal integrator, which is represented by $1/s$, and replace it by a discrete-time approximation in the z-domain. This approximation is then substituted into the s-domain transfer function. Several of these methods will be derived below, and further extensions to multiple integrators will be given later. The operational substitution method avoids the Laplace and z-transforming problems discussed previously and can be more accurate. The cost to the user will be more algebra and a smaller time stepsize.

3.2 FORWARD-LOOKING RECTANGLES (EULER'S METHOD)

We will first attempt to approximate the ideal integrator, $1/s$, in discrete time. This means that we want to approximate the process

$$y(t) = \int_0^t u(\tau)d\tau \tag{3-1}$$

Equivalently, by using the Laplace transform and ignoring the initial condition,

$$Y(s) = \frac{1}{s} U(s) \tag{3-2}$$

as shown in Figure 3.1. It should be remembered from introductory calculus that the integral is just the area under the curve $u(t)$. This idea will be used in the discussion that follows. Many techniques are available for computing the area under the curve, which are collectively referred to as quadrature methods. We will not discuss these further, but the interested reader should refer to Ralston & Rabinowitz (1978) and Lapidus and Seinfeld (1971).

One simple way to approximate the integral above is to replace the integral with a sum, as follows:

$$y_n = \sum_{k=0}^{n-1} u_k T \tag{3-3}$$

This will give an approximation to the output at time n; however, it would be better to have a recursive expression relating the input and output, as discussed in Chapter 2. This will now be found inductively: At time $n = 1$,

$$y_1 = u_0 T$$

At time $n = 2$,

$$y_2 = u_1 T + u_0 T = u_1 T + y_1$$

And at time $n = 3$,

$$y_3 = u_2 T + u_1 T + u_0 T = u_2 T + y_2$$

By studying the subscripts in each of these equations, it is clear that the difference equation representing this process is

$$y_{n+1} = y_n + Tu_n \tag{3-4}$$

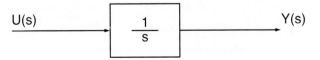

U(s) ⟶ $\dfrac{1}{s}$ ⟶ Y(s)

Figure 3.1 Open loop integration.

Furthermore, it should also be clear that this method is equivalent to that first learned in introductory calculus, that is, approximating the area under a curve by forward-looking rectangles, as shown in Figure 3.2.

To better understand this approximate integrator, we will apply the standard tools of the control engineer. First, we will find a transfer function for the difference equation (assuming the initial condition is zero).

$$Y(z) = \frac{T}{z-1}\, U(z) \qquad (3\text{--}5)$$

By comparison to the original continuous-time system, the method of forward-looking rectangles, often called *Euler's method*, is found by equivalence:

$$\frac{1}{s} \Leftrightarrow \frac{T}{z-1} \text{ or } s \Leftrightarrow \frac{z-1}{T} \qquad (3\text{--}6)$$

For multiple integrals, the approximation becomes

$$\left(\frac{1}{s}\right)^n \Leftrightarrow \left(\frac{T}{z-1}\right)^n \text{ or } s^n \Leftrightarrow \left(\frac{z-1}{T}\right)^n \qquad (3\text{--}7)$$

It will now be instructive to compare the Euler approximation with the ideal integrator, $1/s$. First, consider the poles of each system. The continuous integrator has a pole at

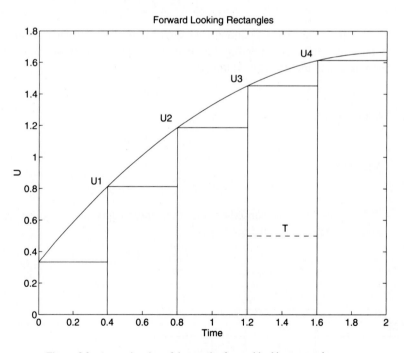

Figure 3.2 Approximation of the area by forward looking rectangles.

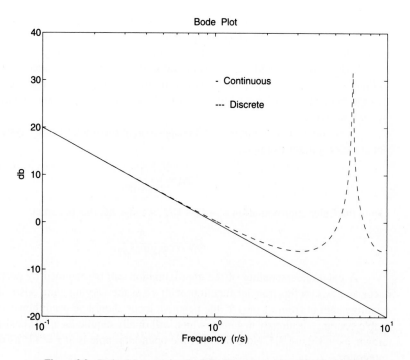

Figure 3.3   Bode plot comparing the Euler integrator with an ideal integrator.

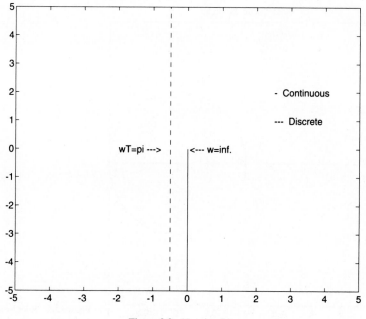

Figure 3.3   Nyquist plot.

$s = 0$. The discrete integrator has a pole at $z = +1$. This is a good thing, as it is known from Chapter 2 that the s-plane origin maps into $+1$ in the z-plane, using the exact mapping $z = e^{sT}$. In fact, it will turn out that it is a requirement for any integrator to have a pole at $z = +1$. Otherwise, it could not be an integrator. Referring to the frequency responses given in Figure 3.3, we can see that for fairly low frequencies the approximations are quite good. At high frequencies $(\omega T = \pi)$, however, we can see that there is some signal amplification by the approximation.

Now consider what happens in the approximation of a first-order system with a pole not at the origin, for example,

$$H(s) = \frac{1}{s - a}$$

Using the Euler approximation $s \rightarrow (z - 1)/T$, we see that this becomes

$$G(z) = \frac{T}{z - 1 - aT}$$

A good understanding of the approximation can be obtained by performing a root locus analysis on this transfer function, with aT as the varying gain. Also, since there is a negative sign in front of the aT, the negative locus should be used; however, assuming a stable system, a is already negative, which will finally require the usual positive locus to be drawn, as in Figure 3.4. Note first that the open-loop pole is at $z = +1$, as required of the integrator.

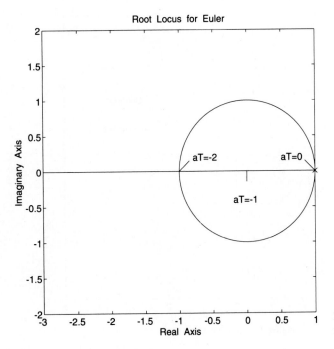

Figure 3.4　Root locus generated by applying Euler's method to a simple system.

It can be seen from the root locus that the simulation will become unstable at $aT = -2$ when the locus leaves the left side of the unit circle. However, it is important to consider accuracy as well as stability. The exactly mapped z-plane pole is $z = e^{aT}$. For a stable system, $aT < 0$, the pole starts at $z = +1$ ($aT = 0$) and goes to $z = 0$ ($aT = -\infty$). The exact pole then represents a monotonically decreasing impulse response from the continuous system. However, the approximation has a pole on the negative real z-axis for $aT < -1$. This corresponds to a high-frequency oscillating signal, which is an inappropriate response from this system. Thus for $aT < -1$, the approximation is not accurate and will be called *detuned*. Thus the useful, or tuned, range of aT for this system is $-1 < aT < 0$. It is also very important to notice that the choice of simulation timestep is now bounded by the pole locations of the system. For this example, and in general for most methods,

$$T_{max} < \frac{1}{|a_{max}|} \tag{3-8}$$

Another example now follows to clarify the situation further.

Example 3.1

Returning to example 2.3, we let $G(s) = 2/(s + 2)$. Then, with Euler substitution of $s = (z - 1)/T$,

$$G(z) = \frac{2T}{z - 1 + 2T}$$

whereas the ZOH model was

$$G_{ZOH}(z) = \frac{1 - e^{-2T}}{z - e^{-2T}}$$

For $T = 1$;

$$G(z) = \frac{2}{z + 1}$$

whereas

$$G_{ZOH}(z) = \frac{0.864}{z - 0.136}$$

The Euler approximation is seen to have its pole at $z = -1$, which is in the left-half z-plane, indicating oscillations in the time response. Also, since the pole is on the unit circle, the simulation is marginally stable. The need for a smaller timestep is suggested.

For $T = 0.1$;

$$G(z) = \frac{0.2}{z - 0.8}$$

whereas

$$G_{ZOH}(z) = \frac{0.181}{z - 0.819}$$

This is a much better approximation, as is observed from the closeness of the two sets of coefficients. Notice also that the approximation is obtained at the cost of only a little algebra, whereas the ZOH approximation requires partial fraction expansions and transform tables.

Some readers may be aware of the δ-operator formulation, as discussed in Middleton and Goodwin (1990), which is very similar to Euler's method. The δ-operator is defined as operating on $x(t)$ via

$$\delta x(t) = \frac{x(t+T) - x(t)}{T} \tag{3-9}$$

When this operator is used in control system design it has the advantage of superior numerical robustness as compared to the z-operator. The δ-operator reduces to the usual differential operator as T goes to zero. With this assumption, the δ-operator allows digital design to be performed in the s-plane as it would for continuous-time design. Design can also be done in the complex δ-plane. More information on this topic can be found in Middleton and Goodwin.

3.3 GENERAL PROCEDURE

Now that one popular substitution method has been derived, the general procedure for any substitution method will be given. Following this, some of the other methods will be derived. A table of methods is given in Table 3.1 for further reference.

Referring to the table, we see that two general types of methods are available. One type has the approximations to higher powers of $1/s$ as the same power of the particular approximation of $1/s$. These methods are typified by Euler's method, backward difference, and Tustin's method and will be referred to as *associative* methods. The other type of method has different approximations to higher powers of $1/s$ than the approximation to $1/s$ raised to that power. These methods are typified by Halijak, Boxer-Thaler, and z-transform methods and will be referred to as *non-associative* methods. It should be noted that these methods can all be derived from one equation, which also gives a measure of the accuracy of the method. This will be done in Chapter 9 after providing the necessary background.

GENERAL PROCEDURE

1. Obtain $G(s)$.
   a. For associative methods there is no need to divide by s^n.
   b. For nonassociative methods, divide through by the highest power of s in the denominator, say, n; that is

$$G(s) \cdot \frac{1/s^n}{1/s^n}$$

The result will be a transfer function in ascending powers of $1/s$. This is done because integration is a low-pass process, whereas approximating s is a high-pass

TABLE 3.1 COMMON FAMILIES OF OPERATIONAL SUBSTITUTION METHODS

Method	$\dfrac{1}{s}$	$\dfrac{1}{s^2}$	$\dfrac{1}{s^3}$	$\dfrac{1}{s^4}$
Euler	$\dfrac{T}{z-1}$	$\dfrac{T^2}{(z-1)^2}$	$\dfrac{T^3}{(z-1)^3}$	$\dfrac{T^4}{(z-1)^4}$
First difference	$\dfrac{Tz}{z-1}$	$\dfrac{T^2z^2}{(z-1)^2}$	$\dfrac{T^3z^3}{(z-1)^3}$	$\dfrac{T^4z^4}{(z-1)^4}$
Tustin	$\dfrac{T(z+1)}{2(z-1)}$	$\dfrac{T^2(z+1)^2}{4(z-1)^2}$	$\dfrac{T^2(z+1)^3}{8(z-1)^3}$	$\dfrac{T^2(z+1)^4}{16(z-1)^4}$
Madwed	$\dfrac{T(z+1)}{2(z-1)}$	$\dfrac{T^2(z^2+4z+1)}{6(z-1)^2}$	$\dfrac{T^3(z^3+11z^2+11z+1)}{24(z-1)^3}$	$\dfrac{T^4(z^4+26z^3+66z^2+26z+1)}{120(z-1)^4}$
Halijak	$\dfrac{Tz}{z-1}$	$\dfrac{T^2z}{(z-1)^2}$	$\dfrac{T^3(z^2+z)}{2(z-1)^3}$	$\dfrac{T^4(z^3+2z^2+z)}{4(z-1)^4}$
Explicit z-transform	$\dfrac{T}{z-1}$	$\dfrac{T^2z}{(z-1)^2}$	$\dfrac{T^3(z^2+z)}{2(z-1)^3}$	$\dfrac{T^4(z^3+4z^2+z)}{6(z-1)^4}$
Implicit z-transform	$\dfrac{Tz}{z-1}$	$\dfrac{T^2z}{(z-1)^2}$	$\dfrac{T^3(z^2+z)}{2(z-1)^3}$	$\dfrac{T^4(z^3+4z^2+z)}{6(z-1)^4}$
Boxer-Thaler	$\dfrac{T(z+1)}{2(z-1)}$	$\dfrac{T^2(z^2+10z+1)}{12(z-1)^2}$	$\dfrac{T^3(z^2+z)}{2(z-1)^3}$	$\dfrac{T^4(z^3+4z^2+z)}{6(z-1)^4}-\dfrac{T^4}{720}$

(noisy) process. For associative methods the algebra gives the same result, and thus approximating s and $1/s$ gives the same discrete transfer function. This will not be the case for nonassociative methods.

3. Substitute expressions for $1/s^n$ from Table 3.1. Control system software can be used for this; Program CC is particularly useful. A matrix method for these substitutions will be presented later.

4. Choose T and evaluate the resulting expressions. The following rule of thumb is suggested: Choose T so that the sampling rate in radians/second is approximately five times the highest natural frequency in the system. Then

$$T = \frac{2\pi}{5\,|s|_{max}} \tag{3-10}$$

If T is chosen any larger than this, important dynamics may be lost. Alternatively, choosing a smaller T can cause significant round-off errors as there are terms, including T^n, that go to zero rapidly as T gets smaller. This will also require more computation to obtain the solution.

The following example should illustrate the procedure.

Example 3.2

1. Let

$$G(s) = \frac{2}{s^2 + 3s + 2}$$

2. Division by s^2 gives

$$G(s) = \frac{\dfrac{2}{s^2}}{1 + \dfrac{3}{s} + \dfrac{2}{s^2}}$$

3. The Halijak method from Table 3.1 gives

$$G(z) = \frac{\dfrac{2T^2z}{(z-1)^2}}{1 + \dfrac{3Tz}{z-1} + \dfrac{2T^2z}{(z-1)^2}}$$

or

$$G(z) = \frac{2T^2z}{(z-1)^2 + 3Tz(z-1) + 2T^2z}$$

or finally

$$G(z) = \frac{2T^2z}{(1+3T)z^2 + (-2-3T+2T^2)z + 1}$$

4. Choose $T = 2(\text{pi})/(5*2) \sim 0.5$. Then

$$G(z) = \frac{0.5z}{2.5z^2 - 3z + 1}$$

A good check on the algebra is to verify that the DC gains of the continuous and discrete systems are the same, using the final-value theorems. Here, $G(s)$, $s = 0$, gives 1, and $G(z)$, $z = +1$, gives 1. Thus, there is a good chance that no error was made.

3.4 OTHER ASSOCIATIVE METHODS

Backward-looking Rectangles (Backward Euler)

Another simple approximation to the integral, in the sense of area under the curve, is obtained by using backward-looking rectangles, as in Figure 3.5. This method is commonly referred to as the *backward Euler method* or the *backward difference method*.

A little analysis shows that the difference equation representing this process is

$$y_{n+1} = y_n + Tu_{n+1} \tag{3-11}$$

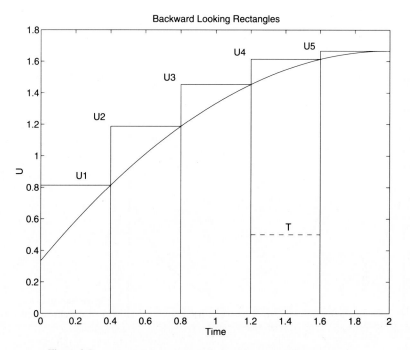

Figure 3.5   Approximation of the area using backward looking rectangles.

Notice that this equation requires the present input, *u,* before calculating the present output, *y.* Again, with *z*-transforms to obtain the transfer function of the approximation, the corresponding substitution method is

$$\frac{1}{s} \Leftrightarrow \frac{Tz}{z-1} \text{ or } s \Leftrightarrow \frac{z-1}{Tz} \tag{3–12}$$

For multiple integrals, this is

$$\left(\frac{1}{s}\right)^{n} \Leftrightarrow \left(\frac{Tz}{z-1}\right)^{n} \text{ or } s^{n} \Leftrightarrow \left(\frac{z-1}{Tz}\right)^{n} \tag{3–13}$$

Since the approximation for higher powers of $1/s$ is just the approximation for $1/s$ raised to the corresponding power, this method is associative. Notice that this approximation also has a pole at $z = +1$, as well as a zero at $z = 0$.

This approximation can also be analyzed by applying it to the simple system considered earlier. Let $G(s) = 1/(s - a)$, $a < 0$ for stability. Then

$$G(z) = \frac{Tz}{(1 - aT)z - 1}$$

The response of this system can be analyzed by again looking at its root locus while (aT) changes, as in Figure 3.6. The locus starts at $z = +1$ $(aT = 0)$ and goes to $z = 0$

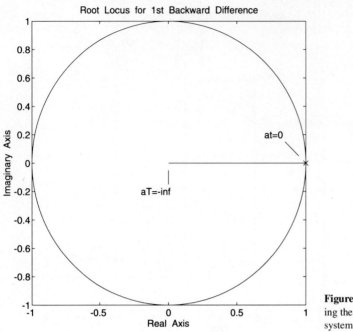

Figure 3.6 Root locus generated by applying the Backward Euler method to a simple system.

$(aT \rightarrow -\infty)$. Hence this simulation will always be stable, regardless of the choice of timestep. It would appear that this method will also always remain accurate since the root locus does not move into the left-half z-plane. Thus this method never becomes severely detuned. A disadvantage of the method, however, is its implicitness, or nonclosed loop realizability. That is, its output at a given time is a function of the input at the same time. This disadvantage will be discussed later.

Example 3.3

Returning to the example $G(s) = 2/(s + 2)$, we get further insight on this method. The approximate system using the backward Euler method is

$$G(z) = \frac{2Tz}{(1 + 2T)z - 1}$$

For $T = 1$,

$$G(z) = \frac{2z}{3z - 1} = \frac{0.667z}{z - 0.333}$$

whereas

$$G_{ZOH}(z) = \frac{0.864}{z - 0.136}$$

This approximation does not appear to be very good, but it is at least stable and gives a monotonically decreasing impulse response. It also has the correct DC gain, as did forward Euler. It will turn out that yielding the correct DC gain will be a requirement for all approximations to the integral. Amazingly enough, this property will be automatically built into the integrators and will require no further consideration by the user. Also notice that now there is a power of z in the numerator, again yielding a not strictly proper system.

Trapezoidal Rule (Tustin's Method)

A third method, which is also based on geometrical considerations, is the trapezoidal rule. Here, rather than using rectangles to approximate the integral, trapezoids are used, as in Figure 3.7. The resulting approximation is considerably better and has amazingly useful properties. This method is referred to by numerical analysts as the *trapezoidal rule*, by control engineers as *Tustin's method*, and by signal processors as the *bilinear transformation*. It should be noted that Tustin's derivation of this method was significantly different but yielded the same result (Tustin, 1947).

Repeated application and summation of the trapezoids gives the difference equation representing this process as

$$y_{n+1} = y_n + \frac{T}{2}(u_{n+1} + u_n) \tag{3-14}$$

The z-transform gives the following approximation for the integral

$$\frac{1}{s} \Leftrightarrow \frac{T}{2}\left(\frac{z+1}{z-1}\right) \text{ or } s \Leftrightarrow \frac{2}{T}\left(\frac{z-1}{z+1}\right) \tag{3-15}$$

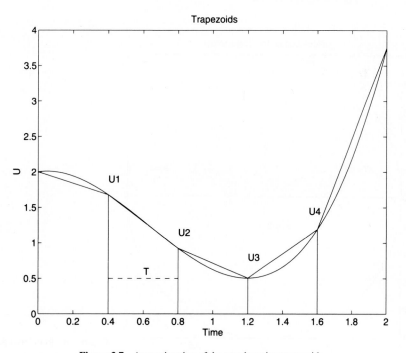

Figure 3.7   Approximation of the area by using trapezoids.

For multiple integrals, the transformation is

$$\left(\frac{1}{s}\right)^n \Leftrightarrow \left[\frac{T}{2}\left(\frac{z+1}{z-1}\right)\right]^n \text{ or } s^n \Leftrightarrow \left[\frac{2}{T}\left(\frac{z-1}{z+1}\right)\right]^n \tag{3-16}$$

and so this method is also an associative method.

Applying this to a first-order system will give further insight into its properties. With $G(s) = 1/(s - a)$, $a < 0$ for stability, the approximate system becomes

$$G(z) = \frac{T(z+1)}{2(z-1) - aT(z+1)}$$

By performing a root locus analysis with aT as the gain, we get the root locus of Figure 3.8. It can be seen that the locus starts at $z = +1$ ($aT = 0$) and goes to $z = -1$ ($aT \rightarrow -\infty$). Thus the simulation will always be stable. With regard to accuracy, it is seen that the locus passes through $z = 0$ when $aT = -2$. Thus for aT more negative than -2, the simulation is detuned, and the response will not resemble that which is expected. Another problem with this method is its non–closed-loop realizability.

Example 3.4

Returning to the earlier example $G(s) = 2/(s + 2)$, we obtain the following discrete system:

$$G(z) = \frac{2T(z+1)}{(2+2T)z - (2-2T)}$$

For $T = 1$ this is $G(z) = \dfrac{0.5(z+1)}{z}$.

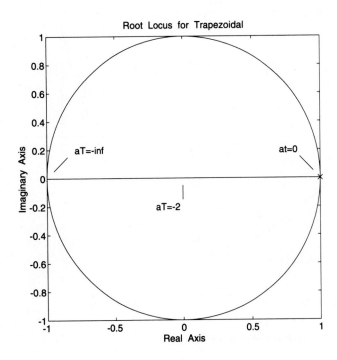

Figure 3.8 Root locus generated by applying the trapezoidal rule to a simple system.

This is clearly not as good as the response of the backward Euler method since the closed-loop pole is at the origin. It is certainly better than the normal Euler method. Although there appears, from this example, to be some shortcomings to this method, it will soon become apparent that it is quite useful.

Other associative methods are possible. In fact all the methods presented in Chapter 4 could be considered to be associative. It should be observed that these methods have the inherent simplicity of using the same discrete function, raised to the appropriate power, throughout the substitution process. The disadvantage corresponding to this simplicity is less accuracy than can be obtained by using different approximations for different powers of s in the continuous-time transfer function.

3.5 DERIVATION OF NONASSOCIATIVE METHODS

The operational substitution methods to be described now are significantly different from those introduced earlier. Those methods had the property that the approximation to $1/s^n$ was equal to the approximation to $1/s$ raised to the nth power. The two methods that follow, as well as Halijak and Madwed, do not have this property. They are based on the idea that a better approximation to $1/s^n$ can be obtained directly, rather than approximating $1/s$ and raising it to the appropriate power.

Boxer-Thaler Method (z-forms)

The derivation of the Boxer-Thaler method is very important since it begins to introduce information from complex variable theory. It will turn out that most of the useful information about simulation methods is obtained from this theory.

Recall the exact s-plane to z-plane mapping, $z = e^{sT}$. Since we are trying to obtain s as some function of z, this equation is rearranged to give us

$$s = \frac{1}{T} \, ln(z) \tag{3–17}$$

This is very nice; however, the useful thing about z-transforms is that they directly represent easily programmable difference equations. The above expression could be substituted directly into an s-plane transfer function; however, it is very difficult to program $ln(z)$. Rather than attempt it, we seek a rational approximation to the $ln(z)$ function. The $ln(z)$ function is difficult to approximate rationally since it has an infinite order branch point at the origin. Several Taylor series expansions are available for this function. The one chosen by Boxer and Thaler (1956) is

$$ln(z) = 2 \left[\left(\frac{z-1}{z+1} \right) + \frac{1}{3} \left(\frac{z-1}{z+1} \right)^n + \ldots + \frac{1}{n} \left(\frac{z-1}{z+1} \right)^3 + \ldots \right], \ n \text{ odd and } z > 0 \tag{3–18}$$

Rosko (1972) suggests substituting $w = (z - 1)/(z + 1)$ to simplify the following algebra. Doing so yields

$$ln(z) = 2\left[w + \frac{1}{3}w^3 + \ldots + \frac{1}{n}w^n + \ldots \right] \qquad (3\text{--}19)$$

Since, as mentioned earlier, $1/s$ is a low-pass function, it is more useful to obtain an approximation to $1/s$. Hence

$$\frac{1}{s} = \frac{T}{ln(z)} = \frac{T}{2}\frac{1}{(w + w^3\!/3 + \ldots)} \qquad (3\text{--}20)$$

By long division this becomes

$$\frac{1}{s} = \frac{T}{2}\left[\frac{1}{w} - \frac{1}{3}w - \frac{4}{45}w^3 - \ldots \right] \qquad (3\text{--}21)$$

Truncating this series after the lowest power of w yields a useful substitution method:

$$\frac{1}{s} \approx \frac{T}{2} \cdot \frac{1}{w} = \frac{T}{2}\left[\frac{z+1}{z-1} \right] \qquad (3\text{--}22)$$

This can be recognized as the previously introduced trapezoidal rule. It should be noticed that truncating the series after more terms yields an approximation with more than one pole. For example, keeping two terms gives

$$\frac{1}{s} \approx \frac{T}{2}\left[\frac{1}{w} - \frac{w}{3} \right] = \frac{T}{3}\frac{(z^2 + 4z + 1)}{(z^2 - 1)} \qquad (3\text{--}23)$$

which should be recognized as Simpson's rule. It will turn out that this method has very little practical utility in the simulation environment, although it is theoretically a more accurate approximation to the integral than all of the methods considered thus far. This will be discussed at length in Chapter 4.

The object of this approach was to obtain better accuracy for multiple integrals. This can be done by raising the above series to the appropriate power. In considering the double integral, the above Taylor series expansion must be squared as follows:

$$\frac{1}{s^2} = \frac{T^2}{4}\left[\frac{1}{w^2} - \frac{2}{3} + \text{higher-order terms in } w^2 \right] \qquad (3\text{--}24)$$

Truncating the series after two terms gives

$$\frac{1}{s^2} \approx \frac{T^2}{4}\left[\frac{3 - 2w^2}{3w^2} \right] = \frac{T^2}{12}\left[\frac{z^2 + 10z + 1}{(z-1)^2} \right] \qquad (3\text{--}25)$$

which is also found in Table 3.1. More on the accuracy of these methods will be given later; however, it should be noted that they are more accurate than any of the other approximations for the double integral. The individual elements of the Boxer-Thaler family are also the most accurate of all the corresponding elements of the other methods in terms of Taylor series truncation error when T is assumed to be small. Unfortunately this does not mean that they are the most accurate when substituted as a family.

Using the above process, we can obtain approximations for the other multiple integrals. These "z-forms" are tabulated up to 12 integrals in Jury (1964).

Z-Transform Substitution Method

The z-transform substitution method is another operational method that is derived fairly easily. The reader is cautioned not to become confused by the similar names. We now have the z-transform method (that is, the usual transform approach for difference equations), the z-forms (of Boxer and Thaler), and now the z-transform substitution method. This method assumes that good approximations to a multiple integral can be obtained by transforming their corresponding impulse response. The approximation then becomes

$$\hat{G}_n(z) = T \cdot Z \left[L^{-1} \left[\frac{1}{s^n} \right] \right] \tag{3–26}$$

It turns out that this expression has a closed form. Rosko (1972) gives the following complicated formula:

$$\hat{G}_n(z) = \lim_{a \to 0} (-1)^{n-1} \frac{\partial^{n-1}}{\partial a^{n-1}} \left(\frac{z}{z - e^{-aT}} \right) \tag{3–27}$$

A much more useful equation is given by Jury (1964):

$$\frac{1}{s^n} \Leftrightarrow \frac{T^n}{(n-1)!} (-1)^{n-1} D^{n-1} \left(\frac{z}{z-1} \right)$$

where $\tag{3–28}$

$$D \equiv z \frac{d}{dz}$$

For example, consider a triple integral:

$$\frac{1}{s^3} \Leftrightarrow \frac{T^3}{2} D^2 \left(\frac{z}{z-1} \right) = \frac{T^3}{2} D \left(z \frac{d}{dz} \left(\frac{z}{z-1} \right) \right)$$

$$= \frac{T^3}{2} z \frac{d}{dz} \left(\frac{-z}{(z-1)^2} \right) \tag{3–29}$$

$$= \frac{T^3}{2} \left(\frac{z^2 + z}{(z-1)^3} \right)$$

It is a fairly simple matter to compute these for any given multiple integral. The first four are found in Table 3.1.

3.6 MATRIX METHODS FOR OPERATIONAL SUBSTITUTION ALGEBRA

This section presents a simple matrix method for operational substitution for the simulation of physical systems. Tables are given in the appendix to this chapter for quick calculation either by hand or by computer.

Operational substitution methods have been widely used for the digital simulation of dynamic systems, particularly by engineers, because of the methods' simplistic nature and easy implementation. One of the problems with these methods is the tedious algebra that is associated with the substitution process. This is certainly frustrating for practicing engineers, who wish to do a quick simulation to check the behavior of a system, since the algebra is difficult to do quickly and accurately. The frustration is compounded for students who are trying to learn the method and who make repeated algebra mistakes in deriving the discrete representation. Clearly the process can be automated in a computer, but this appears to make the straightforward technique become rather magical to the user, and to students in particular. As is shown in what follows, the entire process can be greatly simplified by using the coefficient convolution properties of polynomial multiplication. This reduces the tedious algebra to a simple matrix multiply, which is most easily accomplished via the tables in the appendix. We have found that the use of these tables greatly reduces frustration, tedium, and opportunity for error. Furthermore, showing the derivation of the technique enlightens the student in some of the more interesting properties of polynomials. The now fast and accurate implementation does not hold the mystique of a computer-generated result, although the approach is easily automated.

Table 3.1 contains the usual families used to accomplish operational substitution. As shown earlier, the method works as follows. To start with the Laplace transfer function of a particular linear part of a system, its numerator and denominator are divided by the highest power of the Laplace s-variable to give numerator and denominator polynomials that are functions of $1/s$. Then the discrete transfer function approximating the corresponding power of $1/s$ is directly substituted into the continuous transfer function. Algebra is then performed, and the result is a discrete-time transfer function. This can then be repeated for all of the linear blocks in the system, being careful to maintain closed-loop realizability.

A practical problem that has always existed with these methods has been performing the algebra required to obtain the discrete-time system representation. It should be clear that it is indeed tedious, especially if the variable T is allowed to remain a variable. To somewhat alleviate this problem, matrix methods for performing conformal mappings have been presented in Jury and Chan (1973). These matrix transformations only work for the methods in which higher powers of the $1/s$ approximation correspond to the approximation for $1/s$ raised to that higher power. Examples from Table 3.1 are the first difference and Tustin methods. This is also the case for any application of a linear multistep method such as Euler's method or the Adam's methods. Unfortunately, the other methods given in Table 3.1 are usually considered to be more accurate. Hence this section presents matrix methods for performing operational substitution by using all of the methods in Table 3.1 as well as the second-order Adams-Bashforth method, which is discussed in Chapter 4.

The transformations are based on the fact that polynomials can be multiplied by using Toeplitz matrices. A simplified version of this approach is given by $C(z) = A(z)B(z)$, where $A(z)$ is of degree n, $B(z)$ is of degree m, and thus $C(z)$ will be of degree $n + m$. To perform the multiplication without doing the obviously tedious algebra, the coefficients of $A(z)$ can be put into a row vector. The coefficients of $B(z)$ are then put into a matrix that is conformable with the A-vector; that is, the B-matrix will be $(n + 1)\text{x}(n + m + 1)$. To do this, the first row of the B-matrix will contain the coefficients of $B(z)$ with the appropriate number of zeros at the end. The second row will contain first a zero, then the coefficients of $B(z)$, and then the appropriate number of zeros at the end. Likewise, the third row will contain first two zeros, then the coefficients of $B(z)$, and then more zeros. This process of adding an extra zero at the beginning of each extra row is repeated until the B-matrix is conformable with the A-row vector. Symbolically, this is

$$A(z) = z^n + a_{n-1}z^{n-1} + \ldots + a_1 z + a_0 \Leftrightarrow [1\ a_{n-1}\ a_{n-2} \ldots a_1\ a_0]$$

$$B(z) = z^m + b_{m-1}z^{m-1} + \ldots + b_1 z + b_0 \Leftrightarrow \begin{bmatrix} 1 & b_{n-1} & b_{n-2} & b_{n-3} & b_{n-4} & \cdots \\ 0 & 1 & b_{n-1} & b_{n-2} & b_{n-3} & \cdots \\ 0 & 0 & 1 & b_{n-1} & b_{n-2} & \cdots \\ \vdots & \vdots & & & & \ddots \end{bmatrix}$$

Then

$$[1\ c_{n+m-1}\ c_{n+m-2} \cdots c_1\ c_0] = \tag{3-30}$$

$$[1\ a_{n-1}\ a_{n-2} \ldots a_1\ a_0] \begin{bmatrix} 1 & b_{n-1} & b_{n-2} & b_{n-3} & b_{n-4} & \cdots \\ 0 & 1 & b_{n-1} & b_{n-2} & b_{n-3} & \cdots \\ 0 & 0 & 1 & b_{n-1} & b_{n-2} & \cdots \\ \vdots & \vdots & & & & \ddots \end{bmatrix}$$

Equivalently, the B-matrix could be the row vector that is postmultiplied by the A-matrix. The same result occurs with either order. This should also be recognized as merely the convolution of two sequences.

Now that polynomial multiplication is recognized as the product of two matrices, the approach for doing substitution will be presented. Assume a continuous-time transfer function

$$H(s) = \frac{b_n s^n + \ldots + b_1 s + b_0}{s^n + a_{n-1}s^{n-1} + \ldots + a_0}$$

Dividing through by s^n, we get

$$H(s) = \frac{b_0 s^{-n} + b_1 s^{-n+1} + \ldots + b_{n-1}s^{-1} + b_n}{a_0 s^{-n} + a_1 s^{-n+1} + \ldots + a_{n-1}s^{-1} + 1}$$

Now, using any of the approximations from Table 3.1, we get the following:

$$\hat{H}(z) = \frac{b_0 N_n(z) + b_1(z-1)N_{n-1}(z) + \ldots + b_{n-1}(z-1)^{n-1}N_1(z) + b_n(z-1)^n}{a_0 N_n(z) + a_1(z-1)N_{n-1}(z) + \ldots + a_{n-1}(z-1)^{n-1}N_1(z) + (z-1)^n}$$

where $N_i(z)$ is the numerator of the given approximation for s^{-i}. It should be observed that all the denominators of the methods in Table 3.1 corresponding to s^{-i} are the same, that is, $(z-1)^i$. It can now be recognized that both the numerator and the denominator can be factored in a matrix fashion as

$$\hat{H}(z) = \begin{bmatrix} b_0\ b_1 \ldots b_n \\ a_0\ a_1 \ldots 1 \end{bmatrix} \begin{bmatrix} N_n(z) \\ (z-1)N_{n-1}(z) \\ (z-1)^2 N_{n-2}(z) \\ \cdot \\ \cdot \\ \cdot \\ (z-1)^{n-1}N_1(z) \\ (z-1)^n \end{bmatrix} = H\,Q_n \qquad (3\text{--}31)$$

The coefficients of the entries of the matrix, Q_n, containing the z-terms can now be written as row vectors after performing the required multiplications. The matrix Q_n must be obtained for every possible order of $G(s)$ of interest. In Rosko (1972), it is suggested that substitution becomes dominated by round-off errors for $n > 4$ if T is "small." A family of four matrices will then be created below for each substitution method, although the matrices for higher-order systems could easily be generated. Multiplication of the continuous-time transfer function coefficient matrix, H, with the appropriate approximation matrix, Q_n, will allow the rapid transformation from continuous-time to discrete-time by using the substitution approach, thereby avoiding the tedious algebra. It should also be noted that these matrices become fairly easy to generate by hand, when the tables in the appendix are not available, by remembering the definition of Q_n.

Example 3.5

This example will repeat Example 3.2, using the matrix approach and a different substitution method to demonstrate the utility of the method.

The system is

$$G(s) = \frac{2}{s^2 + 3s + 2}$$

The first operation is to divide by the numerator and denominator by the highest power of s to give

$$G(s) = \frac{2\left(\dfrac{1}{s^2}\right)}{1 + 3\left(\dfrac{1}{s}\right) + 2\left(\dfrac{1}{s^2}\right)}$$

Here the Boxer-Thaler method is to be used and $H = HQ_2$, or

$$\hat{H} = \begin{bmatrix} 2 & 0 & 0 \\ 2 & 3 & 1 \end{bmatrix} \begin{bmatrix} \dfrac{T^2}{12} & \dfrac{10T^2}{12} & \dfrac{T^2}{12} \\ \dfrac{T}{2} & 0 & -\dfrac{T}{2} \\ 1 & -2 & 1 \end{bmatrix}$$

$$\hat{H} = \begin{bmatrix} \dfrac{2T^2}{12} & \dfrac{20T^2}{12} & \dfrac{2T^2}{12} \\[3mm] \dfrac{2T^2}{12} + \dfrac{3T}{2} + 1 & \dfrac{20T^2}{12} - 2 & \dfrac{2T^2}{12} - \dfrac{3T}{2} + 1 \end{bmatrix}$$

This then gives

$$H(z) = \frac{2T^2z^2 + 20T^2z + 2T^2}{(2T^2 + 18T + 12)z^2 + (20T^2 - 24)z + (2T^2 - 18T + 12)}$$

This clearly demonstrates the utility of this method. It may, however be a little easier to substitute the numerical value for T before the matrix multiply when doing this by hand.

This section has presented a fast way of doing the algebra associated with operational substitution simulation methods. Clearly, the tables given in the appendix could be expanded to include any of the linear multistep methods. These tables should allow the tedious algebraic process to be completed with relative ease.

3.7 THE ERROR IN OPERATIONAL SUBSTITUTION

This section addresses the inherent accuracy of the entire substitution process. It will turn out that the accuracy of an entire family is limited by the least accurate member.

Consider the following system:

$$G(s) = \frac{b_0}{s^4 + a_3 s^3 + a_2 s^2 + a_1 s + a_0}$$

By dividing through by s^4, this equation can be represented by the diagram in Figure 3.9. Although this resembles one of the many available canonical forms (Kailath 1980), we are unaware of its general usage. Once in this form, the multiple integral blocks can be replaced by a given operational substitution family, as in Figure 3.10.

The errors incurred by each substitution can effectively be added into the original continuous multiple integrals, as shown in Figure 3.11. It is then seen that the response to each error term is the same as that for the input except for the numerator scaling factor b_0; that is,

$$Y_i(z) = \frac{G(z)}{b_0} E_i(z) \tag{3-32}$$

where $E_i(z)$ represents the error in the approximation for $1/s^i$. It then follows that the error in the simulation is dominated by the least accurate multiple integral approximation. Unfortunately, with Table 3.1 this is always the approximation for $1/s$, and the error is on the order of either T^2 or T^3 for all of the approximations to $1/s$ in Table 3.1. This will be further clarified in Chapter 9. It can thus be concluded that the common operational substitution methods are never better than first- or second-order accurate.

One advantage of operational substitution methods is that for every s-plane pole there

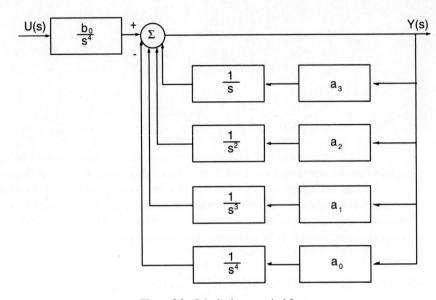

Figure 3.9 Substitution canonical form.

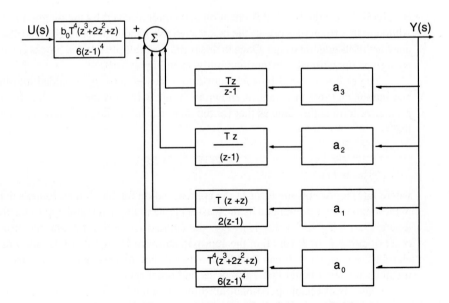

Figure 3.10 Halijak implementation of substitution canonical form.

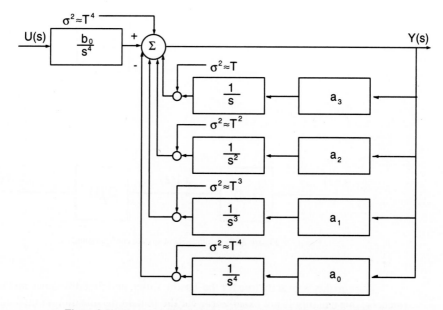

Figure 3.11 Error in typical implementation of substitution canonical form.

is only one z-plane pole. This means that the order of the approximate discrete-time system will be the same as that of the original continuous-time system. This keeps the computational burden of the simulation as small as possible. For the better linear multistep methods, which are described in Chapter 4, there will be several z-plane poles for every s-plane pole. That will be the cost of improved accuracy. Another advantage of operational substitution is the relative ease of obtaining the difference equation, either through the associated algebra or through the matrix method presented in the previous section.

3.8 APPLICATION TO MULTIELEMENT SYSTEMS

One of the motivating factors for not using the ZOH approach is the simulation of multielement systems, which require several inverse transforms. For operational substitution methods, things are a little better. Consider Figure 3.12a. We wish to replace it by Figure 3.12b, which was not possible by using ZOH without degrading the accuracy of the approach. This gives

$$Y_1(z) = G_1(z)U(z)$$

$$Y(z) = G_2(z)Y_1(z)$$

or

$$Y(z) = G_2(z)G_1(z)U(z)$$

Thus, $G(s) = G_1(s)G_2(s)$ and $G(z) = G_1(z)G_2(z)$.

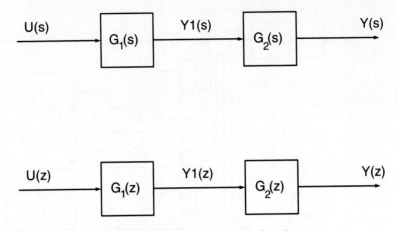

Figure 3.12   Multielement system configuration.

The equality is exactly true for the Tustin, Euler, and first difference methods. Unfortunately, this equality is not exactly true for the remaining substitution methods or for the ZOH method since these methods are not linear in $(1/s)^n$. With the nonassociative methods, substituting and then multiplying transfer functions is not equivalent to multiplying transfer functions and then substituting. It is only equivalent for the three associative methods mentioned above. This is a fairly easy statement to verify. However, using the methods that are not linear in powers of $1/s$ usually gives a more accurate simulation for a large transfer function since the approximations for $1/s^n$ are more accurate than those for $(1/s)^n$.

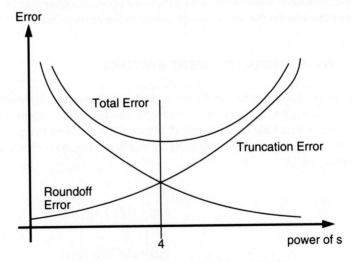

Error as a function of transfer function size.

Figure 3.13   Error as a function of transfer function size.

Multiplying separate continuous-time transfer functions to take advantage of the accuracy of a nonassociative method, however, can yield large round-off errors for large values of n because of the corresponding high powers of T. Rosko (1972) suggests that there is a trade-off, as shown in Figure 3.13, for choosing the size of the original transfer functions in a multielement system. The trade-off is between round-off error for a large n and truncation error for a small n. Rosko suggests that the optimal value of n is about 4.

The implications of all this are that sometime separate blocks need to be multiplied to improve the accuracy of nonassociative methods. Likewise, high-order blocks should always be broken into lower-order blocks in series to reduce round-off error, using either associative or nonassociative methods. For example, four first-order systems should be combined to form one fourth-order system when using nonassociative methods to reduce the truncation error in the approximation. Conversely, a sixteenth-order system should be broken into four fourth-order blocks to prevent the round-off error corresponding to T^{16} terms, using any substitution method.

3.9 SIMULATION OF CLOSED-LOOP SYSTEMS

Closed-loop systems present further problems for simulation. Consider Figure 3.14. Clearly block diagram algebra could be performed on this system to give a single transfer function, which could then be simulated. This is not the approach that should be taken, however, because block diagram algebra is to be avoided and because various signals in the loop may be important. Furthermore, the blocks may be part of a larger simulation that allows various blocks to change as subsystem components are redesigned. Thus, this problem is similar to the multielement problem discussed in an earlier section. It is slightly more complicated, however, since the output is fed back to the input. Using a not strictly proper family, such as Tustin's method (and most of the other substitution methods), can cause some problems in implementation since they are not closed-loop realizable. Rosko (1972) gives the following procedure for simulating a closed-loop system of any complexity.

CLOSED-LOOP SYSTEM PROCEDURE

1. Obtain an approximation to each block in the loop by using the procedure given for open-loop systems.

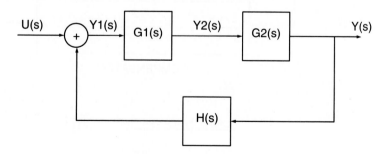

Figure 3.14   A typical closed-loop system.

2. Insert a delay in any feedback loop if (a) the output of its summing junction is to be an observed signal, or (b) the first transfer function after the summing junction is not strictly proper. It should be noted that these two conditions apply most of the time. Adding a delay in the feedback loop is somewhat destabilizing, even for associative methods. If the sampling is done fast enough, however, there is little change in the response. (It should be remembered that these methods were derived when "close enough" was acceptable. By using the methods in the later chapters, these little heuristic arguments will not be needed.)

3. Choose T. The proper choice for T cannot usually be determined since the closed-loop poles of the system may not be known beforehand. Also, the closed-loop behavior of nonassociative methods is unpredictable.

4. Obtain a difference equation for each block in the loop.

5. Program each block separately.

6. If the response does not appear to be reasonable, reduce T and repeat.

Example 3.6

As an example of applying the operational substitution method in a closed-loop environment, consider the system shown in Figure 3.15a. The plant represents a linearized two-degree-of-freedom helicopter model. The plant output Y is the horizontal position, and the plant input U is the rotor thrust angle. The compensator shown in the figure was designed by using a modern control theory method to place the closed-loop eigenvalues at selected locations. Note that both of the transfer functions in the figure are strictly proper.

The first substitutional method used will be Tustin's. Figure 3.15b shows the transfer functions for the plant and compensator with a simulation timestep $T = 0.04$. Each of the transfer functions have equal degrees for the numerator and denominator polynomials. This will always be the case with the Tustin method; therefore, this method is not closed-loop realizable. If we needed to simulate each block separately or to compute the value of the control signal U, we could not do it with these transfer functions. Solving for $U(nT)$ involves knowing

PLANT

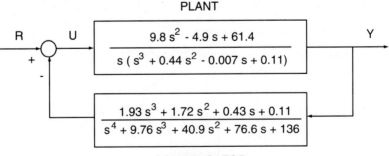

CONTINUOUS - TIME SYSTEM

Figure 3.15a   Continuous-time system.

TUSTIN SUBSTITUTION WITH ADDED DELAY

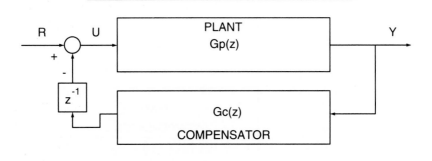

$$Gp(z) = \frac{0.0039\ z^4 - 3.87 \times 10^{-5}\ z^3 - 0.0077\ z^2 + 0.0001\ z + 0.0039}{z^4 - 3.98\ z^3 + 5.95\ z^2 - 3.95\ z + 0.98}$$

$$Gc(z) = \frac{0.0325\ z^4 - 0.064\ z^3 - 0.0011\ z^2 + 0.064\ z - 0.031}{z^4 - 3.62\ z^3 + 4.92\ z^2 - 2.98\ z + 0.677}$$

Figure 3.15b   Tustin substitution.

the value of $Y(nT)$ since the compensator is not strictly proper. However, $Y(nT)$ cannot be computed without knowing the value of $U(nT)$ since the plant is not strictly proper. If only the relationship between the reference input R and the output Y is needed, the block diagram in Figure 3.15b could be reduced to a single block, and the resulting difference equation could be programmed to compute Y. However, it would not be possible to compute U, which would generally be desired.

To obtain a simulation that is closed-loop realizable, a delay of one timestep can be inserted in the feedback loop, as shown in Figure 3.15c. The effect of the delay is to make the compensator's denominator polynomial fifth degree while leaving the compensator's numerator polynomial fourth degree. Therefore, there is a one-step delay between Y and U, so that $U(nT)$ can be computed from $Y[(n-1)T]$. However, this delay is not modeling anything that exists in the continuous-time system; it is an error term in addition to the approximation error of the method. Stability and accuracy of the simulation may suffer because of this extra delay.

The second method considered is Halijak's. Figure 3.15d gives the transfer functions, using the same T as before. It can be seen that the discrete-time plant model is strictly proper with this method. Therefore, the configuration is closed-loop realizable without any additional delay.

To compare the time responses of the Tustin (with added delay) and Halijak methods, a step signal of magnitude 0.1 radians was applied at the signal R. Figure 3.15e shows the response of the continuous-time system to this input. Figure 3.15f shows the errors in the output signal obtained from the two substitution methods. It is clear from those plots that the Halijak simulation has considerably less error than Tustin's method, particularly in the first ten seconds. Most of the large error in Tustin's method is due to the added delay that was necessary to obtain a simulation of the closed-loop system without combining blocks. Figure 3.15g shows the control input signal, U, for the continuous-time system, and Figure 3.15h shows the

TUSTIN SUBSTITUTION

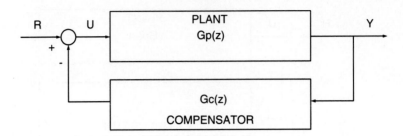

$$Gp(z) = \frac{0.0039\,z^4 - 3.87 \times 10^{-5}\,z^3 - 0.0077\,z^2 + 0.0001\,z + 0.0039}{z^4 - 3.98\,z^3 + 5.95\,z^2 - 3.95\,z + 0.98}$$

$$Gc(z) = \frac{0.0325\,z^4 - 0.064\,z^3 - 0.0011\,z^2 + 0.064\,z - 0.031}{z^4 - 3.62\,z^3 + 4.92\,z^2 - 2.98\,z + 0.677}$$

Figure 3.15c   Tustin substitution with added delay.

HALIJAK SUBSTITUTION

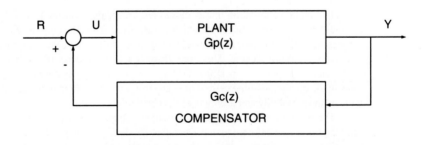

$$Gp(z) = \frac{z\,(0.0153\,z^2 - 0.0307\,z + 0.0156)}{z^4 - 3.98\,z^3 + 5.95\,z^2 - 3.95\,z + 0.98}$$

$$Gc(z) = \frac{z\,(0.0556\,z^3 - 0.165\,z^2 + 0.163\,z - 0.0537)}{z^4 - 3.67\,z^3 + 5.06\,z^2 - 3.11\,z + 0.719}$$

Figure 3.15d   Halijak substitution.

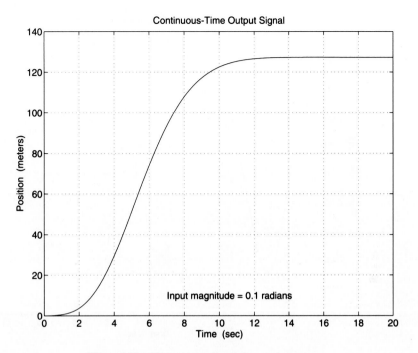

Figure 3.15e   Step response for continuous-time system.

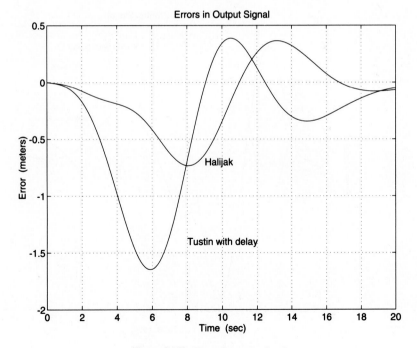

Figure 3.15f   Errors in output signal.

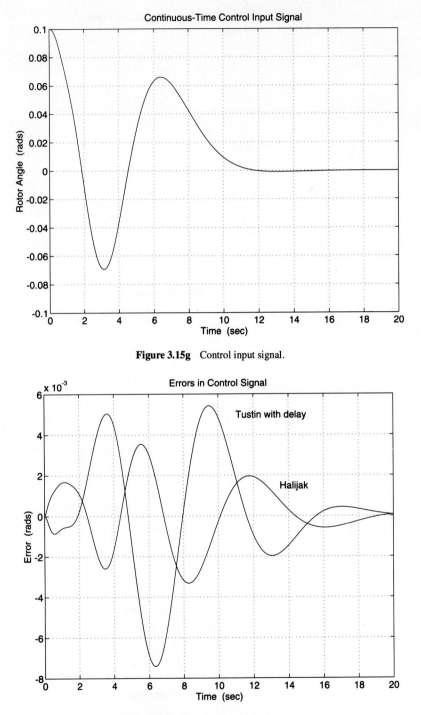

Figure 3.15g   Control input signal.

Figure 3.15h   Errors in control input signal.

errors from the two substitution methods. Once again, Halijak's method has the error with much smaller maximum magnitude. It should also be remembered that Tustin's method without the extra delay would not be able to compute this signal without block diagram manipulation.

3.10 TIME-VARYING SYSTEMS

Some continuous-time systems are considered to have multiplicative parameters that are known functions of time. In this situation, the usual constant coefficient Laplace transform theory does not apply as expected. The reason that the Laplace transform is so useful is that it changes a differential equation in time into an algebraic equation in (s). If a function of time is explicitly involved in the coefficients, this is not the case. Specifically, it transforms a differential equation in time into a differential equation in (s). Thus, the transformation does not really help matters much. For reference, the following Laplace transform pair is included:

$$L\left[t^m \frac{d^n y(t)}{d^n t} \right] \Leftrightarrow (-1)^m \frac{d^m [s^n Y(s)]}{ds^m} \tag{3–33}$$

Assuming zero initial conditions, we can see that if the function of t is a low degree polynomial, say, just t, the resulting differential equation in (s) will only be of first order and may be solvable with little effort.

As an approximation, the standard Laplace transform approach can be used where the transfer function coefficients, which are actually functions of time, are considered to be "frozen" and are thus basically considered time-varying constants. The following example should clarify this technique.

Consider Bessel's equation in time:

$$t^2 \ddot{y} + t\dot{y} + (t^2 - n^2)y = u$$

Let the initial conditions be zero. This can be rewritten as

$$\ddot{y} + \frac{1}{t}\dot{y} + \frac{(t^2 - n^2)}{t^2} = \frac{1}{t^2} u$$

Then the assumption that the time-varying coefficients are "frozen" constants leads to a characteristic equation in the s-variable:

$$\Delta(s) = s^2 + \frac{1}{t} s + \frac{(t^2 - n^2)}{t^2} = 0$$

The stability of the system can now be studied as time changes. For very small values of time, it is seen that the damping ($1/t$) is very large but that there is also a large unstable pole because of the $-n^2/t^2$ term. As time increases, the damping continues to get smaller. Similarly the instability arising from the last term eventually goes away when $t > n$. Finally, as time goes to infinity, the damping gets very small, and the last term, now indicating oscillation frequency, goes to 1. Here the characteristic equation is

$$\Delta(s) \approx s^2 + 1$$

This description does indeed correspond to the response expected from Bessel's equation. After some transient initial growing, followed by damped oscillations, a very lightly damped oscillation persists with radian frequency 1.

The approach just described is very heuristic and caution must be used in evaluating the results. Time-varying systems can be extremely complicated. Unexpected results can occur, particularly when systems are very lightly damped or have periodically varying coefficients. As an example, Mathieu's equation is

$$\ddot{y} + (a + b\cos(t))y = u$$

This system behaves very strangely, and the above analysis leads to somewhat incorrect conclusions (Jordan & Smith, 1979).

3.11 OPERATIONAL METHODS FOR TIME-VARYING SYSTEMS

To simulate a time-varying system using operational methods, the frozen transfer function approach is taken. An example is included to illustrate the procedure. The user is again warned of potential problems when the damping gets small.

Consider the dynamics of a rocket constrained to travel in a vertical line over the launch site. Ignoring, or linearizing, the nonlinearities due to the changing distance from the center of the earth and the aerodynamic drag effects, the linearized but time-varying equations of motion are

$$m(t)\,\ddot{x} + k(t)\,\dot{x} = u - m(t)g$$

where m is the time-varying mass of the rocket due to fuel consumption, k is the aerodynamic drag, g is acceleration due to gravity, and u is the thrust of the solid rocket engine. The mass of the rocket varies as

$$m(t) \;=\; (120000 - 2000\,t - 20000\,g\,(t - 40))kg \text{ for } t < 40 \text{ sec}$$

$$m(t) \;=\; 20000kg \text{ for } t > 40 \text{ sec}$$

The aerodynamic drag is assumed to be a constant, $k = 1000$. The thrust input is assumed to be a constant, $u = 1000000$ Newtons until $t = 40$ and then $u = 0$ afterward. Notice here that acceleration due to gravity resembles a second input and that $g = 10m/s/s$. The procedure for using operational methods for time-varying systems is now presented through this example.

Example 3.7

1. Obtain the frozen transfer function from the given differential equation:

$$X(s,t) = \frac{\dfrac{1}{m(t)}}{s^2 + \dfrac{k}{m(t)}s}\, U(s) - \frac{10}{s^2 + \dfrac{k}{m(t)}s}$$

2. Divide through by the highest power of s:

$$X(s,t) = \frac{\dfrac{1}{m(t)s^2}}{1 + \dfrac{k}{m(t)s}}\, U(s) - \frac{\dfrac{10}{s^2}}{1 + \dfrac{k}{m(t)s}}$$

3. Substitute your favorite operator. If using Halijak,

$$X(z,nT) = \frac{\dfrac{T^2 z}{m(nT)(z-1)^2}}{1 + \dfrac{kTz}{m(nT)(z-1)}}\, U(z) - \frac{\dfrac{10\,T^2 z}{(z-1)^2}}{1 + \dfrac{kTz}{m(nT)(z-1)}}$$

$$X(z,nT) = \frac{\dfrac{T^2 z}{m(nT)}}{(z-1)^2 + \dfrac{kTz(z-1)}{m(nT)}}\, U(z) - \frac{10\,T^2 z}{(z-1)^2 + \dfrac{kTz(z-1)}{m(nT)}}$$

$$X(z,nT) = \frac{\dfrac{T^2 z}{m(nT)}}{\left[1 + \dfrac{kT}{m(nT)}\right]z^2 - \left[2 + \dfrac{kT}{m(nT)}\right]z + 1}\, U(z)$$

$$-\frac{10\,T^2 z}{\left[1 + \dfrac{kT}{m(nT)}\right]z^2 - \left[2 + \dfrac{kT}{m(nT)}\right]z + 1}$$

4. Use an inverse transform to obtain a difference equation:

$$x_{n+2} = \left[\frac{m(nT)}{m(nT)+kT}\right] \cdot \left[\left(2 + \frac{kT}{m(nT)}\right)x_{n+1} - x_n + \frac{T^2}{m(nT)}u_{n+1} - 10T^2\right]$$

Care must be taken in shifting the difference equation forward or backward in time. The coefficients must remain relatively the same; that is,

$$x_{n+1} = \left[\frac{m((n-1)T)}{m((n-1T)+kT}\right] \cdot \left[\left(2 + \frac{kT}{m((n-1)T)}\right)x_n - x_{n-1} + \frac{T^2}{m((n-1)T)}u_n - 10T^2\right]$$

5. Choose T and evaluate the coefficients.

The rule of thumb given earlier is still applicable to the frozen transfer function if the pole positions for all time are used. Even with this rule, T may still need to be determined experimentally.

Perhaps it is fortunate that these systems do not arise very often, as the frozen transfer function approach is questionable. The reader is again reminded that most of the substitution methods originated when close enough was acceptable. It may, or may not, be fortunate that nonlinear systems seem to arise much more often than time-varying systems.

3.12 OPERATIONAL METHODS FOR NONLINEAR SYSTEMS

Nonlinear systems are extremely difficult to deal with. The major problem, from a control theory viewpoint, is that the important superposition property is lost. All linear system theory assumes that superposition holds, which is not the case for nonlinear systems. This implies the loss of the impulse response kernel, and thus the loss of the convolution integral. This then implies the loss of a system transfer function that completely characterizes the system behavior for all inputs, thereby requiring the determination of the system response for every possible input of interest. This is a major motivation for system simulation. Chapter 8 contains significantly more information on nonlinear systems.

Operational methods can be useful for nonlinear systems since it is often possible to put the system into block diagram format by using the following procedure.

1. Separate the given system into linear dynamic blocks and nonlinear static blocks.
2. Obtain a discrete transfer function for each dynamic block.
3. Obtain difference equations for each dynamic block. Obtain a nonlinear static equation for each nonlinear block.
4. Select a time increment, probably experimentally.

Example 3.8 (Rosko, 1972)

Consider the system given in Figure 3.16. Using Tustin's approximation gives the system of Figure 3.17. A simple BASIC program for simulating this system with a sinusoidal input follows:

```
T = 0.1
xs = 0
for k = 1 to N
    u = 2*sin(k*T)
    x = 4*u
    if u < -1 then x = -4
    if u > 1 then x = 4
    y = (1/(2 + .5*T))*((2 - .5*T)*y + T*x + T*xs)
    xs = x
    pset(k,y)
next k
end
```

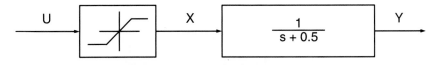

Figure 3.16   A simple open loop nonlinear system.

For closed-loop systems with nonlinearities, the procedure is very similar to that given earlier for closed-loop linear systems, with the appropriate modifications. Basically, the only change in the procedure is that a delay in the feedback loop is required when

1. The first block after the summer is not strictly proper.
2. The first block after the summer is nonlinear.
3. The output of the summer is a simulation response.

An example is shown in Figure 3.18a and b.

3.13 OTHER APPROACHES FOR TRANSFER FUNCTION REPRESENTATIONS

In signal processing, much research has also been performed in the simulation area. Here the motivation is to perform an analog filter design (since designers and control engineers may be more comfortable in the s-plane) and then convert it into a digital filter, using a standard method. Some of the more popular methods are discussed below. For more information, consult Oppenhiem and Schafer (1975) or Rabiner and Gold (1975).

1. Step invariance method.   The idea here is to obtain a discrete system that yields the exact same step response as the original continuous system. We have already encountered such an approximation, the zero-order hold. As it turns out, the step invariance method is indeed equivalent to the ZOH approximation.

2. Mapping of differentials.   The idea here is to map continuous-time integrators into discrete-time integrators. Thus this method is what we have been calling operational substitution.

3. Bilinear transformation with prewarping.   Although the bilinear transformation is equivalent to Tustin's method, a transformation can be added to it to map the

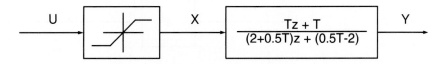

Figure 3.17   Approximation of simple nonlinear system.

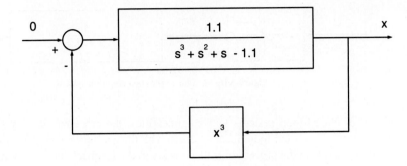

Figure 3.18a Block diagram of the chaotic Cook scroll system.

primary strip in the s-plane into the entire left-half w-plane. This transformation is called prewarping and is performed because the bilinear transformation maps an entire left-half plane into the unit circle. Filters are now designed in the warped w-plane and then mapped back by using the appropriate transformation.

4. Impulse invariance method.   The idea in the impulse invariance method, as in step invariance, is to obtain a discrete system that gives the same impulse response as the original continuous system. The transformation is easier than the ZOH and is given by

$$H(z) = TZ[L^{-1}[G(s)]] \tag{3–34}$$

What this means is that following a partial fraction expansion, terms are mapped as

$$\frac{k_i}{s + p_i} \Longleftrightarrow \frac{k_i z}{z - e^{-p_i T}} \tag{3–35}$$

5. Matched z-transform.   The method z-transform is similar to the impulse invariance method except that both system poles and zeros are mapped by using the transformation

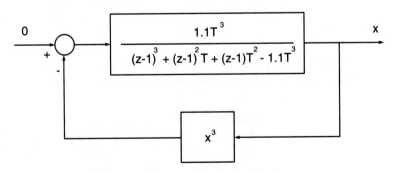

Figure 3.18b Block diagram of the chaotic Cook scroll system, simulated by using Euler's substitution method.

$$s + a \Leftrightarrow 1 - z^{-1} e^{-aT} \qquad (3\text{--}36)$$

Clearly this requires factoring the numerator and denominator of the continuous system before transformation. In addition, it is necessary to map the continuous-time zeros that are at infinity. This is a problem since the entire negative real z-plane axis is effectively at infinity. Two solutions have been widely employed. One is to map the extra zeros into the z-plane origin. The other is to map them into $z = -1$. Both approaches work reasonably well for various systems.

To perform these transformations, the use of some control system software package is suggested. Program CC is especially appealing here since it has many of these transformations built into its CONVERT command. Also, MATLAB has many of these transformations in its c2d command.

Many other methods for performing digital simulation by using a control engineering approach are available in the literature. We have attempted to present the most popular and versatile methods. The reader is encouraged to consult the literature for additional information.

3.14 CONCLUSIONS

All of the methods discussed so far are really based on a transfer function representation. As such, they have been developed and advocated by engineers, particularly in the areas of control systems and signal processing. The methods that follow in the remaining chapters are primarily based on state space representations of systems and have been developed mostly by mathematicians and numerical analysts. As will be seen, the analysis of these methods is more systematic and their behavior is more predictable. However, much of the intuition associated with the transfer function representation is lacking in these state space–based methods. The intent of the presentation will be to put the engineering intuition into these numerical integration methods so that their understanding will be more apparent to the many people who have received considerable engineering training in linear system theory.

APPENDIX: FAMILIES OF SUBSTITUTION OPERATOR MATRICES

First Difference

$$Q_1 = \begin{bmatrix} T & 0 \\ 1 & -1 \end{bmatrix}$$

$$Q_2 = \begin{bmatrix} T^2 & 0 & 0 \\ T & -T & 0 \\ 1 & -2 & 1 \end{bmatrix}$$

$$Q_4 = \begin{bmatrix} T^3 & 0 & 0 & 0 \\ T^2 & -T^2 & 0 & 0 \\ T & -2T & T & 0 \\ 1 & -3 & 3 & -1 \end{bmatrix}$$

$$Q_4 = \begin{bmatrix} T^4 & 0 & 0 & 0 & 0 \\ T^3 & -T^3 & 0 & 0 & 0 \\ T^2 & -2T^2 & T^2 & 0 & 0 \\ T & -3T & 3T & -T & 0 \\ 1 & -4 & 6 & -4 & 1 \end{bmatrix}$$

Tustin

$$Q_1 = \begin{bmatrix} \dfrac{T}{2} & \dfrac{T}{2} \\ 1 & -1 \end{bmatrix}$$

$$Q_2 = \begin{bmatrix} \dfrac{T^2}{4} & \dfrac{2T^2}{4} & \dfrac{T^2}{4} \\ \dfrac{T}{2} & 0 & -\dfrac{T}{2} \\ 1 & -2 & 1 \end{bmatrix}$$

$$Q_3 = \begin{bmatrix} \dfrac{T^3}{8} & \dfrac{3T^3}{8} & \dfrac{3T^3}{8} & \dfrac{T^3}{8} \\ \dfrac{T^2}{4} & \dfrac{T^2}{4} & -\dfrac{T^2}{4} & -\dfrac{T^2}{4} \\ \dfrac{T}{2} & -\dfrac{T}{2} & -\dfrac{T}{2} & \dfrac{T}{2} \\ 1 & -3 & 3 & -1 \end{bmatrix}$$

$$Q_4 = \begin{bmatrix} \dfrac{T^4}{16} & \dfrac{4T^4}{16} & \dfrac{6T^4}{16} & \dfrac{4T^4}{16} & \dfrac{T^4}{16} \\ \dfrac{T^3}{8} & \dfrac{2T^3}{8} & 0 & -\dfrac{2T^3}{8} & -\dfrac{T^3}{8} \\ \dfrac{T^2}{4} & 0 & -\dfrac{2T^2}{4} & 0 & \dfrac{T^2}{4} \\ \dfrac{T}{2} & -T & 0 & T & -\dfrac{T}{2} \\ 1 & -4 & 6 & -4 & 1 \end{bmatrix}$$

Euler (Forward Difference)

$$Q_1 = \begin{bmatrix} 0 & T \\ 1 & -1 \end{bmatrix}$$

$$Q_2 = \begin{bmatrix} 0 & 0 & T^2 \\ 0 & T & -T \\ 1 & -2 & 1 \end{bmatrix}$$

$$Q_3 = \begin{bmatrix} 0 & 0 & 0 & T^3 \\ 0 & 0 & T^2 & -T^2 \\ 0 & T & -2T & T \\ 1 & -3 & 3 & -1 \end{bmatrix}$$

$$Q_4 = \begin{bmatrix} 0 & 0 & 0 & 0 & T^4 \\ 0 & 0 & 0 & T^3 & -T^3 \\ 0 & 0 & T^2 & -2T^2 & T^2 \\ 0 & T & -3T & 3T & -T \\ 1 & -4 & 6 & -4 & 1 \end{bmatrix}$$

Z-Transform Method

$$Q_1 = \begin{bmatrix} T & 0 \\ 1 & -1 \end{bmatrix}$$

$$Q_2 = \begin{bmatrix} T^2 & 0 & 0 \\ T & -T & 0 \\ 1 & -2 & 1 \end{bmatrix}$$

$$Q_3 = \begin{bmatrix} 0 & \dfrac{T^3}{2} & \dfrac{T^3}{2} & 0 \\ 0 & T^2 & -T^2 & 0 \\ T & -2T & T & 0 \\ 1 & -3 & 3 & -1 \end{bmatrix}$$

$$Q_4 = \begin{bmatrix} 0 & \dfrac{T^4}{6} & \dfrac{4T^4}{6} & \dfrac{T^4}{6} & 0 \\ 0 & \dfrac{T^3}{2} & 0 & -\dfrac{T^3}{2} & 0 \\ 0 & T^2 & -2T^2 & T^2 & 0 \\ T & -3T & 3T & -T & 0 \\ 1 & -4 & 6 & -4 & 1 \end{bmatrix}$$

Halijak

$$Q_1 = \begin{bmatrix} T & 0 \\ 1 & -1 \end{bmatrix}$$

$$Q_2 = \begin{bmatrix} T^2 & 0 & 0 \\ T & -T & 0 \\ 1 & -2 & 1 \end{bmatrix}$$

$$Q_3 = \begin{bmatrix} 0 & \dfrac{T^3}{2} & \dfrac{T^3}{2} & 0 \\ 0 & T^2 & -T^2 & 0 \\ T & -2T & T & 0 \\ 1 & -3 & 3 & -1 \end{bmatrix}$$

$$Q_4 = \begin{bmatrix} 0 & \dfrac{T^4}{4} & \dfrac{2T^4}{4} & \dfrac{T^4}{4} & 0 \\ 0 & \dfrac{T^3}{2} & 0 & -\dfrac{T^3}{2} & 0 \\ 0 & T^2 & -2T^2 & T^2 & 0 \\ T & -3T & 3T & -T & 0 \\ 1 & -4 & 6 & -4 & 1 \end{bmatrix}$$

Madwed

$$Q_1 = \begin{bmatrix} \dfrac{T}{2} & \dfrac{T}{2} \\ 1 & -1 \end{bmatrix}$$

$$Q_2 = \begin{bmatrix} \dfrac{T^2}{6} & \dfrac{4T^2}{6} & \dfrac{T^2}{6} \\ \dfrac{T}{2} & 0 & -\dfrac{T}{2} \\ 1 & -2 & 1 \end{bmatrix}$$

$$Q_3 = \begin{bmatrix} \dfrac{T^3}{24} & \dfrac{11T^3}{24} & \dfrac{11T^3}{24} & \dfrac{T^3}{24} \\ \dfrac{T^2}{6} & \dfrac{3T^2}{6} & -\dfrac{3T^2}{6} & -\dfrac{T^2}{6} \\ \dfrac{T}{2} & -\dfrac{T}{2} & -\dfrac{T}{2} & \dfrac{T}{2} \\ 1 & -3 & 3 & -1 \end{bmatrix}$$

$$Q_4 = \begin{bmatrix} \dfrac{T^4}{120} & \dfrac{26T^4}{120} & \dfrac{66T^4}{120} & \dfrac{26T^4}{120} & \dfrac{T^4}{120} \\[2ex] \dfrac{T^3}{24} & \dfrac{10T^3}{24} & 0 & -\dfrac{10T^3}{24} & -\dfrac{T^3}{24} \\[2ex] \dfrac{T^2}{6} & \dfrac{2T^2}{6} & -T^2 & \dfrac{2T^2}{6} & \dfrac{T^2}{6} \\[2ex] \dfrac{T}{2} & -T & 0 & T & -\dfrac{T}{2} \\[2ex] 1 & -4 & 6 & -4 & 1 \end{bmatrix}$$

Boxer-Thaler

$$Q_1 = \begin{bmatrix} \dfrac{T}{2} & \dfrac{T}{2} \\[2ex] 1 & -1 \end{bmatrix}$$

$$Q_2 = \begin{bmatrix} \dfrac{T^2}{12} & \dfrac{10T^2}{12} & \dfrac{T^2}{12} \\[2ex] \dfrac{T}{2} & 0 & -\dfrac{T}{2} \\[2ex] 1 & -2 & 1 \end{bmatrix}$$

$$Q_3 = \begin{bmatrix} 0 & \dfrac{T^3}{2} & \dfrac{T^3}{2} & 0 \\[2ex] \dfrac{T^2}{12} & \dfrac{9T^2}{12} & -\dfrac{9T^2}{12} & -\dfrac{T^2}{12} \\[2ex] \dfrac{T}{2} & -\dfrac{T}{2} & -\dfrac{T}{2} & \dfrac{T}{2} \\[2ex] 1 & -3 & 3 & -1 \end{bmatrix}$$

$$Q_4 = \begin{bmatrix} -\dfrac{T^4}{720} & \dfrac{124T^4}{720} & \dfrac{474T^4}{720} & \dfrac{124T^4}{720} & -\dfrac{T^4}{720} \\[2ex] 0 & \dfrac{T^3}{2} & 0 & -\dfrac{T^3}{2} & 0 \\[2ex] \dfrac{T^2}{12} & \dfrac{8T^2}{12} & -\dfrac{18T^2}{12} & \dfrac{8T^2}{12} & \dfrac{T^2}{12} \\[2ex] \dfrac{T}{2} & -T & 0 & T & -\dfrac{T}{2} \\[2ex] 1 & -4 & 6 & -4 & 1 \end{bmatrix}$$

Adams-Bashforth Second Order

$$Q_1 = \begin{bmatrix} 0 & \dfrac{3T}{2} & -\dfrac{T}{2} \\[2mm] 1 & -1 & 0 \end{bmatrix}$$

$$Q_2 = \begin{bmatrix} 0 & 0 & \dfrac{9T^2}{4} & -\dfrac{6T^2}{4} & \dfrac{T^2}{4} \\[3mm] 0 & \dfrac{3T}{2} & -\dfrac{4T}{2} & \dfrac{T}{2} & 0 \\[3mm] 1 & -2 & 1 & 0 & 0 \end{bmatrix}$$

$$Q_3 = \begin{bmatrix} 0 & 0 & 0 & \dfrac{27T^3}{8} & -\dfrac{27T^3}{8} & \dfrac{9T^3}{8} & -\dfrac{T^3}{8} \\[3mm] 0 & 0 & \dfrac{9T^2}{4} & -\dfrac{15T^2}{4} & \dfrac{7T^2}{4} & -\dfrac{T^2}{4} & 0 \\[3mm] 0 & \dfrac{3T}{2} & -\dfrac{7T}{2} & \dfrac{5T}{2} & -\dfrac{T}{2} & 0 & 0 \\[3mm] 1 & -3 & 3 & -1 & 0 & 0 & 0 \end{bmatrix}$$

$$Q_4 = \begin{bmatrix} 0 & 0 & 0 & 0 & \dfrac{81T^4}{16} & -\dfrac{108T^4}{16} & \dfrac{54T^4}{16} & -\dfrac{12T^4}{16} & \dfrac{T^4}{16} \\[3mm] 0 & 0 & 0 & \dfrac{27T^3}{8} & -\dfrac{54T^3}{8} & \dfrac{36T^3}{8} & -\dfrac{10T^3}{8} & \dfrac{T^3}{8} & 0 \\[3mm] 0 & 0 & \dfrac{9T^2}{4} & -\dfrac{24T^2}{4} & \dfrac{22T^2}{4} & -\dfrac{8T^2}{4} & \dfrac{T^2}{4} & 0 & 0 \\[3mm] 0 & \dfrac{3T}{2} & -\dfrac{10T}{2} & \dfrac{12T}{2} & -\dfrac{6T}{2} & \dfrac{T}{2} & 0 & 0 & 0 \\[3mm] 1 & -4 & 6 & -4 & 1 & 0 & 0 & 0 & 0 \end{bmatrix}$$

PROBLEMS

1. You are given the following system: With the system initially at rest, except $x(0) = 0.2$, simulate the system by using Halijak's substitution method for $T = 0.16$. Discuss.

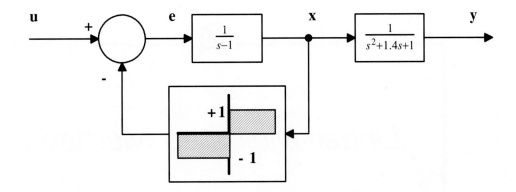

2. What is the Taylor series expansion for e^x.

3. Consider $H(s) = 1/(s^2 + 1)$. Find the exact mapping of the s-plane poles into the z-plane as a function of T. Also, sketch the root locus as a function of T when using the first difference and the Halijak substitution methods. Which is best?

4. Determine a discrete integrator to substitute for s in the transfer function $H(s) = 1/\sinh(\sqrt{s})$. This nonrational transfer function arises in many thermal problems. You should end up with a nonrational function of z. (Hint: Where are the system poles?)

5. Given the second-order system $(s + 1)/(s^2 + s + 10)$
   (a) Find a discrete representation using the z-transform substitution method.
   (b) Plot the z-plane poles and zeros of the discrete representation as a function of T. Compare to what the directly mapped poles and zeros do.
   (c) Map the approximate z-plane poles and zeros back into the s-plane, using the exact mapping as a function of T. Compare to that which the actual poles and zeros do.
   (d) Based on the above, suggest a range for T.
   (e) Analytically determine the response to a unit step input, using your value of T.
   (f) Simulate part e.

4

Linear Multistep Methods

4.1 INTRODUCTION

The substitution methods of the previous chapter are very convenient for quickly discretizing low-order transfer functions. However, it should be pretty clear by now that they generally lack mathematical rigor, particularly when we are attempting to determine stability and accuracy as a function of the simulation timestep. Additionally, although they eliminate much tedious work by eliminating partial fraction expansions, they still require a significant amount of algebra. The methods to be presented in this chapter avoid both of these problems. Solid mathematical analysis tools will be derived where stability and accuracy as a function of simulation timestep will be clearly discernible. In addition, the methods will require little, if any, preliminary algebraic manipulation. The only requirement will be that the system is in a state space representation first. The result will be an easily implementable and very dependable discretization approach. The linear multistep methods are based on discretizing only the integration process, not the entire system. This is in contrast to the methods presented in the previous chapters, where the entire transfer function was discretized. These methods can thus be applied directly to nonlinear and time-varying systems.

4.2 INITIAL-VALUE PROBLEMS

The approach taken for initial-value problems is first to put the given system into a state space form and then simulate the resulting state equations. This is the approach advocated by applied mathematicians and is theoretically more developed than the previous material. Basically we are obtaining the solution to initial-value problems numerically in ordinary differential equations. The development somewhat parallels that of Lambert (1973), which should probably be considered the standard reference on this material. An unique control theory approach is taken throughout this discussion, however.

The general initial-value problem to be solved is defined as follows;

$$\dot{x} = f(x,u,t), \ x(0) \text{ given} \tag{4-1}$$

with any system outputs given by the nonlinear function

$$y = g(x,u,t) \tag{4-2}$$

Here x is an n-dimensional state vector, f is a vector of generally nonlinear derivative functions, u is a vector of time-dependent input functions, and t is time. The form of this system that most control engineers are familiar with is the usual linear state space representation (D'Azzo & Houpis, 1981)

$$\dot{x} = Ax + Bu, \ y = Cx + Du \tag{4-3}$$

where y is a vector of output functions, and the matrices $A, B, C,$ and D are of appropriate dimension and constant for time-invariant systems.

It will be assumed that the system of equation 4–1 obeys a Lipschitz condition; that is, its time derivative has a finite bound, L, during the time period of the simulation. The following is a definition of the Lipschitz constant (Lambert, 1973),

$$L = \sqrt{\lambda \left\{ \left(\frac{\partial f(x,u,t)}{\partial x} \right)^T \left(\frac{\partial f(x,u,t)}{\partial x} \right) \right\}_{max}} \tag{4-4}$$

which is the largest singular value of the system Jacobian matrix over the time period of the simulation. This is also an upper bound on the absolute value of the largest eigenvalue of the system Jacobian matrix in this interval. If L is finite, we can expect the system to have a solution. We will not normally be concerned with considerations of existence and uniqueness, however, as most engineering problems do indeed have a unique solution.

4.3 THE INTEGRATION PROBLEM

The very useful and appealing linear multistep methods are presented here. This chapter and the next will deal with their stability and accuracy. These methods are probably the most useful, accurate, and easy to program of all the methods that will be presented. A fundamental understanding of their properties is also essential to understanding the numerical integration process in general.

First consider the general simulation problem for a state space system. This can be represented diagrammatically, as in Figure 4.1. The problem is to approximate the integral with some discrete dynamic system, as in Figure 4.2, where $H(z)$ is a discrete-time transfer function. For the open-loop situation, the derivative function does not depend on the state; that is, $\dot{x} = f(u,t)$. For the closed-loop situation, the derivative of the state depends on the current state value; that is, $\dot{x} = f(x,u,t)$. The idea is to choose $H(z)$ so that it most accurately approximates an integrator, in both open- and closed-loop situations. It can also be recognized that this is similar to a controls problem, in which in this case the compensator to be designed goes after the system. The problem is further complicated by requiring the output not only to be stable (the usual first controls requirement) but also to be an accurate representation of the original continuous system. Various criteria will be used to achieve this in what follows. Since $H(z)$ is already a z-domain transfer function, it will be studied by using the usual z-transform analysis.

4.4 INTRODUCTION TO LINEAR MULTISTEP METHODS

The general linear k-step method is defined as (Dahlquist, 1956, 1963; Henrici, 1962; Lambert, 1973)

$$\alpha_k x_{n+k} = -\alpha_{k-1}x_{n+k-1} - \ldots - \alpha_0 x_n + T[\beta_k \dot{x}_{n+k} + \beta_{k-1}\dot{x}_{n+k-1} + \ldots + \beta_0\dot{x}_n] \quad (4\text{–}5)$$

or equivalently

$$\sum_{j=0}^{k} \alpha_j x_{n+j} = T \sum_{j=0}^{k} \beta_j \dot{x}_{n+j} \quad (4\text{–}6)$$

where $\dot{x}$ is the generally nonlinear function of x, u, and t given above and is indeed a separate signal. Taking the z-transform of both sides of the above equation, assuming initial conditions are zero, we get as the transfer function for the integrator

$$H(z) = \frac{X(z)}{\dot{X}(z)} = \frac{T(\beta_k z^k + \beta_{k-1}z^{k-1} + \ldots + \beta_1 z + \beta_0)}{(\alpha_k z^k + \alpha_{k-1}z^{k-1} + \ldots + \alpha_1 z + \alpha_0)} = \frac{T\sigma(z)}{\rho(z)} \quad (4\text{–}7)$$

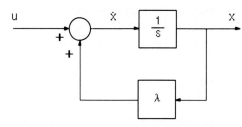

CONTINUOUS - TIME INTEGRATOR
WITH LINEARIZED SYSTEM

Figure 4.1 Exact representation of a dynamic system in feedback form.

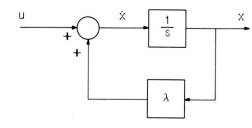

DISCRETE - TIME INTEGRATOR   **Figure 4.2**   The numerical integration prob-
WITH LINEARIZED SYSTEM   lem as a closed-loop system

Here the polynomial $\rho(z)$ is defined as the *first characteristic polynomial* and obviously gives information about the open-loop stability of the integrator, that is, its open-loop poles. The polynomial $\sigma(z)$ is defined as the *second characteristic polynomial* and gives information about the open-loop system zeros. Also, usually $\alpha_k = 1$ to eliminate the arbitrariness of the transfer function, that is, to pin down the coefficients to this reference normalization.

The integration method is called *implicit* if $\beta_k \neq 0$ (Ralston & Rabinowitz, 1978) because the output, x at time $n + k$, is a function of $\dot{x}$ at time $n + k$, which is itself a function of x at time $n + k$. Thus x at time $n + k$ is implicitly a function of itself. This problem was encountered in Chapter 3 when feedback loops required the addition of a delay to have a closed-loop realizable implementation. This ad hoc solution will no longer be acceptable, as discussed later. If $\beta_k = 0$, the method is called *explicit* since the output x at time $n + k$ is explicitly a function of things that have already occurred in the past. Clearly, methods of this type will be preferred because of their computational ease.

It should now be noted that this approach is different from the operational approach (Rosko, 1972). Previously we needed approximations to $(1/s)^n$. Now we need approximations to just $1/s$ since we have a first-order state space system to simulate. However, the approximation to $1/s$ will have k-poles rather than just one pole, which the operational methods had. These extra poles can either improve or degrade accuracy depending on the simulation timestep. This point will be considered at length later.

To illustrate all of the above, we present the following example.

Example 4.1

Consider the trapezoidal rule, $\dfrac{1}{s} \Leftrightarrow \dfrac{T}{2}\left(\dfrac{z+1}{z-1}\right)$, which gives the time domain equation

$$x_{n+1} = x_n + \frac{T}{2}\,[\dot{x}_{n+1} + \dot{x}_n]$$

This is an implicit method, with $k = 1$, that is, an implicit linear one-step method. Then $\beta_0 = 1/2$, $\beta_1 = 1/2$, $\beta_{k>1} = 0$, $\alpha_0 = -1$, $\alpha_1 = 1$, $\alpha_{k>1} = 0$.

One of the great benefits of using linear multistep methods, as opposed to data holds or substitution methods, is the ease of implementation. To demonstrate this, several examples now follow.

Example 4.2

Here the first-order system used in the previous chapters is considered again:

$$G(s) = \frac{2}{s+2}$$

Rather than keeping this transfer function form, it is first necessary to transform the system into state space. For a first-order system, there is only one state, and thus

$$\dot{x} = -2x + 2u$$

This equation is easily simulated by using any linear multistep method. Euler's method is considered first. In linear multistep form, Euler's method is

$$x_{n+1} = x_n + T\dot{x}_n$$

Note here that x is now the system state vector and $\dot{x}$ is the vector of derivatives of the system, or derivative functions.

There are basically two approaches to obtaining the discrete-time system by using linear multistep methods. One is analytical, and the other is more practical. The analytical approach gives a discrete-time state space representation. Here the analytic expression for $\dot{x}$ is substituted directly into the linear multistep method. For a general linear state space system

$$\dot{x} = Ax + Bu$$

This becomes

$$x_{n+1} = x_n + T[Ax_n + Bu_n] = [I + TA]x_n + TBu_n$$

For the given scalar example system this is

$$x_{n+1} = x_n + T[-2x_n + 2u_n] = [1 - 2T]x_n + 2Tu_n$$

Thus we are left with a useful analytical expression for the discrete-time system.

The more practical approach to linear multistep methods is simply to combine the integrator and the derivative function evaluation in a loop. A simple BASIC language code is listed here to demonstrate the comparable simplicity of this method.

```
T = 0.1                          Simulation timestep
x = 0                            I.C.'s
u = 1                            Step input
for n = 1 to 100                 Ten seconds real time
        f = -2*x + 2*u           Function evaluation
        x = x + T*f              Integrate for next x
        pset(n, 200 - 100*x)     Plotting
    next n
    end
```

Several practical points should now be made. One is that an array is not necessary for storing the time vector in the simulation. Many simulations do not use arrays to store x and f, as it usually is not practical to store that much data, particularly for a real-time simulation. Usually it is sufficient to "store" these data in a plot of the simulation output or in a file. Another point of some confusion is whether or not to do the first derivative evaluation inside or outside the loop. In the above program it has been done inside the loop. The alternative, however, requires a rearrangement of the loop. In light of this discussion, the program is given again, using the alternatives discussed.

```
dimension x(101),f(101),u(101)              Initialize arrays
T = 0.1                                     Simulation timestep
x(1) = 0                                    Initial condition
u(1) = 1                                    Initial input
f(1) = -2*x(1) + 2*u(1)                     Initial derivative
for n = 1 to 100                            Ten seconds real time
        u(n + 1) = 1                        Step input
        x(n + 1) = x(n) + T*f(n)            Integrate next x
        f(n + 1) = -2*x(n + 1) + 2*u(n + 1) Get next derivative
next n
end
```

However you decide to program, it is important to be consistent in your approach, so that you may learn to find your errors, and also to be structural, so that all the derivative evaluations go together and all the integrations go together.

Example 4.3

Here a second-order oscillatory system is considered:

$$G(s) = \frac{2}{s^2 + 2s + 2}$$

One state space representation for this system is

$$\dot{x}_1 = x_2$$
$$\dot{x}_2 = -2x_1 - 2x_2 + 2u$$
$$y = x_1$$

Simulating this with Euler's method, we get this BASIC program:

```
T = 0.05                        Simulation timestep
x1 = 0                          Initial condition
x2 = 0                          Initial condition
u = 1                           Step input
for n = 1 to 500                25 seconds real time
        f1 = x2                 Function evaluation
        f2 = -2*x1 - 2*x2 + 2*u Function evaluation
        x1 = x1 + T*f1          Integrate state 1
```

```
                    x2 = x2 + T*f2                Integrate state 2
                    pset(n,200 - 100*x1)          Plot x1
                    pset(n,200 - 100*x2)          Plot x2
              next n
              end
```

Example 4.4

It should now be clear that linear multistep methods are particularly easy to implement. This is demonstrated further by this example, in which a nonlinear, third-order jet engine system given in state space form is simulated by using Euler's method (Hartley & Beale, 1987). The BASIC program for this simulation follows.

```
1 SCREEN 2
2 CLS
20 T  = 0.002
49 REM Initial Conditions
50 P4  =  0.53831
51 PB  =  1.77504
52 N  =  0.54589
90 LINE (0,0)-(0,199)
91 LINE (0,199)-(640,199)
94 REM ** THIRD-ORDER BRENNAN & LEAKE ENGINE MODEL
95 REM P4 = combustor pressure, PB = combustor density, N = rotor
96 REM speed, W3D =  compressor discharge mass flow, T3  =
97 REM compressor discharge temp.
98 REM System's eigenvalues linearized about WFD = 1 are -81, -32,
99 REM -2.8. WFD = fuel input mass rate. WFD = 1.
100 For K  =  1 to 1000
101 REM Algebraic Relations
102 WFD  =  1
103 T3  =  0.64212 + 0.35788 * N * N
104 W3D  =  1.3009*N-0.139825*P4-0.13982*SQR(P4*P4
+0.41688*N*N - 0.0899*P4*N
109 REM Dynamic Relations
110 FP4  =  (0.93586*P4/PB+31.486)*WFD+21.435*W3D*T3
-53.86*P4*P4/PB
120 FPB  =  37.78*W3D-38.448*P4+0.66849*WFD
130 FRN  =  1.258*(P4*P4/PB-W3D*N*N)/N
149 REM Numerical Integration: Euler's Method
150 P4  =  P4 + T * FP4
160 PB  =  PB + T * FPB
170 N  =  N + T * FRN
189 REM Output Graphics
190 PSET(K,199-100*P4)
191 PSET(K,199-100*PB)
192 PSET(K,199-100*N)
199 NEXT K
999 END
```

Notice that almost no off-line work is necessary before implementation once the system equations became available. All that is required beyond choosing the method (Euler's) and the simulation timestep is the programming of the simulation. The programming of linear multistep methods usually proceeds as demonstrated here; that is, variables are initialized outside the main loop, the system equations (derivatives) are evaluated as a block in the loop, the numerical integration method is performed as a block also in the main loop, and then some type of data storage is accomplished. This is the real benefit of linear multistep methods and also the justification of the development of a clear analytical theory.

4.5 DERIVATION OF LINEAR MULTISTEP METHODS WITH TAYLOR SERIES

The question that now arises is this: How does one derive these methods in order to have good accuracy properties? Many texts, such as Lambert (1973) or Lapidus and Seinfeld (1971) go into great detail and give several approaches. We will consider Taylor series expansion exclusively because of its simplicity of solution and numerical insightfulness. First we give a simple introduction to the approach and define some important terms. A more complete presentation then follows.

The Taylor series expansion for x at time $n + 1$ in terms of the function x and its derivatives at time n is

$$x_{n+1} = x_n + T\dot{x}_n + \frac{T^2}{2!}\ddot{x}_n + \ldots + \frac{T^m}{m!}x_n^{(m)} \ldots \tag{4-8}$$

If this series is truncated after two terms on the right side, Euler's method is the result:

$$x_{n+1} = x_n + T\dot{x}_n, \text{ with } \alpha_1 = 1, \alpha_0 = -1, \beta_0 = 1 \tag{4-9}$$

The *local truncation error* (LTE) can now be defined as the error incurred in taking one timestep, assuming no numerical imprecision. For Euler's method this becomes (Lambert, 1973)

$$\text{LTE} = \frac{T^2}{2}\ddot{x}_n + \ldots$$

The leading term in this series is the *principal local truncation error,* and the numerical coefficient in it is usually called the *error constant,* which is 1/2 for Euler's method. Clearly all of this should be minimized. It should be recognized also that for Euler's method

$$\text{LTE} = x_{n+1} - x_n - T\dot{x}_n$$

This definition of the local truncation error assumes that all values used to compute x_{n+1} are exactly equal to their theoretical values and that infinite precision is available for the calculations. Therefore, the only error that is being considered is the approximation of the integration process by $H(z)$.

By using this Taylor series expansion, any method can be derived, as is demonstrated below for a simple integrator. Following this apparently tedious design process, a general formula is given that will greatly simplify integrator design.

We will now derive the most accurate implicit one-step method by using this Taylor series approach. It will clearly demonstrate the procedure involved and prepare the reader for the general solution that follows. An implicit one-step method is generally written

$$\alpha_1 x_{n+1} + \alpha_0 x_n = T(\beta_1 \dot{x}_{n+1} + \beta_0 \dot{x}_n) \tag{4-10}$$

The problem is to determine the coefficients. We already know $\alpha_1 \equiv 1$; thus there are three coefficients left to be determined. These will be found by using Taylor series expansions. First, we set

$$x_{n+1} = x_n + T\dot{x}_n + \frac{T^2}{2!}\ddot{x}_n + \dots$$

and

$$\dot{x}_{n+1} = \dot{x}_n + T\ddot{x}_n + \frac{T^2}{2!}x^{(3)}{}_n + \dots$$

Then, moving everything in the integrator equation to the left side and substituting the above Taylor series expansions, we get

$$\left[x_n + T\dot{x}_n + \frac{T^2}{2}\ddot{x}_n + \frac{T^3}{6}x_n^{(3)} + \dots \right] + \alpha_0 x_n$$

$$- T\beta_1 \left[\dot{x}_n + T\ddot{x}_n + \frac{T^2}{2}x_n^{(3)} + \dots \right] - \beta_0 \dot{x}_n = \text{LTE} \tag{4-11}$$

To minimize the LTE, we want to set as many as possible of the leading terms of this expression to zero. It can also be rewritten as (Lambert, 1973)

$$\text{LTE} = C_0 x_n + C_1 T\dot{x}_n + C_2 T^2 \ddot{x}_n + C_3 T^3 x_n^{(3)} + \dots \cong 0 \tag{4-12}$$

This expression is effectively the performance measure that is to be minimized. Since the higher-order terms have higher powers of T, and T is assumed to be small, the LTE is minimized by setting as many of the leading terms to zero as possible. The C-coefficients can now be written in terms of the α's and β's from inspection:

$$C_0 = 1 + \alpha_0$$
$$C_1 = 1 - \beta_1 - \beta_0$$
$$C_2 = \frac{1}{2} - \beta_1$$
$$C_3 = \frac{1}{6} - \frac{1}{2}\beta_1$$

Since our goal is to minimize the LTE, and since we have only three unknowns, the first three of these equations will be set to zero to solve for the three unknowns. The remaining term will be the leading term of the LTE. Solving for the unknowns, we get

$$\alpha_1 \equiv 1$$

$$C_0 = 1 + \alpha_0 \equiv 0, \text{ gives } \alpha_0 = -1$$

$$C_2 = \frac{1}{2} - \beta_1 \equiv 0, \text{ gives } \beta_1 = \frac{1}{2}$$

$$C_1 = 1 - \beta_1 - \beta_0 \equiv 0, \text{ and } \beta_1 \text{ from above gives } \beta_0 = \frac{1}{2}$$

Thus

$$H(z) = \frac{T}{2}\left(\frac{z+1}{z-1}\right) \tag{4-13}$$

which is easily recognizable as the trapezoidal rule, or Tustin's method. It is interesting that this method has fallen out of four completely different derivations. This fact speaks highly of its general utility. Even more useful properties will follow later.

We can now compute the error constant as

$$C_3 = \frac{1}{6} - \frac{1}{2}\left(\frac{1}{2}\right) = \frac{-1}{12}$$

The principal local truncation error, which is most often given by the leading term in the remaining Taylor series, is

$$\text{LTE} \cong \frac{-T^3 x^{(3)}}{12}$$

This example clearly shows both the process involved in deriving methods and also the tedium involved. Fortunately, a general solution to this problem is available and is given below. These equations will be derived in Chapter 5 based on stability region considerations. Remember,

$$\text{LTE} = \sum_{j=0}^{k} [\alpha_j x_{n+j} - T\beta_j \dot{x}_{n+j}] = C_0 x_n + C_1 T \dot{x}_n + C_2 T^2 \ddot{x}_n \ldots = \sum_{i=0}^{\infty} C_i T^i x_n^{(i)} \tag{4-14}$$

The linear multistep method is defined to have the *order of accuracy*, p, if

$$C_0 = C_1 = C_2 = \ldots = C_p = 0, \text{ and } C_{p+1} \neq 0 \tag{4-15}$$

For example, Euler's method is first-order accurate ($C_2 = 1/2$), and the trapezoidal rule is second-order accurate ($C_3 = -1/12$). The relationships between the C's and the α's and β's are now given and will be referred to as Lambert's equations (Lambert, 1973):

$$C_0 = \alpha_0 + \alpha_1 + \alpha_2 + \ldots + \alpha_k \text{ or equivalently } [\rho(z)]_{z \equiv 1}$$

$$C_1 = \alpha_1 + 2\alpha_2 + \ldots + k\alpha_k - (\beta_0 + \beta_1 + \ldots + \beta_k)$$

$$\text{or equivalently } \left[\frac{d\rho(z)}{dz} - \sigma(z)\right]_{z \equiv 1} \tag{4-16}$$

$$C_q = \frac{1}{q!}(\alpha_1 + 2^q\alpha_2 + \ldots + k^q\alpha_k) - \frac{1}{(q-1)!}(\beta_1 + 2^{q-1}\beta_2 + \ldots + k^{q-1}\beta_k)$$

By using these equations, along with any other constraints, we can derive linear multistep methods with the order of accuracy apparently limited only by the need to have as many coefficients as equations ($2k \geq p + 1$ for explicit methods and $2k \geq p$ for implicit methods) as possible.

4.6 CONTROL THEORY INSIGHT

At this point, it will be enlightening to consider the integration process from a control theory viewpoint (Beale & Hartley, 1987). This will yield some useful insights into the problem.

The integration process can be considered operationally as follows:

$$X(s) = \frac{1}{s} \dot{X}(s) \tag{4-17}$$

We will be able to obtain fundamental properties of the discrete integrator by comparing its impulse response with that of the continuous integrator. Assume a Dirac delta function input of area b is applied at a time equal to zero to the above system; that is,

$$\dot{x}(t) = b\delta(t), \text{ which gives } \dot{X}(s) = b \tag{4-18}$$

It should be noted that the derivative of x is considered to be a signal itself; thus it is not given as $sX(s)$. This will be more useful now. The response of the system for this input is then

$$X(s) = \frac{b}{s} \tag{4-19}$$

One useful error measure used in control theory is the steady state error. This will be compared for the continuous and discrete integrators in what follows. Applying the continuous-time final value theorem, we get

$$x(\infty) = \lim_{s \to 0} sX(s) = \lim_{s \to 0} s\,\frac{b}{s}$$

or

$$x(\infty) = b \tag{4-20}$$

In discrete time, the integration process is given by

$$X(z) = H(z)\dot{X}(z) = \frac{T\sigma(z)}{\rho(z)} \dot{X}(z) \tag{4-21}$$

The derivative input is now considered to be a discrete-time impulse. These are a little different from continuous-time impulses, and care must be taken when comparing the two responses. In discrete time, an impulse of height b is effectively held at that level for the entire sampling period. Thus the area under the discrete impulse, which is how the Dirac delta function is defined, is its height b times its width T, or bT. To give it an area of b, as in the continuous case, its height must be scaled to become b/T. Thus the steady state response to this scaled impulse becomes

$$x_\infty = \lim_{z \to 1} (z - 1) \frac{T\sigma(z)}{\rho(z)} \frac{b}{T} \qquad (4\text{--}22)$$

It is desirable to have this value be b so that the discrete integrator behaves like the continuous integrator. Clearly this will not be possible unless $(z - 1)$ is a factor of $\rho(z)$. Otherwise, the final value would be zero. Thus, the following is a requirement on the coefficients of the integrator.

Requirement 1
For the discrete integrator to behave like the continuous integrator, $(z - 1)$ must be a factor of the denominator; that is,

$$\rho(z) = (z - 1)\rho_f(z) \qquad (4\text{--}23)$$

where $\rho_f(z)$ is the polynomial remaining after $(z - 1)$ is factored out. This requirement has a further meaning. Since $z = +1$ is now a root of $\rho(z)$, this means that

$$\rho(1) = C_0 = 0 \qquad (4\text{--}24)$$

which is identical to the first equation obtained from Lambert's equations and which was also observed for the substitution methods of Chapter 3. This is an obviously desirable property for a discrete integrator.

Requirement 2
A second requirement also comes from the definition of the final value theorem. To apply that theorem, $\rho(z)$ must have all of its roots in a stable configuration. These two requirements together are usually referred to as *zero stability*.

We now have

$$x_\infty = \lim_{z \to 1} (z - 1) \frac{T\sigma(z)}{(z - 1)\rho_f(z)} \frac{b}{T} = \lim_{z \to 1} \frac{b\sigma(z)}{\rho_f(z)} \qquad (4\text{--}25)$$

Thus, for this final value to be b, the following is necessary.

Requirement 3
$$\text{For } x_\infty = b \qquad (4\text{--}26)$$
$$\rho_f(1) = \sigma(1)$$

This is similar to the C_1 equation from the Taylor series approach above. In fact, about a page of algebra will show that these expressions are equivalent, specifically

$$\rho_f(1) = \left[\frac{d\rho(z)}{dz} \right]_{z=1} = \sigma(1) \qquad (4\text{--}27)$$

This requirement is usually referred to as *consistency*.

With these three requirements, we now have the equivalence of the continuous and discrete integrators in steady state. It is fortunate that these are the first two equations that must be solved (set equal to zero) from Lambert's equations. Any further equations set equal to zero will merely improve accuracy, as will be shown in Chapter 5. The three requirements are necessary for a discrete integrator to perform similarly to a continuous integrator. These requirements were also known to the numerical analysts, and a complementary discussion can be found in Henrici (1962).

4.7 INTRODUCTION TO THE ADAMS FAMILY

It is now appropriate to introduce a very popular method, the Adams-Bashforth second-order integrator, or AB-2.

$$H(z) = \frac{\frac{T}{2}(3z-1)}{z(z-1)} \tag{4-28}$$

This method can be obtained from the Taylor series equations as the most accurate two-step explicit method, with one pole constrained to be at the origin. The reader should verify this statement. The most accurate two-step explicit method with no pole constraints is Hermite's method, which is unstable and thus not acceptable:

$$H(z) = \frac{T(4z+2)}{z^2 + 4z - 5} \tag{4-29}$$

In fact, the entire family of Adams-Bashforth methods can be derived as the most accurate k-step explicit methods with $(k-1)$-poles constrained to be at zero. Likewise, the entire family of Adams-Moulton methods can be derived as the most accurate k-step implicit methods with $(k-1)$-poles constrained to be at zero.

Returning to the AB-2, we can see that

$$\alpha_0 = 0,\ \alpha_1 = -1,\ \alpha_2 = 1,\ \beta_0 = -\frac{1}{2},\ \beta_1 = \frac{3}{2},\ \beta_2 = 0$$

Then

$$C_0 = \alpha_0 + \alpha_1 + \alpha_2 = 0$$
$$C_1 = \alpha_1 + 2\alpha_2 - \beta_0 - \beta_1 = 0$$
$$C_2 = \frac{1}{2}(\alpha_1 + 4\alpha_2) - \beta_1 = 0$$
$$C_3 = \frac{1}{6}(\alpha_1 + 8\alpha_2) - \frac{1}{2}\beta_1 = \frac{5}{12}$$

A pole-zero plot is also shown in Figure 4.3. A frequency response is obtained by replacing $z = e^{j\omega T}$, and it is given in Bode and Nyquist forms in Figures 4–4 and 4–5,

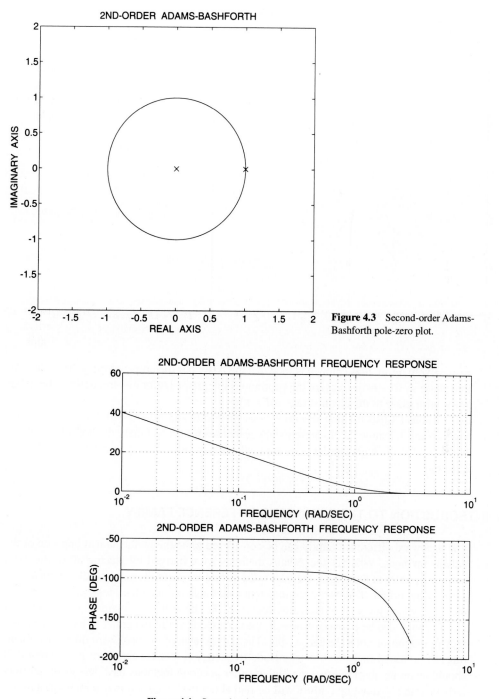

Figure 4.3 Second-order Adams-Bashforth pole-zero plot.

Figure 4.4 Second-order Adams-Bashforth frequency response.

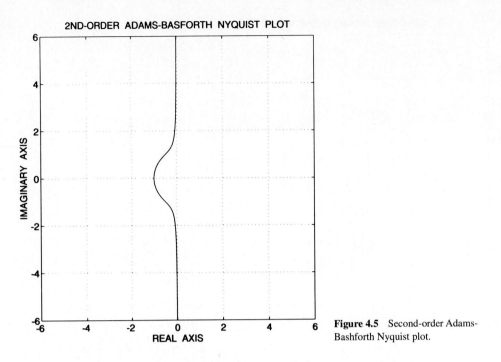

Figure 4.5   Second-order Adams-Bashforth Nyquist plot.

respectively. From these plots it is seen that the AB-2 has the proper slope (-20 db/decade) and phase shift (-90 degrees) of the continuous integrator $1/s$ for low frequencies, but it has some error close to the Nyquist rate ($\omega T \approx \pi$).

Table 4.1 lists the first several Adams-Bashforth and Adams-Moulton methods. Figures 4.6 and 4.7 show the Bode frequency responses for each of these methods. It should be observed that the higher-order accurate methods have less error, particularly near the Nyquist rate.

4.8 INTRODUCTION TO THE BACKWARD DIFFERENCE FAMILY

The backward difference family can be considered to be the mathematical opposite of the Adams methods. Whereas the Adams methods were derived by placing all of the extra (spurious) poles at the origin, the backward difference methods place all of the open-loop zeros at the origin. The backward difference methods are then the most accurate implicit methods, which have all of the zeros constrained to be at the z-domain origin. An explicit family could also be derived; however, these methods are found either to have poor accuracy properties or not to be zero stable. The coefficients of the backward difference methods can be found in Table 4.2. The Bode frequency responses are given in Figure 4.8. It should again be observed that the higher-order accurate methods have less error, particularly near the Nyquist rate. More will be said about backward difference methods and implicit methods in general later.

TABLE 4.1   NUMERICAL INTEGRATION ALGORITHMS

	Adams-Bashforth Predictors	
Order of Method— Number of Steps	Algorithm	Error
1–1	$x_{n+1} = x_n + Tf_n$ (Euler's method)	$\dfrac{T^2 x^{(2)}}{2}$
2–2	$x_{n+2} = x_{n+1} + \dfrac{T}{2}[3f_{n+1} - f_n]$	$\dfrac{5T^3 x^{(3)}}{12}$
3–3	$x_{n+3} = x_{n+2} + \dfrac{T}{12}[23f_{n+2} - 16f_{n+1} + 5f_n]$	$\dfrac{3T^4 x^{(4)}}{8}$
4–4	$x_{n+4} = x_{n+3} + \dfrac{T}{24}[55f_{n+3} - 59f_{n+2} + 37f_{n+1} - 9f_n]$	$\dfrac{251T^5 x^{(5)}}{720}$

	Adams-Moulton Correctors	
Order of Method— Number of Steps	Algorithm	Error
1–1	$x_{n+1} = x_n + Tf_{n+1}$	$\dfrac{-T^2 x^{(2)}}{2}$
2–1	$x_{n+1} = x_n + \dfrac{T}{2}[f_{n+1} + f_n]$	$\dfrac{-T^3 x^{(3)}}{12}$
3–2	$x_{n+2} = x_{n+1} + \dfrac{T}{12}[5f_{n+2} + 8f_{n+1} - f_n]$	$\dfrac{-T^4 x^{(4)}}{24}$
4–3	$x_{n+3} = x_{n+2} + \dfrac{T}{24}[9f_{n+3} + 19f_{n+2} - 5f_{n+1} + f_n]$	$\dfrac{-19T^5 x^{(5)}}{720}$

4.9 PROPERTIES OF LINEAR MULTISTEP METHODS

With the above insight, it is now appropriate to give the following definitions (see Lambert, 1973, or Henrici, 1962, for more information).

1. A linear multistep method is said to be *consistent* if it has order, $p \geq 1$. From the definition of order, this implies that both $C_0 = C_1 = 0$, as was similarly stated earlier.

2. A linear multistep method is said to be *zero-stable* if all of its poles are in a stable configuration. This means that $\rho(z)$ cannot have any roots outside the unit circle or repeated roots on the unit circle. Keep in mind that these two definitions stem from the requirements for open-loop integration, that is, when $\dot{x}$ is not a function of x.

3. The necessary and sufficient conditions for a linear multistep method to be *convergent* are that it be consistent and zero-stable (Dahlquist, 1956).

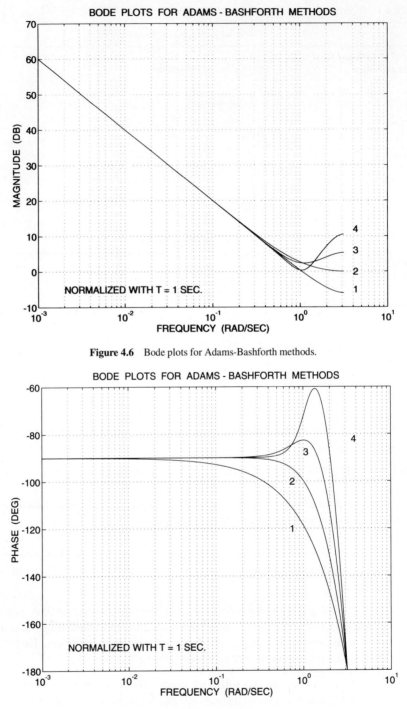

Figure 4.6   Bode plots for Adams-Bashforth methods.

Figure 4.6   (continued)

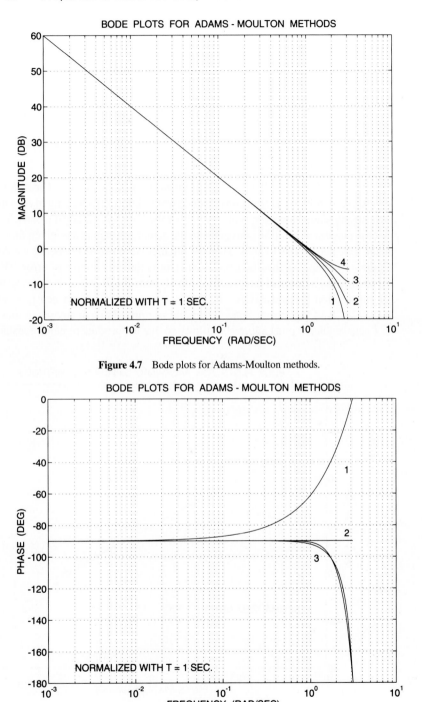

Figure 4.7   Bode plots for Adams-Moulton methods.

Figure 4.7   (continued)

TABLE 4.2   COEFFICIENTS FOR BACKWARD DIFFERENCE METHODS

k	1	2	3	4
β_k	1	2/3	6/11	12/25
α_4				1
α_3			1	−48/25
α_2		1	−18/11	36/25
α_1	1	−4/3	9/11	−16/25
α_0	−1	1/3	−2/11	3/25

A brief digression into what divergence means is appropriate here. Clearly, from the above impulse response discussion, convergence would imply errors going to zero in infinite time, or in steady state. However, that is not what the numerical analyst means. The idea here is that the numerical solution converges to the actual solution at any given time point ($t = nT$) as the timestep T goes to zero. Thus the smaller we make T, the better we do, as shown in Figure 4.9, assuming that the truncation error is the only source of error. This will be true for most of the numerical integration methods considered, which is opposed to the situation for substitution methods. Here, the round-off problems are not nearly as severe, mainly because of working in state space. It should be noticed that at any given time point $t = nT$; as T gets smaller, n must approach infinity so that the product remains constant. This seems

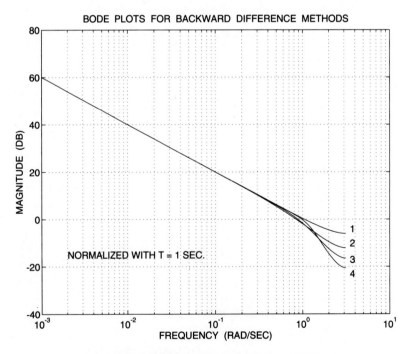

Figure 4.8   Bode plots for backward difference methods.

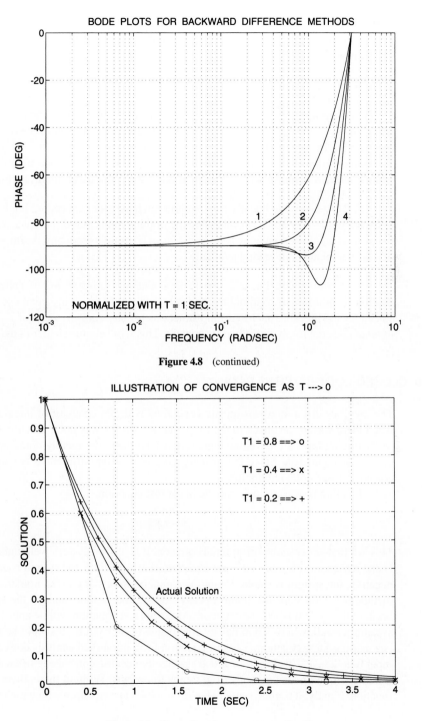

Figure 4.8   (continued)

Figure 4.9   Illustration of convergence as $T \to 0$.

to imply the use of the final-value theorem, even though the basic idea of when the approximation is converging is different. It is not totally clear to us how these two ideas are completely resolved, but it has been a great topic for debate.

　　4. A linear k-step method is said to be *maximal* if it is of order $2k$, for an implicit method, or of order $2k - 1$ for an explicit method. This statement should be quite clear since for an implicit method there are $2k + 2$ unknowns in the integrator. The leading term in the denominator is defined to be unity, which leaves $2k + 1$ coefficients to be determined. This can be done by setting the first $2k + 1$ C-coefficients from Lambert's equations equal to zero, which implies a maximum order for the method to be $2k$ (reduce all of this by 1 for an explicit method). Unfortunately, the situation is not as good as it might seem.

　　5. No zero-stable linear k-step method can have order exceeding $k + 1$ when k is odd or exceeding $k + 2$ when k is even. Here the implication is that if the order were increased, the method would no longer be zero-stable, as in Hermite's predictor. A simple justification of this statement is not readily available.

　　6. A zero-stable linear k-step method that has order $k + 2$ is called an optimal method. Unfortunately, optimal methods turn out not to be of any great use in simulation since they are unstable in closed-loop situations for any nonzero value of T. Thus the term *optimal* does not imply useful in this context. This will become clearer in the next section.

4.10　CLOSED-LOOP INTEGRATION

The previous discussion was mostly concerned with the behavior of integration methods when there was no feedback in the loop, that is, when $\dot{x}$ was not a function of x. Thus we really did not have a differential equation to solve but were merely doing open-loop integration of input data. Since the primary goal is to solve differential equations, we now consider closed-loop integration.

　　It is first necessary to define the linear test equation,

$$\dot{x} = \lambda x + u \qquad\qquad (4\text{--}30)$$

This is a first-order scalar differential equation. It is obviously very well understood, which is one reason it is chosen as the test equation. If λ is positive, the system is unstable. If λ is negative, the system is stable. Using this equation to test any given numerical integration method is a good standard. It allows easy study of the method, and not the system, simply because the system is well understood. Also, this test equation will be applicable to vector systems as well, where λ will be any eigenvalue of the system. It should be noted that this is a modified test equation. The test equation from numerical analysis does not contain an input function, u. This has been added essentially because control engineers are more comfortable with transfer functions from inputs to outputs. Without the u, there is no transfer function. Finally, it is good sense to have a standard test equation with which to compare all integration methods. Otherwise, we would have to test every method on every possible system, and it would be difficult to make general statements about integration methods.

This system can be represented in block diagram form, as in Figure 4.10. The transfer function for this system is easily found to be

$$G(s) = \frac{X(s)}{U(s)} = \frac{1}{s - \lambda} \tag{4–31}$$

The discrete approximation of this system is found by replacing the integrator with its discrete equivalent, as shown in Figure 4.11. The closed-loop transfer function should now be found for this system. First

$$\dot{X}(z) = \lambda X(z) + U(z) \tag{4–32}$$

and

$$X(z) = H(z)\dot{X}(z) = \frac{T\sigma(z)}{\rho(z)} \dot{X}(z) \tag{4–33}$$

Substitution gives

$$X(z) = \frac{\lambda T\sigma(z)}{\rho(z)} X(z) + \frac{T\sigma(z)}{\rho(z)} U(z) \tag{4–34}$$

The remaining algebra yields

$$G(z) = \frac{X(z)}{U(z)} = \frac{T\sigma(z)}{\rho(z) - \lambda T\sigma(z)} \tag{4–35}$$

This is the closed-loop equation for the integration process. Thus we have effectively reformulated the closed-loop integration problem as a control systems problem. It has the characteristic equation given by

$$\Delta(z) = \rho(z) - \lambda T\sigma(z) = 0 \tag{4–36}$$

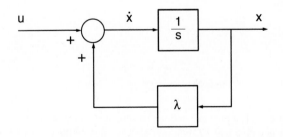

CONTINUOUS - TIME INTEGRATOR
WITH LINEARIZED SYSTEM

Figure 4.10 Continuous-time integrator with linearized system.

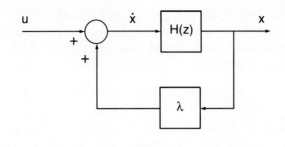

DISCRETE - TIME INTEGRATOR
WITH LINEARIZED SYSTEM

Figure 4.11 Discrete-time integrator with
linearized system.

This resembles the standard root locus problem from classical controls (Thaler, 1989). The closed-loop poles of the system move from the open-loop integrator poles, given by the roots of $\rho(z)$, to the open-loop integrator zeros, given by the roots of $\sigma(z)$, as the gain changes. The loop gain is given by $-\lambda T$, which is the product of the continuous system pole and the timestep. Although the negative sign would indicate the use of a negative locus, this is not the case as we will be primarily concerned with stable systems. For a stable system, λ is necessarily negative, and a positive locus is then used since the double negatives cancel. Hence all of the normal root locus rules for positive gain can be used to create a root locus for the integrator, assuming a stable continuous-time system. It is interesting, and expected from control theory, that the closed-loop information can be directly obtained from the open-loop information. It should be noticed that the designer has some choice in the closed-loop pole locations. Given the continuous system poles, T can be chosen to give a desired response. It will turn out that the smaller T is, the more accurate the simulation will be, and necessarily, the longer it will take to run. Thus the designer must always make some choices between accuracy and computation time. Several examples will now be given to demonstrate this discussion.

Example 4.5

$$AB\text{–}2: H(z) = \frac{\dfrac{T}{2}(3z-1)}{z(z-1)}$$

This system has a pole at $z = +1$, a pole at $z = 0$, and a zero at $z = +1/3$, as shown in the root locus of Figure 4.12. For negative real λT, one root always stays on the positive real z-axis. This will be defined later to be the *principal root*, as it is always closest to $z = e^{\lambda T}$. There is also another root, on the negative real z-axis, which will also be defined later to be a *spurious root*, as it is never the closest to $z = e^{\lambda T}$. Here the spurious root leaves the unit circle when the gain $\lambda T = -1$. Hence it should be remembered that usually T must be based on the given λ to pre-

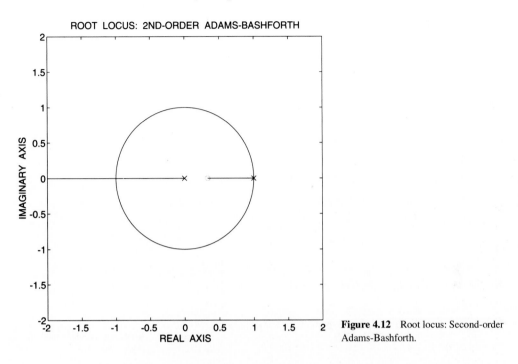

Figure 4.12 Root locus: Second-order Adams-Bashforth.

serve not only stability but also accuracy. There are some methods, usually implicit, that have unconditionally stable properties. The following important statement should now be made.

Explicit Method Property

All explicit methods become unstable for some negative values of λT since they will always have at least one zero at infinity, each of which attracts a closed-loop pole.

Example 4.6

Euler's method: $H(z) = \dfrac{T}{z - 1}$

This integrator has a pole at $z = +1$. Its root locus is given in Figure 4.13. As expected from the earlier discussions, this integrator becomes unstable when real $\lambda T = -2$.

Example 4.7

Backward Euler: $H(z) = \dfrac{Tz}{z - 1}$

This integrator has a pole at $z = +1$ and a zero at $z = 0$. It root locus is given in Figure 4.14. It can be seen that this system is stable for all values of negative λT.

Example 4.8

Trapezoidal rule: $H(z) = \dfrac{T}{2}\left(\dfrac{z + 1}{z - 1}\right)$

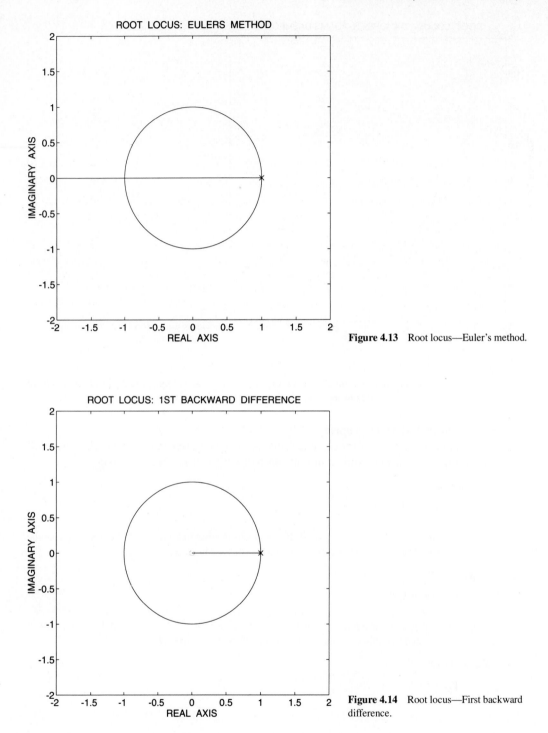

Figure 4.13 Root locus—Euler's method.

Figure 4.14 Root locus—First backward difference.

This system has a pole at $z = +1$ and a zero at $z = -1$. Its root locus is given in Figure 4.15. This method is also unconditionally stable for a stable continuous system.

Example 4.9

Simpson's rule: $x_{n+2} = x_n + \dfrac{T}{3}(\dot{x}_{n+2} + 4\dot{x}_{n+1} + \dot{x}_n)$

with transfer function,

$$H(z) = \frac{\dfrac{T}{3}(z^2 + 4z + 1)}{z^2 - 1}$$

Most people have been introduced to Simpson's rule in an introductory calculus course. It is touted as better than many other methods for numerical integration. What this really means is that it is a better method for doing open-loop integration of data. In fact, Simpson's rule is both optimal and maximal. If a differential equation is involved, however, it does not perform very well, as the root locus analysis now shows. It has a pole at $z = +1$ and at $z = -1$. Two zeros are located at $z = -2 \pm \sqrt{3}$, or approximately $z = -0.268$, $z = -3.732$. The root locus is given in Figure 4.16. It is seen that the spurious root starting at $z = -1$ immediately leaves the unit circle for any negative nonzero gain. Likewise, when simulating an unstable system, the principal root at $z = +1$ leaves the unit circle along the negative locus for any nonzero positive value of gain. Thus the only time that Simpson's rule is stable is when the gain is zero, which corresponds to open-loop integration.

ROOT LOCUS: TRAPEZOIDAL METHOD

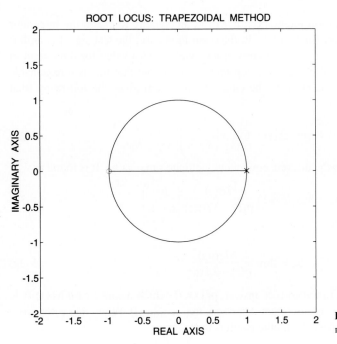

Figure 4.15 Root locus—Trapezoidal method.

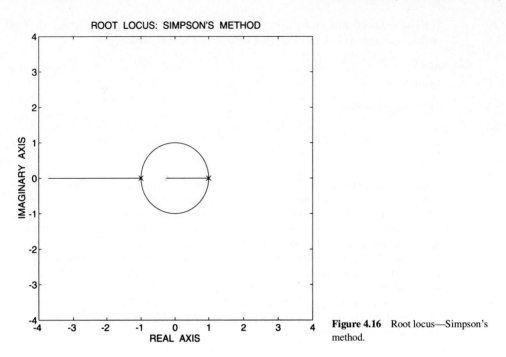

Figure 4.16 Root locus—Simpson's method.

4.11 CONTROL THEORY REVISITED

Just as was done for open-loop integration, it will be interesting to check the integrator coefficients based on correct final values. In the open-loop case, the test signal is an impulse. Here, an impulse is not practical since it will yield a zero value for the output in steady state. The next logical choice is a step input. It turns out that this is a reasonable choice. Applying a step of height b to the closed-loop system gives the following final value:

$$x(\infty) = \lim_{s \to 0} sX(s) = \lim_{s \to 0} s\left(\frac{1}{s - \lambda}\right)\left(\frac{b}{s}\right) = \frac{-b}{\lambda} \tag{4–37}$$

It is desirable to have the discrete equivalent yield the same value. It is found as

$$x_\infty = \lim_{z \to 1} (z - 1)\frac{T\sigma(z)}{\rho(z) - \lambda T\sigma(z)}\left(\frac{bz}{z - 1}\right) \tag{4–38}$$

or

$$x_\infty = \lim_{z \to 1} \frac{bTz\sigma(z)}{\rho(z) - \lambda T\sigma(z)} \tag{4–39}$$

Clearly, to force this to the correct answer, $\rho(1) = 0$, which means that it has $(z - 1)$ as a factor. This is the same requirement as before; that is, $C_0 = 0$. When this is the case, notice that the $T\sigma(z)$ terms cancel, and the result is then

$$x_\infty = \frac{-b}{\lambda} \qquad (4\text{--}40)$$

which is the desired result. Thus it would appear in closed-loop integration that the only requirement for convergence is $C_0 = 0$. In fact, it has been shown (Hartley, 1984) that improved accuracy can be obtained if this is the only constraint; however to do this it is required that significant information about the system must be used to determine the coefficients.

Further insight into this situation can be found from applying a ramp as the test input. If $\dot{x}$ is chosen as the output, however, it is clear that it will go to infinity for the test equation. A better choice for the output is the error signal, which for the given configuration is $\dot{x}$. The transfer function from u to $\dot{x}$ is found as

$$G_e(s) = \frac{\dot{X}(s)}{U(s)} = \frac{s}{s - \lambda} \qquad (4\text{--}41)$$

Notice here that there is now a zero at $s = 0$. The steady state value of this error is seen to be zero when the input is an impulse or a step. It has a nonzero value, however, for a ramp input. This is found to be

$$\dot{x}(\infty) = \lim_{s \to 0} s\left(\frac{s}{s - \lambda}\right)\left(\frac{b}{s^2}\right) = \frac{-b}{\lambda} \qquad (4\text{--}42)$$

For the discrete system, the transfer function from the input, u, to the error signal, $\dot{x}$, is found as

$$G_e(z) = \frac{1}{1 - \dfrac{\lambda T \sigma(z)}{\rho(z)}} = \frac{\rho(z)}{\rho(z) - \lambda T \sigma(z)} \qquad (4\text{--}43)$$

The steady state value of the error for a ramp input is then

$$\dot{x}_\infty = \lim_{z \to 1} (z - 1)\left(\frac{\rho(z)}{\rho(z) - \lambda T \sigma(z)}\right)\left(\frac{bTz}{(z - 1)^2}\right)$$

$$= \lim_{z \to 1} \left(\frac{bTz\rho(z)}{\rho(z) - \lambda T \sigma(z)}\right)\left(\frac{1}{z - 1}\right) \qquad (4\text{--}44)$$

Clearly, to keep this from becoming infinity, it is necessary that $(z - 1)$ be a factor of $\rho(z)$. By setting

$$\rho(z) = (z - 1)\rho_f(z) \qquad (4\text{--}45)$$

we get the following:

$$\dot{x}_\infty = \lim_{z \to 1} \frac{bTz\rho_f(z)}{(z - 1)\rho_f(z) - \lambda T \sigma(z)} = \lim_{z \to 1} \frac{bT\rho_f(z)}{-\lambda T \sigma(z)} \qquad (4\text{--}46)$$

Thus, to have the correct final value,

$$\rho_f(1) = \sigma(1) \qquad (4\text{--}47)$$

which was the requirement from the open-loop case; that is $C_1 = 0$. Then

$$\dot{x}_\infty = \frac{-b}{\lambda} \qquad (4\text{-}48)$$

which is the desired result.

Thus, using steady state analysis, we have shown that the same requirements on the integrator coefficients are in the closed-loop case as were found in the open-loop case. It would seem that further requirements could be found by using higher-order inputs; however, it is not presently clear how to do so with this approach. It will turn out that sinusoids are a good choice; however, another equivalent and more elegant approach will be used.

4.12 WEAK STABILITY THEORY (CLOSED-LOOP INTEGRATION)

What numerical analysts refer to as *weak stability theory,* control engineers would call *closed-loop stability*. Thus what will be discussed now is the stability and accuracy of the integration process when $\dot{x}$ is a function of x. It is clear from earlier discussions that closed-loop stability is a function of both the continuous system poles and the integration timestep, as well as the integrator coefficients. The following definitions should now be made (see Lambert, 1973, or Henrici, 1962, for more information).

1. A linear multistep method is said to be *absolutely stable* for a given value of loop gain, λT, if all the roots of the closed-loop characteristic equation, $\rho(z) - \lambda T \sigma(z)$, are inside the z-plane unit circle. This is equivalent to bounded-input–bounded-output stability.

To clarify, consider the AB-2 again, where the root locus is shown in Figure 4.12. The following formal definitions should be made.

2. The *principal root* is the one, for a given λT, that is closest to the exact mapping of the s-plane root, $z = e^{\lambda T}$.

3. The *spurious roots* are all the remaining roots that are not the principal root.

Clearly the AB-2 method has one principal root and one spurious root. The spurious root leaves the unit circle when $\lambda T = -1$. Thus the AB-2 method is absolutely stable for purely real λT between 0 and -1.

For control engineers, absolute stability is of fundamental importance, and a great number of tools are available for studying it, including the root locus discussed above; the Routh-Hurwitz, Schur-Cohn, and Jury tests; and the frequency response techniques including Bode, Nyquist, and Nichols forms (Phillips & Nagle, 1990). We will put these to great use later, particularly the frequency response approaches. In addition to absolute stability, we are also concerned with accuracy. The above definitions have given little information about the accuracy of a simulation. Numerical analysts give the following definition.

4. A linear multistep method is said to be *relatively stable* for a given λT if the principal root is larger than (or equal to) each of the spurious roots in magnitude (Lambert, 1973).

This should not be confused with the usual control engineering definition of relative stability, which is related to a measure of stability robustness. This definition says that the simulation will be accurate as long as the principal root is the largest. This should be fairly obvious; however, the following terminology borrowed from the communications area seems to shed further light on this issue. A linear multistep method has k-poles in both open and closed loop for a scalar continuous system. These can be considered to be signal sources because each of the poles provides one term in the output expression obtained by partial fraction expansion. The signal that we are concerned about receiving is the one associated with the principal root, that is, the one closest to $z = e^{\lambda T}$. All of the spurious roots are effectively noise sources. Since the magnitude of the roots indicates how fast the signals decay, and thus which ones are dominant, we would always like the principal root to be the largest. When T is chosen so that the spurious roots are larger than the principal roots, the signal is effectively lost in the noise. We clearly would like the *signal-to-noise-ratio* (SNR) (Schwartz, 1970) to be greater than 1. It should also be clear from this discussion that linear multistep methods are not always more accurate than the other methods discussed earlier. They are theoretically more accurate as $T \rightarrow 0$. When larger values of T are used, some care must be taken to make sure that both accuracy and stability are maintained. It should be observed that the AB-2 method is relatively stable for purely real λT between zero and approximately −0.6.

5. A linear multistep method is said to be *tuned* if the principal root is close to the exact z-plane mapping of the continuous system pole, $z = e^{\lambda T}$. The farther the principal root is away from the exact mapping, the more *detuned* the method is.

This is very much like tuning your car radio late at night in the middle of nowhere. You try to tune in on the signal you want, $z = e^{\lambda T}$; however, not only are there noise sources, but also the station can be hard to find. If T is too large, you may get a signal, but it may not be recognizable. This is the case for the trapezoidal rule whose root locus is given in Figure 4.17. This method is absolutely stable for all real negative λT. There is only one root, so it must be the principal root. There are no spurious roots, and hence no spurious noise sources; the method is always relatively stable. The only source of error, then, is detuning. The following heuristic definition is now included.

6. A linear multistep method is said to be *severely detuned* if the principal root is in such a location that the system response will be qualitatively different from that which is expected.

Although not very quantitative, this is still a useful concept. Consider the trapezoidal rule again. When the principal root moves onto the negative real z-axis, the response will be oscillatory, rather than monotonic as it should be. Here the method will be severely detuned. It will only be tuned for λT between zero and −2. Returning to the AB-2, it becomes slightly detuned for large T but never severely detuned; its problem is inaccuracy because of the spurious root/noise source. Recall that the Adams methods all have their spurious open-loop poles located at the z-plane origin.

Remember that tuning only deals with the principal root location. (Is the signal that we are receiving what we expect it to be?) Relative stability deals with the relative locations of the principal and spurious roots. (Is our signal lost in the noise?) There will usually be some of each of these errors in any given simulation. Furthermore, for a system of equa-

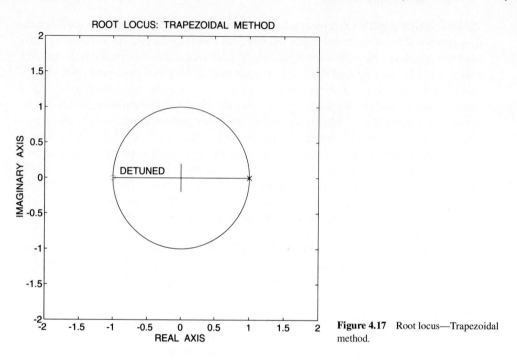

Figure 4.17   Root locus—Trapezoidal method.

tions (vector x), all of the roots associated with each continuous eigenvalue must be considered simultaneously. Thus relative stability must be considered with respect to other principal roots in a system as well.

Discussion of the Root Locus Method

The root locus method used extensively above for illustration is a very powerful tool, with which most control engineers are familiar. Its biggest problem arises when one is studying simulation stability for systems of equations:

$$\dot{x} = Ax + Bu$$

A root locus using the integrator poles and zeros must be done for every eigenvalue of the A matrix. It is somewhat amazing that this process works; however, it does indeed give all of the closed-loop poles of the integration process. This can easily be seen, however, by expanding the original system, using a partial fraction expansion that effectively puts each term in parallel with all the others. A simulation of each term of the partial fraction expansion can then be performed for the given eigenvalue, λ. The resulting root locus process is fairly tedious, and usually one only worries about the largest negative eigenvalue of A, as this will often be sufficient for guaranteeing that the smaller λT-products will also be inside the unit circle. The process is further complicated if the eigenvalues of the continuous system are complex, that is, have an imaginary component. Since the characteristic equation is

$$\rho(z) - \lambda T \sigma(z) = 0 \qquad (4\text{--}49)$$

a complex value of λT gives a complex gain. All of the root locus rules that are normally used consider only real gains. Thus we are left without our basic tool, unless some automated procedure is used. This is easily done in the MATLAB environment, and a representative plot is given in Figure 4.18. For that matter, a plot of all of the closed-loop poles is easily obtained with MATLAB, and comparisons are easily made.

Alternatively, a root locus that uses only real gains can still be constructed by multiplying the above equation by its complex conjugate and then performing the root locus on the result. Unfortunately, the usual root locus rules cannot be used because of the gain and gain squared terms in the equation. The resulting equation is given here for usage with a root-finding package rather than for the more versatile MATLAB package, which can use complex gains:

$$\rho^2(z) - 2rT \cos(\theta)\, \sigma(z)\rho(z) + r^2 T^2 \sigma^2(z) = 0 \qquad (4\text{--}50)$$

Here r is the magnitude of the complex pole, θ is the angle of the complex pole, and rT should be considered the gain. With this equation it is also possible to obtain lines of constant damping (r constant, θ varying) or lines of constant settling (θ constant, r varying), for the given integrator and timestep, in the z-plane. A simple manipulation of this equation will give the lines of constant damping and constant natural frequency.

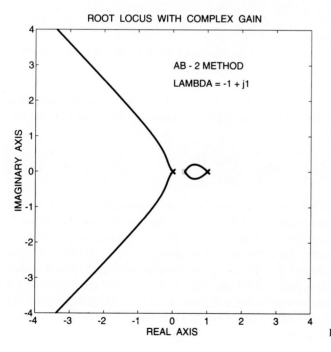

Figure 4.18 Root locus with complex gain.

4.13 INTRODUCTION TO STABILITY REGIONS

Although the root locus method was easily automated, it does leave the user with some loss of intuition when complex gains/eigenvalues are used. To remedy this loss of intuition, the very powerful stability region method for determining stability is introduced here. This is derived by starting with the closed-loop characteristic equation,

$$\rho(z) - \lambda T \sigma(z) = 0 \tag{4–51}$$

This equation can be rearranged to solve for the loop gain,

$$\lambda T = \frac{\rho(z)}{\sigma(z)} \tag{4–52}$$

It is now possible to solve for a stability boundary in a new complex λT-plane by mapping the stability boundary from the z-plane. This is done by replacing $z = e^{j\omega T} = \cos(\omega T) + j\sin(\omega T)$ in the above equation, or equivalently, $z^n = e^{jn\omega T} = \cos(n\omega T) + j\sin(n\omega T)$, rather than $z^n = [\cos(\omega T) + j\sin(\omega T)]^n$; this simple realization can save much effort. Now map the entire boundary in the z-plane into the λT-plane by letting ωT vary from zero to 2π, going around the circle in a counterclockwise direction. Once this is done for a given integrator, T can be chosen so that all of the λT-products for a particular stable continuous system are inside this stability region. The resulting simulation with this value of T will then necessarily be stable if the original system is stable. This bears further emphasis.

Stability Region Property

If all of the λT-products for a given system lie inside the stability region for the given integrator, all of the closed-loop poles of the simulation will lie inside the unit circle in the z-plane.

The stability boundaries, or boundary loci, for several common integrators are given below. It should be noticed that the stability region is effectively the inverse of the frequency response scaled by T:

$$H(j\omega) = \frac{T\sigma(e^{j\omega T})}{\rho(e^{j\omega T})} \tag{4–53}$$

In fact, the stability boundary for a given integrator is its inverse Nyquist plot. Because of its nice properties, the inverse Nyquist plot is often used by control engineers to check closed-loop stability (D'Azzo & Houpis, 1981) as the loop compensation changes. We will take advantage of these properties later. The inverse Nyquist plot for a continuous integrator is given in Figure 4.19. Notice that this goes straight up the axis and to infinity when $\omega \to \infty$. It then returns up the negative imaginary axis and reaches the origin when $\omega = 0$ again. Thus the stability region for a numerical integrator should resemble this Nyquist plot for a continuous integrator. It should also be noticed that the stability region is a conformal mapping of the complex z-plane into the complex λT-plane.

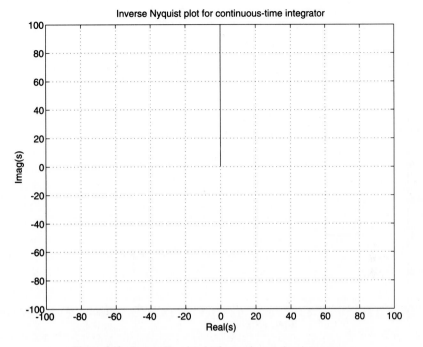

Figure 4.19   Inverse Nyquist plot for continuous-time integrator.

Example 4.10: Euler's method: $\lambda T = z - 1$

This example is particularly easy. The stability region is just a shifted unit circle, as in Figure 4.20. Choosing T so that λT is always within this region will guarantee that the simulation will remain stable. Of course, this assumes that the original system is stable. If it were unstable, we would want the principal roots associated with the unstable poles to be outside the unit circle or stability region, depending on the plane that we are using. One problem with using Euler's method, which can now be made clear from its stability region, is in simulating continuous systems with very lightly damped complex poles. These would be located very close to the imaginary axis, away from the origin in the λT-plane. For Euler's method, the stability region is a circle and thus never approaches the imaginary λT-axis at any point other than the origin. It is thus nearly impossible to have the stability region encircle these undamped complex poles, unless very small timesteps are used. If the poles are purely imaginary, Euler's method will always give an unstable simulation since the stability region can never include points on the imaginary λT-axis away from the origin. The recommendation is that if there are very lightly damped poles in the system, some method other than Euler's should be used.

Example 4.11. Trapezoidal rule: $\lambda T = \dfrac{2(z-1)}{z+1}$

The stability region is given in Figure 4.21. This method has the remarkable property that the entire left-half λT-plane is the stability region for the method. This implies that the simulation of any stable system, using any timestep, will be stable if simulated with the trapezoidal rule. For this reason, the trapezoidal rule, by way of bilinear substitution, has become an extremely

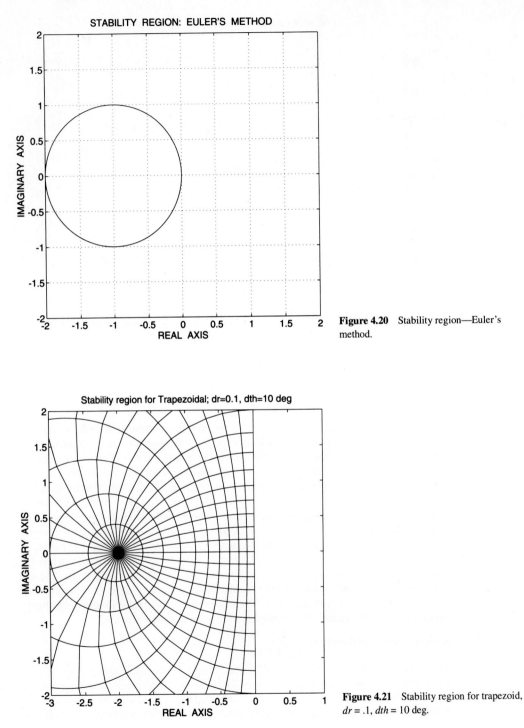

Figure 4.20 Stability region—Euler's method.

Figure 4.21 Stability region for trapezoid, *dr* = .1, *dth* = 10 deg.

popular method in the control and communications area. Some caution must be used, however, as this rule says nothing of accuracy. Also this method is implicit, which makes it somewhat less useful, as will be discussed later.

Example 4.12. Backward Euler: $\lambda T = \dfrac{z-1}{z}$

The division by z here has a very profound effect on the otherwise shifted unit circle previously encountered. In Figure 4.22 notice that the encirclement is clockwise, whereas we used a counterclockwise direction in the z-plane. This means that everything outside of this circle is stable. From a stability point of view, this is even better than the trapezoidal rule. However, it is not as nice as it first seems. Because this method can make the simulation of even unstable systems stable, clearly it can then be a severely detuned method if λT is chosen inside the stability region in the right-half λT-plane ($\lambda T > 2$). This method is also implicit.

Example 4.13. AB-2: $\lambda T = \dfrac{2z(z-1)}{3z-1}$

This mapping is given in Figure 4.23. It is seen that the stability region of the AB-2 method goes significantly farther up the imaginary λT-axis than that of Euler's method. It still does not lie exactly on the imaginary axis and should not be used for purely conservative oscillations. It is usually sufficient for most engineering systems, however. Notice also that it only goes out to -1 on the real λT-axis, which is only half as far as that for Euler's method. Thus we have gained some accuracy, particularly up the imaginary axis, but have lost some stability for large negative λT. Stability regions for third- and fourth-order Adams-Bashforth methods are given in Figures 4.24 and 4.25, respectively.

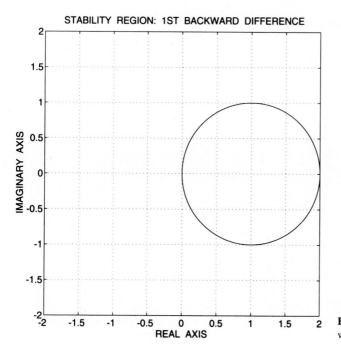

STABILITY REGION: 1ST BACKWARD DIFFERENCE

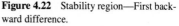

Figure 4.22 Stability region—First backward difference.

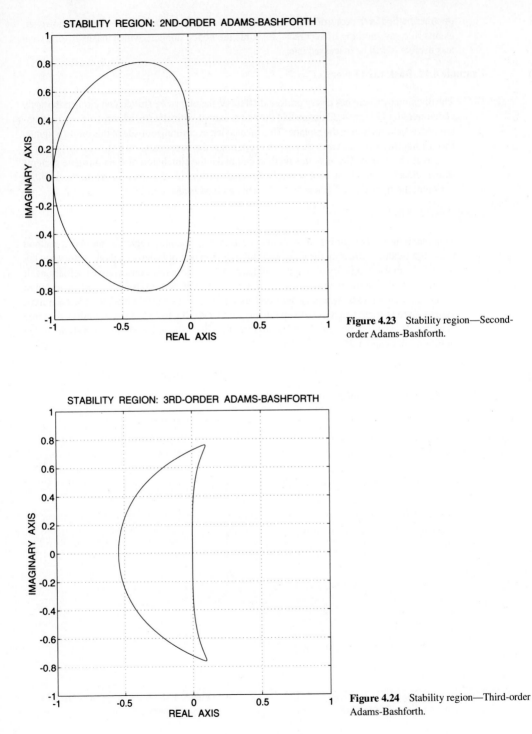

Figure 4.23 Stability region—Second-order Adams-Bashforth.

Figure 4.24 Stability region—Third-order Adams-Bashforth.

STABILITY REGION: 4TH-ORDER ADAMS-BASHFORTH

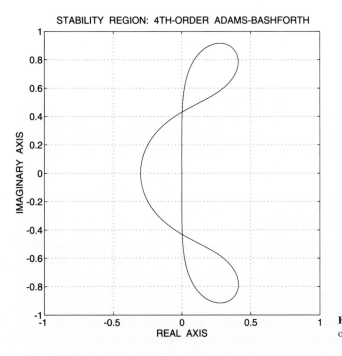

Figure 4.25 Stability region—Fourth-order Adams-Bashforth.

Now consider the other Adams-Bashforth methods. The higher the order, the smaller the stability region. Thus accuracy is traded for stability and computation time. Also notice that the stability regions go up the axis better for higher-order accuracy methods, which should be expected. Notice that the AB-3 includes a little "ear" in the right-half plane. The stability region for this method goes up just to the right of the imaginary λT-axis. Thus this method imposes a little more stability on a simulation than the real system has for lightly damped systems. For undamped conservative systems, the simulations become slightly damped with this method. The AB-4 stability region goes right up the imaginary axis and does very well for conservative systems. It has interesting clockwise loops that lie in the right-half λT-plane. These loops are unstable regions because of the direction of encirclement. They continue to enlarge for higher-order Adams-Bashforth methods.

Example 4.14. Simpson's rule: $\lambda T = \dfrac{3(z^2 - 1)}{z^2 + 4z + 1}$

Simpson's rule does not have a visible stability region. It is basically the origin of the λT-plane corresponding to open-loop integration, or $\lambda T = 0$. It turns out that all optimal methods suffer from the same problem.

We will finish this section with one more definition and an observation.

Definition

A numerical integration method is said to be A-*stable* if its stability region contains the entire left-half λT-plane.

This is clearly a severe constraint on a method. However, as seen above, both the trapezoidal rule and the backward Euler method are A-stable methods. The following observation can then be made.

Observation

No explicit linear multistep method can be A-stable.

Any explicit method will have a pole excess of at least 1, implying a zero at infinity, and follows from the explicit method property. The resulting root locus must then always leave the unit circle, even for complex gains. Thus explicit methods will always have a finite stability region

4.14 THE $\hat{s}$ PLANE

It is recognized that some people probably are still not comfortable with the z-plane and do not understand intuitively what the pole locations mean there. To address this problem and to give even further insight, we introduce another complex plane. This plane will be a mapping of the z-plane pole locations, which are the real pole locations that the simulation uses, back into a continuous s-like-plane. The original system being simulated has s-plane poles. The poles that are now obtained are the equivalents of the z-plane poles that the integrator is using and will be placed in what will be called the $\hat{s}$-plane. The technique used to obtain these $\hat{s}$-pole locations is now given. First, choose T to get all of the λT-products inside the stability region of the integrator being used. Once this is done, substitute these products into the characteristic equation $\rho(z) - \lambda T \sigma(z) = 0$ to get all of the z-plane poles corresponding to each of these products. These z-plane poles are then mapped back into the $\hat{s}$-plane by

$$\hat{s} = \frac{1}{T} \, ln(z) \qquad (4\text{--}54)$$

where it should be remembered that

$$ln(re^{j\theta}) = ln(r) + j\theta + j2n\pi, \; n = \text{all integers} \qquad (4\text{--}55)$$

Recall that with a k-step integrator, each s-plane pole results in k z-plane poles. Therefore, for a nth-order system, there will be kn poles in the z-plane and kn poles in the $\hat{s}$-plane. Putting the $\hat{s}$-plane pole locations and the s-plane pole locations on the same plot will give a good indication of the tuning errors in the simulation as well as the magnitude of the spurious noise. An example is included below.

4.15 A DESIGN EXAMPLE

To illustrate the use of stability regions in choosing the simulation timestep, a design example is included here. The problem is to simulate digitally the operation of a third-order Butterworth filter (see D'Azzo & Houpis, 1981; Kailath, 1980; or Van Valkenburg, 1974, for more information on the optimality of Butterworth forms). Basically, it is a low-pass filter that minimizes the squared error in response to a unit step. The purpose of the simulation could be either to study the properties of the continuous filter before it is constructed or to replace the continuous filter with a discrete equivalent. Whichever, the continuous filter is given below:

$$G(s) = \frac{Y(s)}{U(s)} = \frac{1}{s^3 + 2s^2 + 2s + 1}$$

The first task is to convert this system to state space form. This is the suggested approach for numerical integration methods. It should be noted that linear multistep methods can be used as substitution methods, as in Chapter 3. The state space system in phase variable form is

$$\dot{x}_1 = x_2$$
$$\dot{x}_2 = x_3$$
$$\dot{x}_3 = -x_1 - 2x_2 - 2x_3 + u$$
$$y = x_1$$

or

$$\dot{x} = Ax + Bu, \, y = Cx$$

where $A = \begin{bmatrix} 0 & 1 & 0 \\ 0 & 0 & 1 \\ -1 & -2 & -2 \end{bmatrix}$, $B = \begin{bmatrix} 0 \\ 0 \\ 1 \end{bmatrix}$, $C = [1 \ 0 \ 0]$.

The characteristic equation of the A-matrix is that the denominator of $G(s)$ gives the three system poles, $\lambda = -1, -0.5 \pm j0.866$. One good reason to use this as an example is that the magnitudes of all of these roots are unity. The configuration is given in Figure 4.26. To simulate this, the AB-3 is chosen because it has good accuracy up the λT-axis. The AB-2 would also work well; however, Euler's method is probably not very good for this system, although usable. The AB-3 is given as

$$x_{n+3} = x_{n+2} + \frac{T}{12}(23\dot{x}_{n+2} - 16\dot{x}_{n+1} + 5\dot{x}_n)$$

Then

$$\lambda T = \frac{\rho(z)}{\sigma(z)} = \frac{12\,(z-1)z^2}{23z^2 - 16z + 5}$$

We let $z = e^{j\omega T}$ to get the stability region in Figure 4.27. To get a feel for where the left part of the region crosses the real λT-axis, it is sufficient to substitute $z = -1$ into the above equation. This gives

$$\lambda T = \left[\frac{\rho(z)}{\sigma(z)}\right]_{z=-1} = \left[\frac{12(z-1)z^2}{23z^2 - 16z + 5}\right]_{z=-1} = \frac{-24}{44} = -0.5454$$

Now we combine the system with the stability region. Although the following is easily done by sight, the complete graphical construction will be given for instruction. The

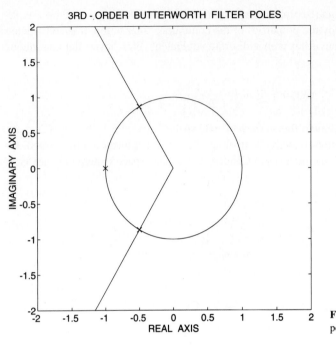

Figure 4.26 Third-order Butterworth filter poles.

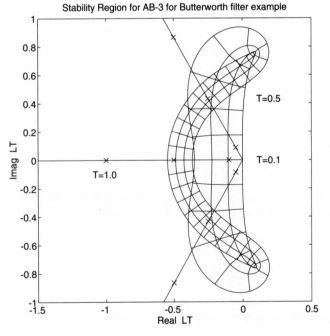

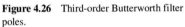

Figure 4.27 Stability region for AB-3 for Butterworth filter example.

λT-product will lie on three lines radiating from $\lambda T = 0$. These should be placed on the stability region. One lies on the negative real axis, and the other two are at 60-degree angles from the negative real axis (see Figure 4.27). First choose $T = 1$, for illustration. The resulting λT-products are clearly outside the stability region. Since we know that the left-most part of the stability region is $\lambda T = -0.5454$, let $T = 0.5$. From the figure it is seen that the resulting λT-products all lie just inside the stability region. Since these are near the left side of the stability region, the resulting simulation is probably not going to be very accurate, as will be shown later. Generally, the closer the λT-products are to the origin, the better the simulation will be.

We have not yet said how the stability region is related to relative stability, and we will do so later. However, the following is a pretty good rule of thumb: The relative stability region is roughly half the size (radially) of the absolute stability region. Thus a much better choice of timestep would be $T = 0.1$, and $T = 0.25$ would be just at the boundary of relative instability, or *SNR* = 1. A program for simulating this system is given below. Notice that the integration methods, being three-step methods, require three starting values at times $n = 0$, 1, and 2. These can be obtained by using Euler's method, followed by AB-2, or Runge-Kutta methods (which will be discussed in Chapter 6). In practice, since this is to act like a real-time digital filter, which operates continuously, the initial starting values can be assumed to be zero since they have no effect after several timesteps. A MATLAB program for performing this simulation is now given.

```
%MATLAB simulation of Butterworth filter using AB-3
T = 0.1
y = [0];
x = [0 0 0]';
f = x;
fs = f;
fss = f;
u = 1;
for n = 1:150,
    f(1) = x(2);
    f(2) = x(3);
    f(3) = -x(1)-2*x(2)-2*x(3)+u;
    x = x+(T/12)*(23*f-16*fs+5*fss);
    fss = fs;
    fs = f;
    y = [y x(1)];
end
plot(0:0.1:15,y)
title('Step response of Butterworth filter using AB-3, T = 0.1 .')
xlabel('Time')
ylabel('Filter output, y = x1')
```

To clarify further, the step response, the z-plane poles, and the $\hat{s}$-plane poles using the AB-3 are given in Figures 4.28 and 4.29 for the cases in which $T = 0.1$ and $T = 0.5$, respec-

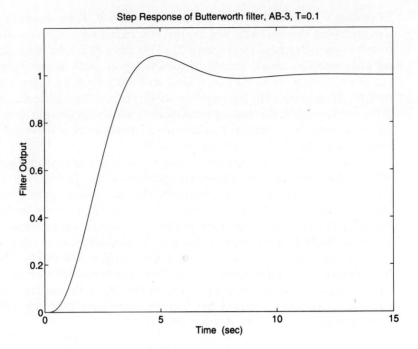

Figure 4.28   Step response of Butterworth filter using AB-3, $T = 0.1$.

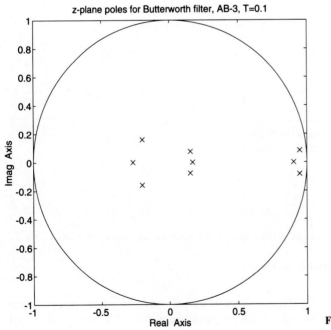

Figure 4.28   (continued)

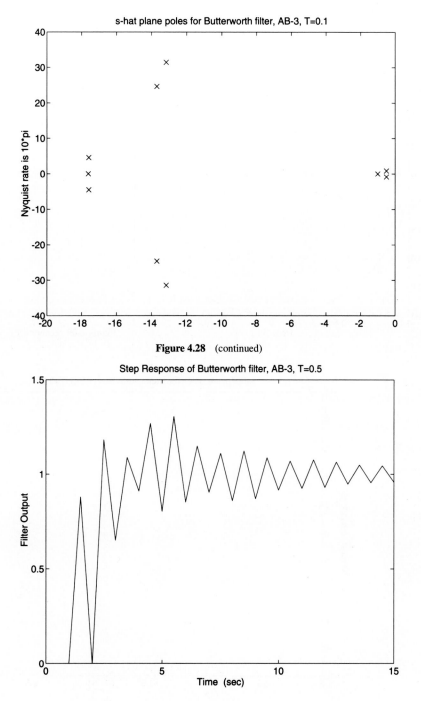

Figure 4.28   (continued)

Figure 4.29   Step response of Butterworth filter using AB-3, $T = 0.5$.

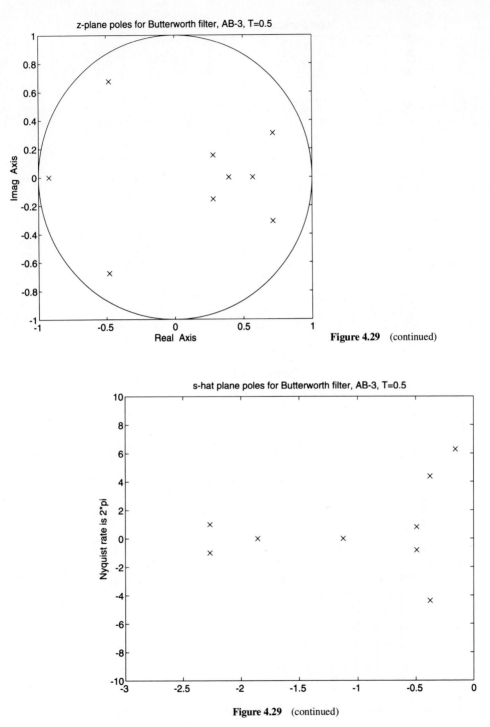

Figure 4.29   (continued)

Figure 4.29   (continued)

tively. Notice that when $T = 0.5$, the z-plane contains spurious poles near the unit circle in the left half of the $z =$ plane. In the $\hat{s}$-plane, these correspond to very lightly damped poles occurring at high frequencies, including two at plus and minus the Nyquist rate. These lightly damped spurious poles manifest themselves in the simulation by high frequency, lightly damped ringing, or noise. When $T = 0.1$, the simulation is dominated by the three principal roots of the simulation lying near $z = +1$. In the $\hat{s}$-plane, these three principal roots lie near the origin. The highly damped spurious roots lie far to the left in the $\hat{s}$-plane and have very little effect on the simulation.

4.16 CHARACTERISTICS OF IMPLICIT METHODS

So far we have said very little about the implementation of implicit methods. As stated earlier, the problem with implicit methods is that they require the solution at a given point in time in order to calculate that same solution. This is generally not very practical. We now discuss how to implement implicit methods by themselves and, later, predictor-corrector methods.

The good properties of implicit methods are clear. They can always give better accuracy for the given number of steps in the method, their error constants are usually smaller for a given order of accuracy, and their stability regions are usually much larger than those for explicit methods. Although all of these factors point in the favor of implicit methods, they are much more difficult to implement.

Examples of Implementing Implicit Methods

When applying implicit methods to linear systems, there is some salvation however. This comes at the cost of doing some algebra. Consider the following linear system, which is often encountered in control systems:

$$\dot{x} = Ax + Bu \tag{4–56}$$

Recall that the trapezoidal rule, an implicit method, is

$$x_{n+1} = x_n + \frac{T}{2} [\dot{x}_{n+1} + \dot{x}_n] \tag{4–57}$$

With an explicit method, this, along with the differential equation, is programmed and inserted into a loop. This cannot be done here since $\dot{x}$ at the present time depends on x at the present time, which depends on $\dot{x}$ at the present time, and so on. This dilemma is solved through the necessary process of performing some algebra before programming. First, insert the differential equation into the trapezoidal rule:

$$x_{n+1} = x_n + \frac{T}{2} [Ax_{n+1} + Bu_{n+1} + Ax_n + Bu_n] \tag{4–58}$$

A little manipulation gives

$$\left[I - \frac{T}{2}A\right]x_{n+1} = \left[I + \frac{T}{2}A\right]x_n + \frac{T}{2}B[u_{n+1} + u_n] \tag{4-59}$$

Then, solving for the output, we get

$$x_{n+1} = \left[I - \frac{T}{2}A\right]^{-1}\left\{\left[I + \frac{T}{2}A\right]x_n + \frac{T}{2}B[u_{n+1} + u_n]\right\} \tag{4-60}$$

This is an equation that can be programmed in a loop and used to simulate a system. The cost of implementing this implicit method has thus been some algebra and a matrix inverse. Any implicit method can always be used as above for a linear system. It should be noted that the output at the present time still depends on the input at the present time.

The above algebraic method can sometimes be used for nonlinear systems as well. Consider the following example.

Example 4.15

Let the differential equation be

$$\dot{x} = -x^2, \, x_0 = 1$$

The trapezoidal rule gives the following difference equation:

$$x_{n+1} = x_n + \frac{T}{2}[-x_{n+1}^2 - x_n^2]$$

This equation must now be solved for x at time $n + 1$. Thus

$$x_{n+1} + \frac{T}{2}x_{n+1}^2 = x_n - \frac{T}{2}x_n^2$$

or

$$x_{n+1}^2 + \frac{2}{T}x_{n+1} + x_n^2 - \frac{2}{T}x_n = 0$$

This quadratic equation is what must be solved at every timestep. Note that it is possible to solve it since at time $n + 1$, x at time n is already known. With $T = 1$ and using the initial condition, we get the following equation:

$$x_1^2 + 2x_1 - 1 = 0$$

with solutions $x_1 = 0.4142$ or -2.4142. Which one is the correct answer? It is hard to tell in general. For this problem, we know that this system has its equilibrium point at the origin. It is then impossible for the sign of the solution to change. The correct answer must be $x = 0.4142$. For a more unknown system, it is extremely difficult to choose. Furthermore, this answer is still not the exact solution to the differential equation. The truncation error of the trapezoidal rule is built into it. It is also interesting to note that the other solution is not totally spurious. If the timestep is made negative, the signs on the solution change, giving $x = 2.4142, -0.4142$. Hence the solution $x = 2.4142$ would be along the trajectory in negative time. The above procedure, including the solution of the quadratic, can now be repeated as far as necessary into the future to obtain the solution to the differential equation.

More often, the above algebraic procedure cannot be used for nonlinear systems because of the inability to solve the resulting algebraic equation and of choosing the correct solution. Many numerical techniques are available for solving nonlinear algebraic equations. The interested reader should consult Ralston and Rabinowitz (1978). A method called fixed point iteration is often used for these problems. First, a good guess is made for x at time $n + 1$. Then the equations are iterated until the solution converges. If the guess was good enough and the timestep small enough, the iteration converges to the correct value. This will be applied to the same nonlinear example considered above:

$$\dot{x} = -x^2,\ x_0 = 1$$

The trapezoidal rule results in the following difference equation:

$$x_{n+1} = x_n + \frac{T}{2}[-x_{n+1}^2 - x_n^2]$$

This equation must now be solved for x at time $n = 1$. Let the first guess at time 1 be $x = 0$. This is a really bad guess, but it will do for the illustration. Then, with $T = 1$,

$$x_{1,1} = x_{0,\infty} + \frac{1}{2}[\dot{x}_{1,0} + \dot{x}_{0,\infty}] = 1 + 0.5(0 - 1) = 0.5$$

The second subscript denotes the iteration number at a given timestep. Infinity corresponds to a converged iteration, which can be a large number but not really infinity. Also notice that the next best guess is 0.5. This is much closer to the actual value of 0.4142 found above than zero was. The next iteration gives

$$\dot{x}_{1,1} = -x_{1,1}^2 = -0.25$$

$$x_{1,2} = x_{0,\infty} + \frac{1}{2}[\dot{x}_{1,1} + \dot{x}_{0,\infty}] = 1 + 0.5(-0.25 - 1) = 0.375$$

Continuing yields

$$\dot{x}_{1,2} = -x_{1,2}^2 = -0.141$$

$$x_{1,3} = x_{0,\infty} + \frac{1}{2}[\dot{x}_{1,2} + \dot{x}_{0,\infty}] = 1 + 0.5(-0.141 - 1) = 0.4295$$

and

$$\dot{x}_{1,3} = -x_{1,3}^2 = -0.185$$

$$x_{1,4} = x_{0,\infty} + \frac{1}{2}[\dot{x}_{1,3} + \dot{x}_{0,\infty}] = 1 + 0.5(-0.185 - 1) = 0.408$$

Thus it can be seen that the iteration is indeed converging to the correct trapezoidal rule solution of 0.4142. Again it should be remembered that the trapezoidal rule truncation error is built into this solution, which is not the exact solution of the differential equation.

Furthermore, this process must be iterated to convergence at each point in time. Thus it can be seen this process is very time consuming.

Iteration Convergence Criterion

The convergence rate of the iteration process depends on the given system, the chosen integration method, the chosen timestep, and the accuracy of the initial guess. The dependence on the timestep and integration method will now be shown. Any implicit method can be written as

$$x_{n+k,j+1} = \text{(some function of known } x\text{'s)} + T\beta_k f(x_{n+k,j}, u_{n+k}, n+k)$$
$$+ \text{(some function of known } \dot{x}\text{'s)}$$

If we assume that $f(x,u,n)$ has a Lipschitz bound L, this is just an iteration on x, with j being the iterate. This can then be written as

$$|x_{n+k,j+1}| \leq TL|\beta_k| \, |x_{n+k,j}| \tag{4-61}$$

Since this is just a difference equation in j, it will be stable, and thus converge, if the pole is less than 1 in magnitude, or

$$TL|\beta_k| \leq 1 \tag{4-62}$$

This equation can then be manipulated to give a bound on the timestep (Lambert, 1973):

$$T \leq \frac{1}{L \, |\beta_k|} \tag{4-63}$$

Thus it can be concluded that even implicit methods when iterated have a bound on their timestep that is usually more restrictive than the usual stability region bound. Conversely, if an algebraic solution can be found, the stability region can be used as usual.

For Example 4.15, $L = 1$ from the slope at the initial condition, and $\beta_k = 1/2$. Thus the maximum timestep that can be used with the trapezoidal rule implemented with fixed point iteration for this system is $T = 2$.

4.17 INTRODUCTION TO PREDICTOR-CORRECTOR METHODS

The previous section considered an iterative procedure for implicit methods. This approach requires an initial guess of the solution at each timestep. The question that arises is, How can we make the best possible guess? It would seem that using an explicit method, of the same order as the implicit method, to make the guess might be a good choice. It turns out that this is the case. Usually the explicit method is called a *predictor*. Likewise, the implicit method is called a *corrector*. Used together in this way, the combination is called a *predictor-corrector method*. First we discuss how to implement these methods and then present a detailed derivation of the closed-loop transfer function. Finally, we give the absolute stability and weak stability properties of these methods along with techniques for finding stability regions.

Normally, for a simulation using explicit methods, or predictors, the procedure is to evaluate (E) the derivative function, then predict (P) the next value. This process, PE, is repeated indefinitely or until the end of the simulation. Using a corrector modifies this procedure. Following the predict (P) and evaluate (E), there is a correct (C), followed by another evaluate (E). The process is then PECE (Lambert, 1973, also discusses the PEC-only mode and shows that it is reducible to an explicit method, and hence it will not be pursued here). The recursive approach of the last section allowed multiple iterations of the corrector. In the current notation this would be PECECECE . . . or $P(EC)^m E$, where m is the number of iterations of the corrector. Some authors, such as Lambert, state that correcting to convergence gives the process the properties of the corrector. This is not exactly true because the convergence of the iterations depends on the choice of timestep. This will be shown more completely later via the stability regions of the $P(EC)^m E$ process.

In choosing specific methods for predictor-corrector pairs, the following guidelines are useful. Do not choose the order of the predictor higher than that of the corrector. If so, the local truncation error will still approach that of the corrector, which is greater than that of the predictor. Hence, we would not want to correct, as the correction could reduce the accuracy of the prediction. If the order of the predictor is lower than the order of the corrector, Lambert (1973) states that the order of the local truncation error approximately improves by 1 for every correction made until it approaches that of the corrector. It should be clear that the order of the predictor should be equal to the order of the corrector; otherwise unnecessary computations must be done.

A simple example of implementing this procedure follows. It should be clear that the complexity of the program has not significantly increased.

Example 4.16

Here the chaotic Duffing double-scroll (Hartley, 1988) system is simulated. The equations are

$$\dot{x} = 10y - 10(0.2857x^3 - 0.1429x)$$
$$\dot{y} = x - y + z$$
$$\dot{z} = -14.28y$$

We will use the forward Euler–backward Euler predictor-corrector pair. The BASIC program for this simulation is

```
T = 0.01
x = 0.7
y = 0
z = -0.7
for n = 1 to 100000000
    fx = -10*y - 10*(-0.1429*x + 0.2857*x^3)

    fy = x - y + z                              Derivative evaluations
    fz = -14.28*y
    xp = x + T*fx
```

```
        yp = y + T*fy                              Predict next states
        zp = z + T*fz
        fxp = -10*yp - 10*(-0.1429*xp + 0.2857*xp^3)
        fyp = xp - yp + zp                         Predict derivatives
        fzp = -14.28*yp
        x = x + T*fxp
        y = y + T*fyp                              Correct next states
        z = z + T*fzp
        pset(100 + 50*x,100 - 50*y)                Phase plane plot
    next n
end
```

4.18 WEAK STABILITY OF PREDICTOR-CORRECTOR METHODS

Weak stability for predictor-corrector methods is considerably more complicated than that for either separately, mainly because of the much more complicated closed-loop transfer function of the process. This will now be derived below.

The following definitions must be made initially. Let the first and second characteristic polynomials of the predictor and corrector be

$$\rho_p(z) = \sum_{j=0}^{k_p} \alpha_{pj} z^j, \; \sigma_p(z) = \sum_{j=0}^{k_{p-1}} \beta_{pj} z^j \tag{4-64}$$

and

$$\rho_c(z) = \sum_{j=0}^{k_c} \alpha_{cj} z^j, \; \sigma_c(z) = \sum_{j=0}^{k_c} \beta_{cj} z^j \tag{4-65}$$

Then Figure 4.30 gives the block diagram of the predictor-corrector pair implemented for the linear test equation in the PECE mode.

The closed-loop transfer function, $\dfrac{X_c(z)}{U(z)}$, will now be obtained through block diagram algebra. The algebra and the result are simplified if the characteristic polynomials above are each multiplied by enough powers of z to make $k = k_c = k_p$. If we start with the predictor equation, its input values all depend on outputs from past corrections (obviously we would not use the predicted values since we have better, corrected ones). Then

$$z^k X_p = [z^k - \rho_p] X_c + T\sigma_p \dot{X}_c, \text{ but } \dot{X}_c = \lambda X_c + U \tag{4-66}$$

where the function of z notation has been dropped for convenience. This gives the predicted output

$$X_p = \frac{z^k - \rho_p}{z^k} X_c + \frac{\lambda T\sigma_p}{z^k} X_c + \frac{T\sigma_p}{z^k} U \tag{4-67}$$

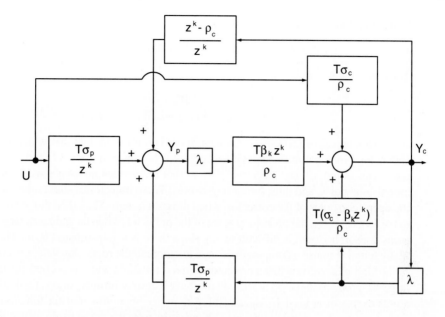

Figure 4.30 Block diagram of PECE mode.

This is the predict step. It is followed by the evaluate step,

$$\dot{X}_p = \lambda X_p + U \tag{4-68}$$

Then the correction is done:

$$\rho_c X_c = T\beta_k z^k \dot{X}_p + T(\sigma_c - \beta_k z^k)\,\dot{X}_c \tag{4-69}$$

or

$$X_c = \frac{T\beta_k z^k}{\rho_c}(\lambda X_p + U) + \frac{T(\sigma_c - \beta_k z^k)}{\rho_c}(\lambda X_c + U) \tag{4-70}$$

Combining the two terms in U, we get

$$X_c = \frac{\lambda T\beta_k z^k}{\rho_c}X_p + \frac{\lambda T(\sigma_c - \beta_k z^k)}{\rho_c}X_c + \frac{T\sigma_c}{\rho_c}U \tag{4-71}$$

To get the closed-loop transfer function, it is now necessary to substitute the predicted value of X into this equation:

$$\rho_c X_c = \lambda T\beta_k z^k[(z^k - \rho_p)X_c + \lambda T\sigma_p X_c + T\sigma_p U] + T\sigma_c U + \lambda T(\sigma_c - \beta_k z^k)X_c \tag{4-72}$$

Moving all of the corrected X's to the left, we get

$$[\rho_c - \lambda T\beta_k(z^k - \rho_p) - (\lambda T)^2\sigma_p\beta_k - \lambda T(\sigma_c - \beta_k z^k)]\,X_c = [\lambda T^2\beta_k\sigma_p + T\sigma_c]\,U \tag{4-73}$$

Cancelling like terms and rearranging, we find

$$[(\rho_c - \lambda T\sigma_c) + \lambda T\beta_k\rho_p - (\lambda T)^2\beta_k\sigma_p]\, X_c = T\,[\lambda T\beta_k\sigma_p + \sigma_c]\, U \qquad (4\text{–}74)$$

The transfer function is then

$$\frac{X_c(z)}{U(z)} = \frac{T\,[\sigma_c + \lambda T\beta_k\sigma_p]}{(\rho_c - \lambda T\sigma_c) + \lambda T\beta_k(\rho_p - \lambda T\sigma_p)} \qquad (4\text{–}75)$$

This transfer function is somewhat more complicated than that which we normally encounter. Note that in the denominator there is both a gain term, λT, and a gain-squared term. This is unfortunate since our normal root locus rules no longer work. There are k-closed-loop poles, and thus, k-root trajectories. These trajectories start at the roots of $\rho_c(z)$, the open-loop poles of the corrector, when the gain is zero. They then move around and go to the roots of $\sigma_p(z)$, the open-loop zeros of the predictor, when the gain gets large. What the locus does in between is difficult to say since there is a gain-squared term. The only way really to tell is to use a computer package to compute the roots. Anyway, the consequence of the high-gain zeros is that a predictor-corrector method will always become unstable for large λT-products since $\sigma_p(z)$ will always have a zero at infinity, or $\rho_c(z)$ will always have a pole excess of at least 1. Another thing to notice about this transfer function is that the zeros now move around as the gain changes. In fact, the zeros begin at the zeros of the corrector and go to the zeros of the predictor as the gain increases. Some examples of the analysis of predictor-corrector methods will be given later.

Although not derived here, the characteristic polynomial for the P(EC)^{m}E mode is (see Lambert, 1973, p. 97)

$$\Delta(z) = [\rho_c(z) - \lambda T\sigma_c(z)] + M_m[\rho_p(z) - \lambda T\sigma_p(z)] \qquad (4\text{–}76)$$

where the nonlinear gain term is

$$M_m = \frac{(\lambda T\beta_k)^m(1 - \lambda T\beta_k)}{[1 - (\lambda T\beta_k)^m]} \qquad (4\text{–}77)$$

Lambert suggests that as m gets large with $\lambda T\beta_k < 1$, M goes to zero, and thus the characteristic equation becomes that of the corrector. This appears to follow correctly; however, it is not true. The term never really goes to zero, and the corrector never really takes over because the iterations of the corrector also work to determine stability. This will be clearer when the stability regions are presented.

4.19 AN EXAMPLE PREDICTOR-CORRECTOR STABILITY STUDY

As an example of the above analysis, consider the forward Euler–backward Euler predictor-corrector pair. Then

$$\rho_p = z - 1,\ \sigma_p = 1,\ \rho_c = z - 1,\ \sigma_c = z,\ \text{thus } \beta_1 = 1$$

From the transfer function equation given above, the closed-loop transfer function becomes

$$H(z) = \frac{X_c}{U} = \frac{T[z + \lambda T]}{[z - 1 - \lambda Tz] + \lambda T[z - 1 - \lambda T]}$$

or

$$H(z) = \frac{T[z + \lambda T]}{z - 1 - \lambda T - (\lambda T)^2}$$

The closed-loop pole is thus at $z = 1 + \lambda T + (\lambda T)^2$, and the closed-loop zero is at $z = -\lambda T$. The root locus is given in Figure 4.31. It should be noticed that the closed-loop pole trajectory starts at $z = +1$, moves to the left, stops, and then moves to the right through $z = +1$ and then outside the unit circle. The place where the trajectory turns around is of some interest since we would say that the method would probably be detuned for gains larger than this. The value of gain when this happens is easily found as follows:

$$\frac{dz}{d\lambda T} \equiv 0 = 1 + 2\,\lambda T, \text{ or } \lambda T = -\frac{1}{2}$$

Plugging this back into the pole equation, we get $z = +3/4$. The value of gain where the root leaves the unit circle at $z = +1$ is also of interest. This is found by setting the expression for the roots equal to $+1$:

$$z = 1 + \lambda T + (\lambda T)^2 = +1$$

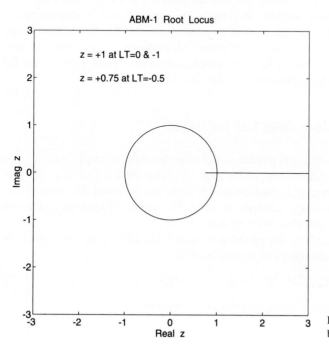

ABM-1 Root Locus

z = +1 at LT=0 & -1

z = +0.75 at LT=-0.5

Figure 4.31 Root locus: Forward Euler–backward Euler pair.

This equation is then solved for λT. The result is $\lambda T = 0, -1$. Thus the largest negative real λT-product for stability is -1. It is interesting to observe that a zero is always equal to a pole at $z = +1$ for some nonzero value of λT, for all predictor-corrector methods. For this method, it occurs for $\lambda T = -1$.

Also, for this method, it is very easy to determine the truncation error. Basically, the z-plane pole should be

$$z = e^{\lambda T} = 1 + \lambda T + \frac{(\lambda T)^2}{2} + \frac{(\lambda T)^3}{6} + \dots \qquad (4\text{--}78)$$

From the closed-loop pole position, it is clear that the leading error term is $(\lambda T)^2/2$ by subtracting the theoretically correct closed-loop pole position from the actual closed-loop pole position. Another interesting thing about this method, and many others to follow, is that instability occurs on the right side of the unit circle. In a simulation, this would manifest itself as a slow ramping off to infinity. Previously, instability was usually due to a pole on the left side yielding high-frequency oscillations. It is important to know the sources of instability for a given method so that they can be recognized easily in implementations.

A stability region can also be found for this method. The characteristic equation can be rewritten as $(\lambda T)^2 + \lambda T + (1 - z) = 0$. Solving for the gain, we get

$$\lambda T = -\frac{1}{2} \pm \sqrt{z - 3/4}$$

To get the λT-plane stability region, $z = re^{j\theta}$ is substituted into the right side of this equation. This is given in Figure 4.32. Notice that now the mapping is 1 to 2, rather than k to 1 as with regular linear multistep methods. In general, the mappings will be many to many, and thus very complicated. It is important to recognize here that for this method, every point in the z-plane has two images in the λT-plane. For example, $z = +1$ maps into both $\lambda T = 0$ and -1. This makes sense, as $\lambda T = -1$ corresponds to the z-plane pole leaving the unit circle. Also notice that the z-plane origin maps into $\lambda T = -0.5 \pm j\sqrt{3/4}$.

4.20 MODIFIERS IN PREDICTOR-CORRECTOR METHODS

Modifiers have often been used with predictor-corrector methods to improve accuracy at a modest cost and also to give an estimate of the local truncation error at each step. The idea is that since we know the local truncation error of both the predictor and the corrector, the leading term of each can be used to eliminate the error. This improved accuracy is obtained at a very modest computational cost. We now derive modifiers.

Assume first that the order of the predictor is equal to the order of the corrector. Then let the principal local truncation error of the predictor be

$$C_{P,p+1}T^{p+1}x^{(p+1)} = x_{\text{true}} - x_P + O(T^{p+2}) \qquad (4\text{--}79)$$

Likewise, let the principal local truncation error of the corrector be

$$C_{C,p+1}T^{p+1}x^{(p+1)} = x_{\text{true}} - x_C + O(T^{p+2}) \qquad (4\text{--}80)$$

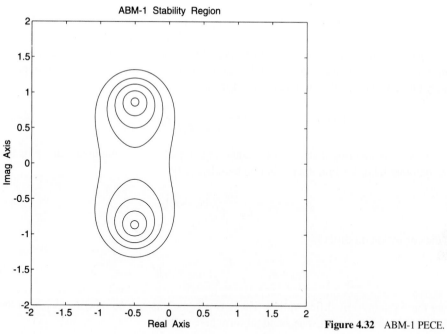

Figure 4.32   ABM-1 PECE.

Recall that the C's are the corresponding error constants. Subtraction of the second equation from the first gives a relationship between the computed values and their local truncation errors. If we drop the subscript $p + 1$ on the error constants, we find the relationship

$$(C_P - C_C)T^{p+1}x^{(p+1)} = x_C - x_P + O(T^{p+2}) \qquad (4\text{–}81)$$

An approximation to the local truncation error of the corrector can be found by multiplying the above by C_C. This gives

$$C_C [C_P - C_C]T^{p+1}x^{(p+1)} = C_C [x_C - x_P] + \text{ignored higher-order terms} \qquad (4\text{–}82)$$

Then the approximation is

$$\text{LTE}_{C,n} = C_C T^{p+1}x^{(p+1)} \cong \frac{C_C}{C_P - C_C} [x_C - x_P]_n \qquad (4\text{–}83)$$

Notice here that the subscript n has been added to denote the present timestep. Likewise an approximation to the local truncation error of the predictor can also be found. Repeating the above by multiplying by C_P, we get

$$C_P [C_P - C_C] T^{p+1}x^{(p+1)} = C_P [x_C - x_P] + \text{ignored higher-order terms} \qquad (4\text{–}84)$$

Then the approximation becomes

$$\text{LTE}_{P,n} = C_P T^{p+1} x^{(p+1)} \cong \frac{C_P}{C_P - C_C} [x_C - x_P]_n \tag{4-85}$$

Again the subscript n has been added, except now the LTE depends on the corrected value, which has not been computed yet. It is here assumed that the present error is approximately the same size as that at the previous timestep. Hence

$$\text{LTE}_{P,n} = C_P T^{p+1} x^{(p+1)} \cong \frac{C_P}{C_P - C_C} [x_C - x_P]_{n-1} \tag{4-86}$$

Now, remembering that the true value of x is equal to the computed value plus the corresponding LTE, we find the prediction modifier to be

$$\hat{x}_{P,n} = x_{P,n} + \frac{C_P}{C_P - C_C} [x_C - x_P]_{n-1} \tag{4-87}$$

and the correction modifier is

$$\hat{x}_{C,n} = x_{C,n} + \frac{C_C}{C_P - C_C} [x_C - x_P]_n \tag{4-88}$$

where the ^ refers to modified values. These two equations can now be inserted into the usual predict-correct modes. The standard form here will be

PM(EC)^{m}ME

It should be noted that the corrector modification is done after correcting to convergence, or as far as required. In general this technique is called a Milne's device (the user should be warned, however, that the method referred to as Milne's method should not be used, because of instability, and is not related to this).

To more clearly demonstrate this process, consider now the midpoint-trapezoidal predictor-corrector pair:

$$x_{n+2} = x_n + 2T\dot{x}_{n+1}, \quad x_{n+2} = x_{n+1} + \frac{T}{2} [\dot{x}_{n+2} + \dot{x}_{n+1}]$$

These methods are both second-order accurate methods with

$$C_{P,2} = \frac{1}{3}, \text{ and } C_{C,2} = \frac{-1}{12}$$

It should also be noticed that the explicit midpoint rule is unstable if used by itself on any stable system. Thus it follows that neither the predictor nor the corrector need be stable separately. Stability is still possible when they are used together. The modifier coefficients can now be computed:

$$\frac{C_P}{C_P - C_C} = \frac{4}{5}, \text{ and } \frac{C_C}{C_P - C_C} = \frac{-1}{5}$$

The PMECME mode for this system follows:

$$\text{P: } x_{P,n+2} = x_{CM,n} + 2T\dot{x}_{CM,n+1}$$

$$\text{M: } x_{PM,n+2} = x_{P,n+2} + \frac{4}{5}[x_{C,n+1} - x_{P,n+1}]$$

$$\text{E: } \dot{x}_{PM,n+2} = F(x_{PM,n+2})$$

$$\text{C: } x_{C,n+2} = x_{CM,n+1} + \frac{T}{2}[\dot{x}_{PM,n+2} + \dot{x}_{CM,n+1}]$$

$$\text{M: } x_{CM,n+2} = x_{C,n+2} - \frac{1}{5}[x_{C,n+2} - x_{P,n+2}]$$

$$\text{E: } \dot{x}_{CM,n+2} = F(x_{CM,n+2})$$

It should be noted that the modifiers use unmodified values on the right side. We hope that the general procedure follows directly from this example, but to clarify further, we present a more detailed example.

Example 4.17

Here the chaotic Duffing double-scroll (Hartley, 1988) system is simulated as in Example 4.8. We will use the forward Euler–backward Euler predictor-corrector pair with modifiers. The BASIC program for this simulation follows:

```
T = 0.01
x = 0.7
y = 0
z = -0.7
for n = 1 to 100000000
        fx = -10*y - 10*(-0.1429*x + 0.2857*x^3)
        fy = x - y + z                              Derivative evaluations
        fz = -14.28*y
        xp = x + T*fx
        yp = y + T*fy                               Predict next states
        zp = z + T*fz
        xh = xp + 0.5*ex
        yh = yp + 0.5*ey                            Modify prediction
        zh = zp + 0.5*ez
        fxp = -10*yh - 10*(-0.1429*xh +
        0.2857*xh^3)                                Predict derivatives
        fyp = xh - yh + zh
        fzp = -14.28*yh
        xc = x + T*fxp                              Correct next states
        yc = y + T*fyp
        zc = z + T*fzp
        ex = xc - xp
        ey = yc - yp
        ez = zc - zp                                Compute prediction error
        x = xc - 0.5*ex
        y = yc - 0.5*ey                             Modify correction
```

```
            z = zc - 0.5*ez
            pset(100 + 50*x,100 - 50*y)                  Phase plane plot
        next n
        end
```

4.21 WEAK STABILITY FOR PREDICTOR-CORRECTOR METHODS WITH MODIFIERS

These methods are indeed very complicated to analyze in an input-output form. It is important to do so, however, to understand the process and its accuracy and stability properties.

The extremely tedious algebra involved in the PMECME mode results in the following transfer function [Marinchek (1992)]:

$$\frac{\hat{X}_c(z)}{U(z)} = \frac{\begin{aligned}&T[z^k[\sigma_c(z) + M_c\sigma_c(z) - M_c\sigma_p(z)]\\ &+ \lambda T\beta_k z^k \sigma_p(z)[1 + M_c - M_p z^{-1}]]\end{aligned}}{\begin{aligned}&z^k(1 + M_c)[\rho_c(z) - \lambda T\sigma_c(z)] - M_c z^k[1 - \lambda T\sigma_p(z)]\\ &+ \lambda T z^k \beta_k(1 + M_c - M_p z^{-1})[\rho_p(z) - \lambda T\sigma_p(z)]\end{aligned}} \tag{4-89}$$

Notice that when the modifier terms go to zero, the usual PECE transfer function results.

It should also be noted that the modifiers look very similar to the standard form for filters, estimators, and identifiers, that being

$$\text{New guess} = \text{old guess} + \text{gain} * \text{error} \tag{4-90}$$

Much more will be done with modifiers in the next chapter, where their stability regions will receive considerable study.

4.22 VARIABLE TIMESTEP AND VARIABLE ORDER METHODS

Historically, modifiers have been used not only to provide greater accuracy but also to reduce the stepsize when the error, $E = e_n = (x_C - x_P)_n$, becomes large. For example, a simulation might contain the statements

$$\text{If } E > \text{MAX then } T = T/2 \text{ (to maintain accuracy)}$$
$$\text{If } E < \text{MIN then } T = 2T \text{ (to increase speed)}$$

Variable Timestep

Some effort has been given to devising more elaborate step control policies than these. Basically there are four major considerations when using a variable stepsize method.

1. Error control: The timestep must be reduced until the error is less than the maximum allowed in any given timestep. Clearly, this requires restarting the method whenever the timestep is reduced.

2. Speed control: As long as the errors are staying small, the timestep should be increased to increase the speed of the simulation. There should be maximum T bounds, however, as discussed next.
3. Stability control: The timestep must be chosen at each point in time so that numerical stability and accuracy are maintained. This is fairly easy for linear systems; however, this is not so clear for nonlinear systems.
4. Convergence control: If the method is correcting to convergence, the timestep must be chosen small enough so that the corrector still converges but large enough so that it converges fairly rapidly at each point in time.

Many specialized methods are available for varying stepsize and they will not be discussed here. The interested reader should consult Shampine and Gordon (1975). The Runge-Kutta methods of Chapter 6 are much more appropriate for implementations of variable stepsize since they do not need to be restarted with k initial conditions, and no knowledge of the error constants is required.

Variable Order Methods

It turns out that varying the order of accuracy is also possible as a means of creating adaptive methods that maintain speed and accuracy. The book by Shampine and Gordon (1975) is a good reference for this material; however, a particularly enlightening historical illustration from Lambert (1973) is repeated here.

To begin this section it is important to define forward and backward differences.

First forward difference: $\Delta f_n = f_{n+1} - f_n$

Second forward difference: $\Delta^2 f_n = \Delta(\Delta f_n) = \Delta(f_{n+1} - f_n) = f_{n+2} - 2f_{n+1} + f_n$ and so on.

First backward difference: $\nabla f_n = f_n - f_{n-1}$

Second backward difference: $\nabla^2 f_n = \nabla(\nabla f_n) = \nabla(f_n - f_{n-1}) = f_n - 2f_{n-1} + f_{n-2}$ and so on.

Usually these operators are used to create a table for hand calculation. For example, the forward difference operator could be applied to $f = sin(x)$ as below.

x_n (degrees)	$f_n(x_n) = sin(x_n)$	Δf_n	$\Delta^2 f_n$	$\Delta^3 f_n$	$\Delta^4 f_n$
0	0				
		.174			
10	0.174		−.006		
		.168		−.004	
20	0.342		−.010		−.001
		.158		−.005	
30	0.500		−.015		
		.143			
40	0.643				

Before the advent of computing machinery, such tables were used in conjunction with difference operator expansions, such as that of Newton-Gregory, to simulate a system. For example, the Adams-Bashforth methods can be expressed as

$$x_{n+1} = x_n + T\left[1 + \frac{1}{2}\nabla + \frac{5}{12}\nabla^2 + \frac{3}{8}\nabla^3 + \frac{251}{720}\nabla^4 + \frac{95}{288}\nabla^5 + \dots\right]\dot{x}_n \qquad (4\text{--}91)$$

and the Adams-Moulton methods can be expressed as

$$x_{n+1} = x_n + T\left[1 - \frac{1}{2}\nabla - \frac{1}{12}\nabla^2 - \frac{1}{24}\nabla^3 - \frac{19}{720}\nabla^4 - \frac{3}{160}\nabla^5 + \dots\right]\dot{x}_{n+1} \qquad (4\text{--}92)$$

By truncating these series after j terms, we get a j step Adams method (try it). Consequently, a difference table could be created by hand for $\dot{x}$. Whenever some high-order difference became higher than a prescribed maximum value, the higher differences could be added directly to the lower-order result. In this way variable order methods were available before computing machinery.

4.23 CONCLUSIONS

Linear multistep methods are extremely versatile and easy to use. Furthermore, a rigorous analytical theory is available for completely understanding how any given method will perform on any given linearized system. Their performance on nonlinear systems will be considered in Chapter 8. Additionally, the methods developed here will be extended to higher-order integrals in Chapter 9, where it will be shown that all of the substitution methods of the previous chapter may be derived from a single equation representing a linear k-step approximation to m-integrals. Before continuing this discussion, the next chapter will attempt to provide more insight into stability regions. As these are probably the most important tools available to the simulation designer in choosing a method and a timestep, it is extremely important to have a clear understanding of the information they contain.

PROBLEMS

1. Use the two-step Adams-Bashforth method as an operational substitutional method in the following system.

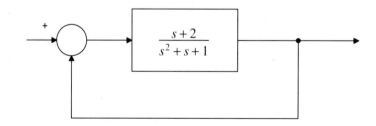

Make an intelligent choice for the timestep and simulate for the following input.

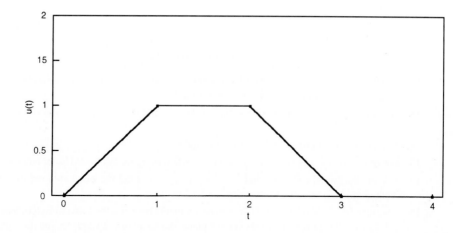

2. Consider the ABM-2 predictor-corrector method, which uses the two-step Adams-Bashforth predictor and the trapezoidal rule as corrector. Using it in the PECE mode on the system $x' = \lambda x + u$, determine the pole and zero loci for λ negative. Indicate the interval of absolute and relative stability and comment on the general utility of the method.

3. You are to derive an explicit two-step method that is first-order accurate. A specific system to be simulated in real time is $x' = -100x + u$. You are constrained by the hardware to use a sampling time of $T = 0.1$ seconds. Force both of the closed-loop poles to be at the origin and indicate where the principal root should be. Comment on the effectiveness of this method versus others. Assume the system is at rest and $u(t)$ = unit step at $t = 0$; simulate the system by using this method for $t = 0 \rightarrow 1$ seconds.

4. The following equation describes the operation of a DC motor: $x'' + x' = u$. Obtain expressions for the simulated output x_n when $u = \delta(t)$ by using the four-step Adams-Bashforth method and the two-step Adams-Moulton method. Compare these with the expression for the exact results. Let $T = 0.2$ above.

5. Obtain the steady-state response for the system $x'' + 0.05x' + x^3 = 0.014 \cos(t) + 0.005$. Include both a phase plane plot and the time response (with the input sinusoid). Choose $x(0) = -0.598$, $x'(0) = -0.0088$.

6. Simulate the system $x' = 100(\sin(5t) - x)$, $x(0) = 1$, using the second-order Adams-Bashforth method with $T = 0.0099$ second for the $t = 0 \rightarrow 5$ seconds, and discuss.

7. We want to implement a variable stepsize algorithm based on the AB-2, trapezoidal predictor-corrector method. The system to be simulated will typically be of the form $\dot{x} = -10(1.2 + \sin(t))x + u$ with $|\delta u/\delta t| < 1$. Discuss the approach, but do not program.

8. Given the second-order system $(1/((s+5)(s+1)))$
   a. Obtain a discrete representation by using the AB-2 method.
   b. Plot the z-plane poles versus T and compare the locations with where they should be.
   c. Map the discrete poles back into the $\hat{s}$-plane and compare.
   d. Suggest a range for T and discuss.

9. Below is an approximate model for a helicopter near hover. Simulate it by using at least a two-step method. Assume the system is at rest and is excited with a square wave input. Make an intelligent choice for the integration timestep

$$
\begin{bmatrix} \dot{q} \\ \dot{\theta} \\ \dot{u} \end{bmatrix} = \begin{bmatrix} -0.4 & 0 & -0.01 \\ 1 & 0 & 0 \\ -1.4 & 9.8 & -0.02 \end{bmatrix} \begin{bmatrix} q \\ \theta \\ u \end{bmatrix} + \begin{bmatrix} 6.3 \\ 0 \\ 9.8 \end{bmatrix} \delta
$$

10. Use the AB-2 method as a substitution operator for $H(s) = 100s/(s + 50)$. Determine the discrete transfer function and suggest a good choice for T.

11. Hermite's predictor is $H(z) = (4z + 2)/(z^2 + 4z - 5)$, and it is the most accurate two-step explicit method. Discuss its usefulness.

12. What is the only optimal and maximal method?

13. For $H(s) = 1/(s - 10)$, is $T = 1$ a good choice when using the backward Euler method?

14. Use the backward Euler method for $x' = -x^2$, $x_0 = 1$. Find the exact solution of the implicit method one step into the future when $T = 10$.

15. Consider $H(s) = 1/(s^2 + 2s + 2)$. It is to be simulated by using the forward Euler method with $T = 1$. Locate the appropriate poles in the s-plane, the λT-plane, the z-plane, and the $\hat{s}$-plane.

16. Find the modifier equations for the forward Euler–backward Euler predictor-corrector pair.

17. Design an integrator that is at least first-order accurate and has all its spurious roots at the origin and its principle roots and the exact mapping of λT when $\lambda T = -2$. Plot a root locus for real λT and the stability region for the resulting method.

18. The "second-open integration" linear multistep method is $x_{k+3} = x_k + (3T/2)(x'_{k+2} + x'_{k+1})$. Determine its order for accuracy, error constant, and applicability to the simulation of dynamic systems.

19. Plot the stability region of the trapezoidal rule. Indicate regions of absolute stability, relative stability, and detuning.

20. Study the usefulness of the delayed Trapezoidal rule

$$
H(z) = \frac{\dfrac{T}{z}(z + 1)}{2^m (z - 1)}
$$

as m increases from 0.

5

Stability Regions Revisited

5.1 INTRODUCTION

The last chapter introduced linear multistep methods and the concept of stability region in the complex λT-plane. The stability region was then shown to be a very useful device for determining the stability of a given integration method for a given linearized system. This chapter expands the discussion on stability regions and demonstrates how they can be used to enhance the understanding of how an integrator works.

5.2 MORE ABOUT FREQUENCY RESPONSES

Frequency response techniques are very useful for control engineers and, consequently, will be very useful for the study of simulation methods. Some background material on the frequency response technique is given here for later use.

Consider the control configuration shown in Figure 5.1 and discussed in Maciejowski, (1989), and Horowitz (1963).

The signal $r(t)$ is a desired reference input; the signal $d(t)$ is some generally unknown disturbance acting on the plant and includes plant uncertainty; the signal $m(t)$ is the error incurred in measuring the plant output, $y(t)$; and $u(t)$ is the control signal applied to the plant. The closed-loop transfer function is computed to be

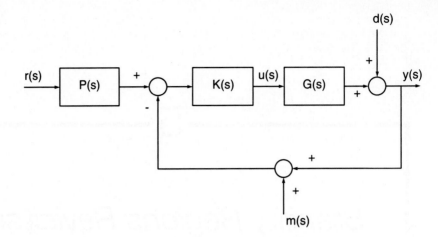

Figure 5.1   Standard feedback configuration for robust control system design.

$$y(s) = S(s)d(s) + H(s)r(s) - T(s)m(s) \qquad (5\text{--}1)$$

Here

$$S(s) = \frac{1}{1 + G(s)K(s)} \qquad (5\text{--}2)$$

is known as the *sensitivity* and indicates sensitivity of $y(t)$ to disturbances; the equation

$$T(s) = \frac{G(s)K(s)}{1 + G(s)K(S)} = S(s)G(s)K(s) \qquad (5\text{--}3)$$

is known as the *complementary sensitivity* and indicates sensitivity of $y(t)$ to measurement errors; and

$$H(s) = \frac{P(s)G(s)K(s)}{1 + G(s)K(s)} = T(s)P(s) = S(s)G(s)K(s)P(s) \qquad (5\text{--}4)$$

is the closed-loop transfer function indicating sensitivity of $y(t)$ to the reference signal. Often the prefilter $P(s) = I$, and then $H(s) = T(s)$. Also note that

$$T(s) + S(s) = 1 \qquad (5\text{--}5)$$

This equation indicates that there is a trade-off between reducing the effects of disturbances and reducing the effects of measurement noise. The magnitude of $T(s)$ is also an indication of the control effort required, $u(t)$. The trade-off is usually managed by making $T(j\omega)$ small at one set of frequencies and then making $S(j\omega)$ small at another set of frequencies. Most often, the measurement noise $m(t)$ will have a very large bandwidth and $T(j\omega)$ will be made small at high frequencies. Correspondingly, the disturbance signals are usually more of a low-frequency nature, and thus $S(j\omega)$ is made small at low frequencies. This trade-off is usually done by making the *loop gain, $L(s) = G(s)K(s)$*, large at low frequencies and small at high frequencies.

To study stability of the closed-loop system, control engineers often use the Nyquist plot. This is a polar plot of the loop gain, $L(j\omega)$. Equivalently, the Bode plot or Nichols plot could be used, but these are not as useful for simulation. The *Nyquist stability criterion* is usually given as follows:

For stability, the number of counterclockwise encirclements of the -1 point by $L(j\omega)$ must be equal to the number of right-half plane poles of $L(s)$.

For the simulation problem, the number of right-half plane poles of $L(s)$ will usually be zero, as we are trying to determine the simulation stability of a stable continuous-time system. The Nyquist plot should then avoid encirclements of the minus 1 point completely.

It should be pointed out that one of the most useful benefits of using the frequency response technique is that closed-loop information can be obtained directly from open-loop information, that being the polar frequency plot. Not only can stability be obtained, but also the frequency response plots of $T(s)$ and $S(s)$ can be obtained directly from the frequency response plot of $L(s)$. However, this does require some additional contours to be added to the usual Nyquist plane. We will first consider where the lines of constant magnitude and phase of $T(j\omega)$ lie in the Nyquist plane. This is a very common addition to the Nyquist plane, and its derivation can be found in D'Azzo and Houpis (1981) and Maciejowski (1989). The contours are given in Figures 5.2 and 5.3.

It should be observed that the contour of unity magnitude is a vertical line lying on $Re[L(j\omega)] = -0.5$. Everything to the left of this line amplifies closed-loop sensitivity to measurement errors, with -1 being infinite amplification. Everything to the right of this line attenuates measurement noise, with zero amplification at the origin.

Of additional interest are the lines of constant magnitude and phase of $S(j\omega)$. These are not well known but are perhaps more useful, as $S(s)$ also represents the transfer function from the reference input $r(t)$ to the error signal coming out of the summing junction. The derivation is now given. First

$$S(s) = \frac{1}{1 + L(s)} \tag{5-6}$$

The frequency response of this function can be found from the frequency response of the loop gain, $L(s)$. Let $L(j\omega) = U(j\omega) + jV(j\omega)$. Then

$$S(j\omega) = \frac{1}{1 + U(j\omega) + jV(j\omega)} \cdot \frac{1 + U(j\omega) - jV(j\omega)}{1 + U(j\omega) - jV(j\omega)}$$

$$= \frac{1 + U(j\omega) - jV(j\omega)}{[1 + U(j\omega)]^2 + [V(j\omega)]^2} \tag{5-7}$$

It follows that

$$|S(j\omega)| = \sqrt{\frac{1}{[1 + U(j\omega)]^2 + [V(j\omega)]^2}} \tag{5-8}$$

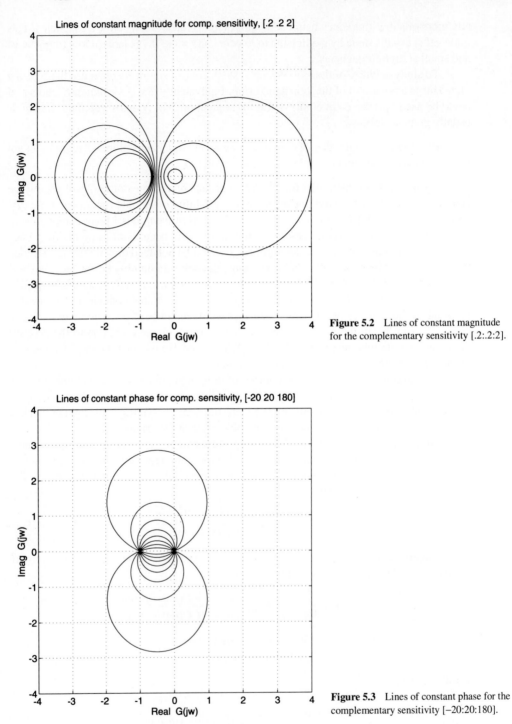

Figure 5.2 Lines of constant magnitude for the complementary sensitivity [.2:.2:2].

Figure 5.3 Lines of constant phase for the complementary sensitivity [−20:20:180].

Thus lines of constant magnitude of the sensitivity function are given by concentric circles in the Nyquist plane centered at −1. The magnitude of $S(j\omega)$ on a given circle equals the reciprocal of the radius of the given circle. Hence the circle of sensitivity magnitude equal to 1 is given by the circle of radius 1 centered at −1. Inside this circle, the sensitivity increases as −1 is approached. Outside the circle, the sensitivity falls to zero as the radius goes to infinity. The phase of the sensitivity function can be found as

$$\angle S(j\omega) = \tan^{-1}\left(\frac{-V(j\omega)}{1 + U(j\omega)}\right) \tag{5–9}$$

This then gives the phase as radial lines centered at −1. The phase is zero along the right side of the real axis from −1. It increases positively in a counterclockwise direction and increases negatively in a clockwise direction from this line. The plot of the sensitivity function is shown in Figures 5.4 and 5.5.

Given the Nyquist plot of the loop gain, it is now possible to read the closed-loop sensitivity and complementary sensitivity directly off the Nyquist plane with the above contours superimposed. This is done in Figure 5.6 for the Butterworth filter example considered in Chapter 4.

As such, the above information would be useful for the simulation problem if a given integrator's open-loop frequency response were used. However, as shown earlier, the stability region for the integrator also happens to be the reciprocal of the frequency response.

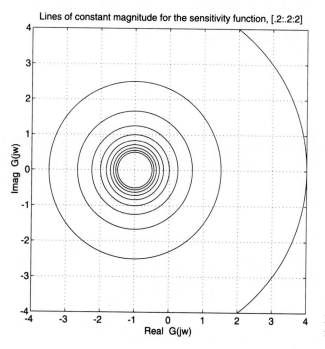

Figure 5.4 Lines of constant magnitude for the sensitivity function [.2:.2:2].

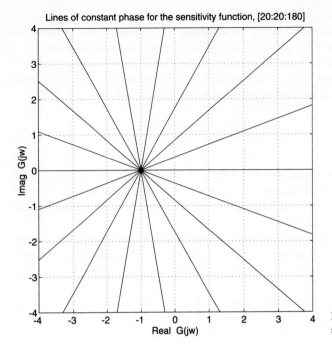

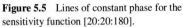

Figure 5.5 Lines of constant phase for the sensitivity function [20:20:180].

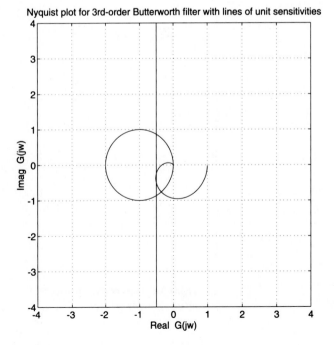

Figure 5.6 Nyquist plot for third-order Butterworth filter with lines of unity sensitivities.

Thus it will be of considerable interest to look at inverse Nyquist plots. A good discussion of this approach can be found in D'Azzo and Houpis (1981). First consider

$$\frac{y(s)}{m(s)} = T(s) = \frac{L(s)}{1 + L(s)} = re^{j\theta} \tag{5-10}$$

Then

$$\frac{m(s)}{y(s)} = \frac{1}{T(s)} = \frac{1}{L(s)} + 1 = \frac{1}{r}e^{-j\theta} \tag{5-11}$$

The lines of constant magnitude and phase of $T(j\omega)$ in the inverse polar, or inverse Nyquist, plot are found from this equation. These contours are very similar in shape to those of $S(s)$ in the regular Nyquist plot. Basically, lines of constant magnitude are circles centered around −1 with the magnitude varying inversely with the radius. The circle of radius 1 corresponds to a magnitude of 1, the circle of infinite radius corresponds to a magnitude of zero, and the circle of radius zero corresponds to a magnitude of infinity. The lines of constant phase are straight lines emanating from the minus 1 point. The phase is zero along the real axis to the right of −1, and increases negatively above this line in a counterclockwise direction. The figure representing this is given in Figure 5.7. This nice radial form is one of the reasons that the inverse polar plot is used.

The reciprocal of the sensitivity function can also be found. Remember that

$$\frac{y(s)}{d(s)} = \frac{1}{1 + L(s)} \tag{5-12}$$

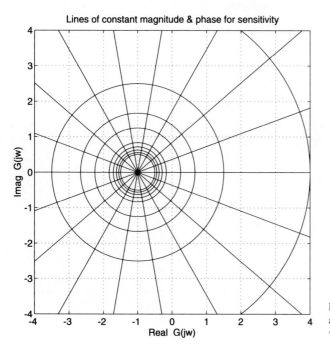

Figure 5.7 Lines of constant magnitude and phase for the sensitivity function and inverse complementary sensitivity.

Then

$$\frac{d(s)}{y(s)} = 1 + L(s) \tag{5–13}$$

which is also radial in $L(j\omega)$. However, we are plotting $1/L(j\omega)$ in this plane and we do not really have L. Thus, this equation is not so nice; rewriting it into a more convenient form, we get

$$\frac{d(j\omega)}{y(j\omega)} = 1 + \frac{1}{1/L(j\omega)} \tag{5–14}$$

It turns out, then, that the contours of $S(s)$ in the inverse plane are similar to those of $T(s)$ in the regular plane. This is shown in Figure 5.8.

These two figures will be of use for investigating the accuracy and stability of the numerical integration process.

5.3 STABILITY REGIONS AND THE INVERSE NYQUIST PLOT

In the last section, it was shown that the open-loop frequency response of a feedback control system contains a great deal of information. If the plant in the control loop is considered to be the integrator, a lot of information about the integration process can be obtained from its frequency response. Interpreting this frequency response in the usual Nyquist plane

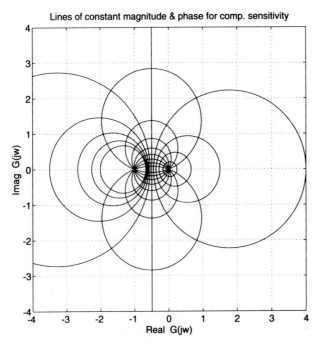

Figure 5.8 Lines of constant magnitude and phase for the complementary sensitivity and inverse sensitivity.

would be very useful. However, interpreting it in the inverse Nyquist plane is even more useful since the reciprocal of the integrator's frequency response also coincides with its stability region (within a scale factor of T). Thus the inverse of the frequency response of the integrator will prove to be of great use for predicting accuracy and stability. Remember that with the inverse Nyquist plot, the -1 point should be encircled rather than avoided.

For the integrator, the closed-loop transfer function is

$$\frac{X(z)}{U(z)} = \frac{T\sigma(z)}{\rho(z) - \lambda T\sigma(z)} \tag{5-15}$$

Its reciprocal is

$$\frac{U(z)}{X(z)} = \frac{\rho(z) - \lambda T\sigma(z)}{T\sigma(z)} = \frac{\rho(z)}{T\sigma(z)} - \lambda \tag{5-16}$$

This last expression would meet our needs better if it was multiplied through by T to make it look more like the stability region equation. The practical effect is that we can work with a normalized frequency response and change the constant λT at will. The resulting closed-loop gains read from the inverse Nyquist plot must then be scaled by $1/T$. This is considered to be much easier than changing the size of the stability region. The resulting equation is

$$\frac{TU(z)}{X(z)} = \frac{\rho(z)}{\sigma(z)} - \lambda T \tag{5-17}$$

Three interpretations of these last two equations will now be given.

1. This form is slightly different from the usual inverse Nyquist form, where the gain is usually unity; here it is λT. In this form, the stability region must encircle the λT point rather than -1, which was already known from the stability region concepts discussed earlier.
2. Equivalently, and alternatively, the more proper use of the inverse Nyquist plot (which is not quite as easy) is to move the stability region by the compensator's frequency response. In this case, it is simply the constant λT. Consequently, the integrator stability region must still encircle the origin given this offset. It should be remembered that the constant λT can generally be complex, corresponding to the system being simulated having complex eigenvalues.
3. Another way to interpret this information is to return to the first equation, offset the stability region by λ, and then scale the size of the stability region by T until the origin is encircled.

These three equivalent approaches are shown in Figures 5.9, 5.10, and 5.11.

Now we will consider how to interpret these plots in terms of the closed-loop frequency responses. From the previous section, we know that the constant closed-loop frequency response magnitude curves are circles centered around the Nyquist point, -1. We can now center these around our Nyquist points, which are λT, for all λ. This is done for the

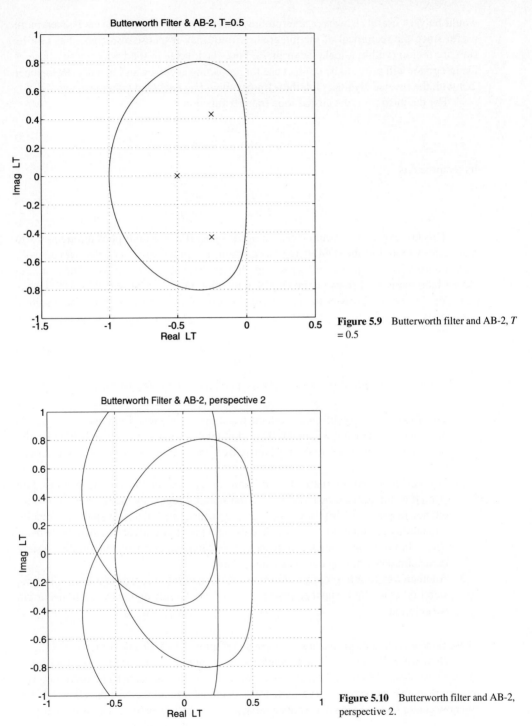

Figure 5.9   Butterworth filter and AB-2, $T = 0.5$

Figure 5.10   Butterworth filter and AB-2, perspective 2.

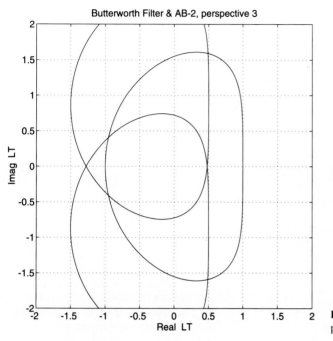

Figure 5.11 Butterworth filter and AB-2, perspective 3, $T = 0.5$.

two methods in Figures 5.9 and 5.10. It can be observed that the second method is closer in nature to the usual inverse Nyquist plot definition. However, it is suggested that the first method is probably more appropriate for the simulation problem since it coincides more closely with the stability region concept.

With the first method, it is now possible to determine the closed-loop frequency response corresponding to each pole of the system. It should be noted that the inverse Nyquist plot for the continuous system is a vertical line going up from the origin, that is, $s = j\omega$. This is what the stability region should aspire to. In terms of the closed-loop frequency response, with λ equal to a negative real constant, $T(s)$ then rolls off monotonically. The Bode response looks like a low-pass system, with a break at $\omega = \lambda$. Likewise, the sensitivity function $S(s)$ acts like a high-pass system, with a break at $\omega = \lambda$; see Figure 5.12.

To see how a discrete integrator would behave, consider the AB-2 stability region applied to the Butterworth filter problem earlier. For low frequencies, the behavior of $T(z)$ and $S(z)$ is very similar to the actual continuous system. However, for high frequencies, the stability region breaks into the left-half plane. Thus, for $T(z)$, it does not continue to roll off for higher frequencies, as shown in Figures 5.13 and 5.14. This means that some higher frequencies will propagate through the system with insufficient attenuation and considerable phase shift. If the timestep is chosen small enough, this effect is insignificant. For larger timesteps however, there can even be amplification of these high-frequency signals. It should be clear that the timestep should always be chosen so that high frequencies are not amplified by the closed-loop system. If this were the case, the simulation would be relatively unstable since the noise sources would be larger than the desired signal.

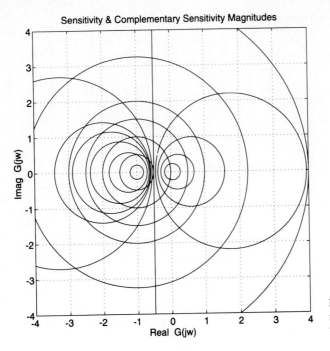

Figure 5.12 Superposition of constant magnitude lines of inverse sensitivity and inverse complementary sensitivity.

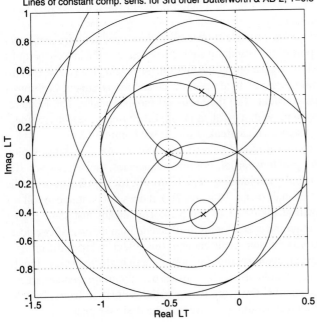

Figure 5.13 Lines of constant complementary sensitivity for third-order Butterworth filter and AB-2, $T = .5$.

Lines of constant comp. sens. for 3rd order Butterworth & AB-2, T=0.5, perspective 2

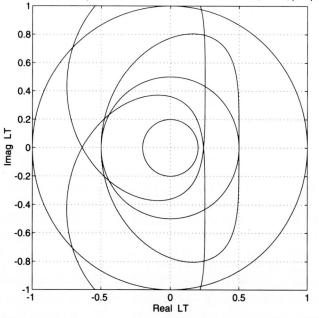

Figure 5.14 Lines of constant complementary sensitivity for third-order Butterworth filter and AB-2, $T = .5$, perspective 2.

The sensitivity function, which is here equivalent to $\dot{x}(z)/u(z)$, behaves in a similar manner. After leveling out at unity magnitude, the stability region breaks into the left-half plane. When this occurs, the sensitivity gets amplified significantly. Equivalently, the derivative in the loop, $\dot{x}$, also becomes amplified at those frequencies and significant error is incurred. Thus the sensitivity function also gives some indication of the error involved in the simulation process. Whenever the stability region breaks into the left-half plane, beyond the vertical line at $-1/2$, some amplification of high-frequency signals will occur. Consequently, this will occur for all explicit methods since their stability regions must be finite. Nonetheless, as long as T remains small, the high-frequency amplification in $\dot{x}$ will be minimal.

5.4 INSIDE THE LAMBDA-T-PLANE

Much can be learned about stability and accuracy from the stability region boundary. However, much more can be learned by considering the entire stability region, that is, the conformal mapping of the entire z-plane, through the integrator, into the integrator's λT-plane. From this, we are able to determine where all of the z-plane poles are located for a given λT-product. Furthermore, relative stability of the simulation can be determined directly by inspection of the inside of the given λT-plane, and thus we may more easily control spurious noise sources. This is a very powerful method, but it can also be difficult to visualize.

Remember that the boundary of the stability region is given by

$$\lambda T = \left[\frac{\rho(z)}{\sigma(z)}\right]_{z=e^{j\theta}} \tag{5-18}$$

Since we are now attempting to map the entire z-plane rather than just the unit circle, this must be modified to

$$\lambda T = \left[\frac{\rho(z)}{\sigma(z)}\right]_{z=re^{j\theta}} \tag{5-19}$$

where now different z-plane radii can be used. The process of mapping the entire plane is done in two steps. First the radius is held constant and the angle θ is varied from zero to 2π. The result will give the mapping of a given z-plane radius circle. This is repeated as many times as required. Then, the angle θ is held constant while the radius is varied from 1 to some smaller number. This task is performed easily with standard control system software packages such as those given in MATLAB in Appendix C (Chicatelli, 1989). The result is a conformal map of the z-plane into the λT-plane. Here angles are necessarily preserved. Alternatively letting $z = x + jy$ would map a z-plane rectangular grid into the λT-plane, but this is not as useful for the simulation problem. Here, we have essentially mapped the rectangular s-plane into the circular z-plane, where the integration is done, and then into the λT-plane of the given integrator, where analysis can be done. Thus the λT-plane should look like the s-plane for the given choice of T, with the understanding that it will obviously be distorted for larger values of T. Restating this, the λT-products should all be placed in the square region of the λT-plane, which is undistorted. The less distortion from squareness there is, the more accurate the resulting simulation will be because the useful region of the λT-plane resembles the s-plane. This is made much clearer by the following examples.

First consider Euler's method:

$$\lambda T = [z - 1]_{z=re^{j\theta}} = re^{j\theta} - 1 \tag{5-20}$$

The entire λT-plane for Euler's method is then very easy to plot. It is just the z-plane itself shifted to the left by 1. It is given in Figure 5.15 for several values of r and θ and superimposed on the s-plane, which contains lines of constant real part and lines of constant imaginary part.

Now, the idea is to choose T so that the λT-product is in the undistorted part of the stability region. For Euler's method, this is somewhere near the origin in the λT-plane. The farther away from the origin this product moves, the more distorted the plane becomes, with respect to the square s-plane.

To further demonstrate the use of the λT-plane, it is possible to read the z-plane pole locations directly from the locations of the λT-product. This is particularly easy for Euler's method. Generally, all we must do is determine which r-line and which θ-line pass through the given λT-plane point. This then gives the z-plane pole location. To demonstrate, recall the Butterwoth filter example from Chapter 4. The values of λ are $\lambda = -1, -0.5 \pm j0.866$.

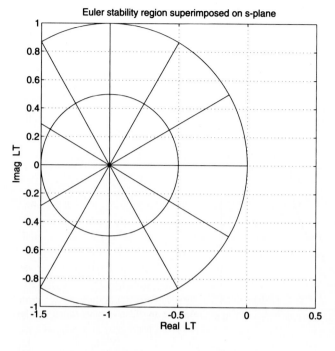

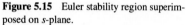

Figure 5.15 Euler stability region superimposed on *s*-plane.

Plotting lines on the Euler λT-plane with the angles of these poles, we derive Figure 5.16. There it is seen that choosing $T = 1$ gives $\lambda T = -1, -0.5 \pm j0.866$, and thus the corresponding *z*-plane poles are z = 0, $0.5 \pm j0.866$. If *T* is reduced to $T = 0.5$, then $\lambda T = -0.5, 0.5 \pm j0.433$, with the corresponding *z*-plane poles $z = 0.5, 0.75 \pm j0.433$.

Next consider the stability region for the AB-2 method. Shown in Figure 5.17 is the mapping of the radii $r = 1, 0.9, 0.8, 0.7,$ and 0.6. Remember that the contours taken in the *z*-plane are chosen to be counterclockwise. The mappings given above are also counterclockwise. Figure 5.18 gives the mapping of the radius $r = 0.5$. Notice that this closes on itself in a clockwise direction. Since this radius is not the stability boundary, this is not necessarily bad. However, it should also be noticed that every point on this contour lies in a part of the previous plot, which has associated with it a larger radius. What this means is that every point in this λ*T*-plane corresponds to two points in the *z*-plane. This is a little confusing but not surprising. The AB-2 is a two-step method and thus has two *z*-plane poles for every *s*-plane pole. This is a two-to-one mapping. All linear *k*-step methods will possess *k*-to-1 mappings, and thus each point in the λ*T*-plane corresponds to *k* points in the *z*-plane. Now it will be possible to determine the two *z*-plane poles from the given λ*T*-product for the AB-2 method. Continuing on to radii smaller than $r = 0.5$ gives larger clockwise enclosures in the λ*T*-plane. This should be interpreted as follows. Suppose λ*T* is chosen to be -0.1. This lies approximately on the $r = 0.9$ contour. Thus there will be a pole at $z \approx 0.9$. But also, underneath this point, on what is called the second Riemann sheet, is another contour of approximate radius, $r = 0.2$. Thus there will also be a *z*-plane pole on the $r \approx 0.2$ circle.

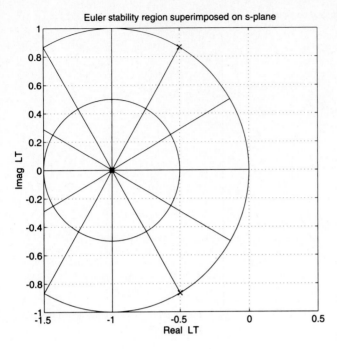

Figure 5.16   Euler stability region superimposed on *s*-plane.

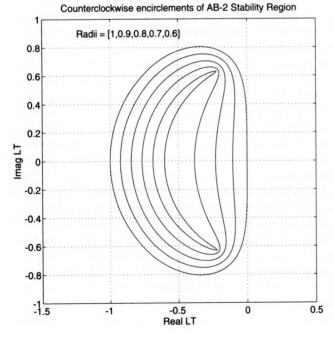

Figure 5.17   Counterclockwise encirclements of AB-2 stability region [1 .9 .8 .7 .6].

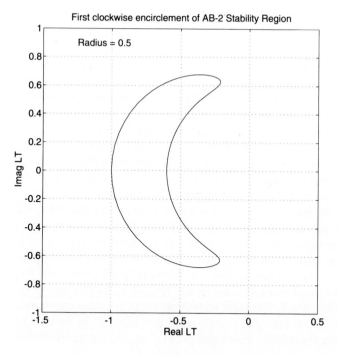

Figure 5.18 plot titled "First clockwise encirclement of AB-2 Stability Region", with "Radius = 0.5" label, y-axis "Imag LT" and x-axis "Real LT".

Figure 5.18 First clockwise encirclements of AB-2 stability region $r = 0.5$.

By looking at the lines of constant angle, we see that this other z-plane pole is at $z \approx -0.2$ and is a spurious pole. Now, we choose λT to be -0.5. Two radial contours and two angle contours also pass through this point. One set is approximately $r = 0.4$ and $\theta = 0$. The other is approximately $r = 0.6$ and $\theta = \pi$. Thus the principal root is at $z \approx 0.4$ and the spurious root is at $z \approx -0.6$. It should be observed that the larger radius, $r = 0.6$, passed through this point on the primary Reimann sheet but that here the equivalent s-plane was so distorted that the damped natural frequency was π. Thus it can be seen that as long as the large radii encirclements possess a smaller angle at a given point, the principal root will be larger than the spurious roots. Alternatively, if the large radii encirclements possess an angle larger than any smaller radii encirclement, the principal root will be smaller than some spurious root. Basically this occurs when the smaller radii encirclements begin to go in a clockwise direction. Inside, or usually to the right of where this occurs, the simulation will be relatively stable. Outside of this range, the simulation will be dominated by spurious noise.

Since there are multiple Riemann sheets, there will necessarily be branch points and branch cuts. The branch cuts can be made anywhere but are here defined to be along the lines where any given radius contour crosses itself. Then necessarily, the branch points will occur where an infinitesimally small turn is made by the contour, that is, where

$$\frac{d\lambda T}{dz} = \frac{\sigma(z)\rho'(z) - \rho(z)\sigma'(z)}{\sigma^2(z)} \equiv 0 \qquad (5\text{--}21)$$

Once the numerator of this equation is solved for z, this point is mapped into the λT-plane through the stability region equation. Branch cuts always connect branch points.

Although the location of the branch cuts is arbitrary, we will choose them to occur wherever a given constant z-plane radius line crosses over itself in the λT-plane. For AB-2, this occurs with $r \approx 0.57$. We can get only from one Riemann sheet to another through a branch cut. The two Riemann sheets for the AB-2 are given in Figures 5.19 and 5.20. Notice also that this derivative is identically +1 at the image of $z = +1$, which is $\lambda T = 0$. For the AB-2 these branch points are at $z = 0.3333 +/- j0.4714$. These correspond to the points $\lambda T = -0.2222 +/- j0.6285$. Usually inside the first branch cut will be the region of relative stability. Returning to the AB-2, we can see that near the λT-plane origin, the primary Riemann sheet looks like a good, undistorted copy of the s-plane. Moving out farther than 0.5 in any direction in the left-half λT-plane gives a well-distorted version of the s-plane, on either Riemann sheet. Thus the user can get an accurate idea of how good a simulation will be by looking at the λT-product locations in the given integrator's λT-plane.

We will now consider the trapezoidal rule. This will be much easier than the AB-2 since it is a one-step method. It is included in the discussion since it has been shown to be so useful. The entire stability region for the trapezoidal rule is shown in Figure 5.21. It should be noted again that the entire left-half λT-plane is within the stability boundary for this method. Referring back to the previous section, we should get good closed-loop frequency response and sensitivity properties since the inverse Nyquist locus goes straight up the imaginary axis. Clearly the method has good properties. There are some drawbacks, however. It should be observed that the z-plane origin maps into $\lambda T = -2$. To the right of this, and not too far up the axis, the stability region gives a fairly undistorted representation

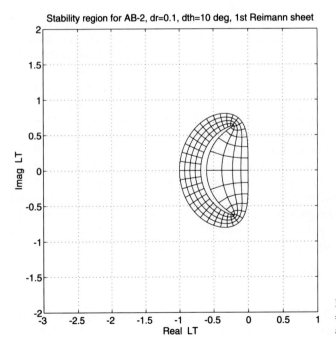

Figure 5.19 Stability region for AB-2, *dr* = 0.1, *dth* = 10 degree .
a. First Reimann sheet.

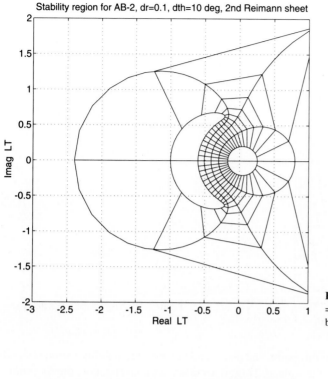

Figure 5.19   Stability region for AB-2, *dr* = 0.1, *dth* = 10 degree
b. Second Reimann sheet.

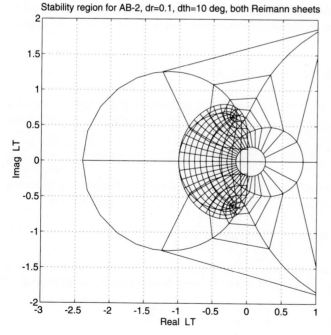

Figure 5.20   Stability region for AB-2, *dr* = 0.1, *dth* = 10 degrees, both Reimann sheets.

STABILITY REGION: TRAPEZOIDAL

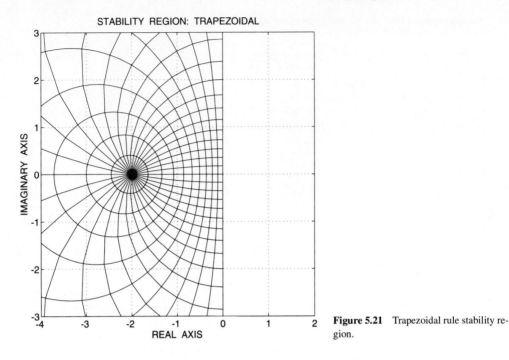

Figure 5.21 Trapezoidal rule stability region.

of the *s*-plane. To the left of this point, and above and below it, the stability region is significantly distorted relative to the *s*-plane. Thus, if accuracy is a consideration, the λT-product for the trapezoidal rule should not exceed 2 in magnitude. If it does, the simulation will be significantly detuned because of the loss of squareness, that is, distortion of the λT-plane.

5.5 A ROBUSTNESS ANALYSIS OF THE SIMULATION PROCESS

To understand completely the simulation process and integrator stability regions, we must understand how the error is propagated in a simulation. To accomplish this, the linear test equation is used:

$$\dot{x} = \lambda x + u \tag{5–22}$$

where negative real λT is assumed, as the analysis for complex λT is a straightforward extension. By Laplace transforming this equation, we get after some manipulation

$$\dot{X}(s) = \frac{s}{s - \lambda}\, U(s) \text{ and } X(s) = \frac{1}{s}\, \dot{X}(s) \tag{5–23}$$

This somewhat unusual representation for the linear test equation has been chosen because the signals $x(nT)$ and $\dot{x}(nT)$ are required to compute the error and its derivative,

$$E(z) = X(z) - \hat{X}(z) \text{ and } \dot{E}(z) = \dot{X}(z) - \hat{\dot{X}}(z) \tag{5–24}$$

where the approximate values from the simulation are

$$\hat{X}(z) = \lambda \hat{X}(z) + U(z) \quad \text{and} \quad \hat{X}(z) = TH(z)\hat{\dot{X}}(z) \tag{5--25}$$

Since we know the actual response for the linear test equation, it is these errors relative to the true solution that we will try to understand and control. To clarify this, we present the block diagram in Figure 5.22. Notice that the continuous-time solution given above is used in block diagram format so that x and its derivative are known and are driving the simulation, along with u. Also notice that the simulation loop has been configured so that it contains both the approximate simulation hatted variables and the error variables.

Several combinations for input and output are possible in this figure. Based on observer design, it is probably appropriate to consider the error variables as outputs with the intent of controlling their magnitude. Then, with x and $\dot{x}$ as the inputs, two important transfer functions are available:

$$\frac{E(z)}{X(z)} = \frac{1}{1 - \lambda TH(z)} \quad \text{and} \quad \frac{\dot{E}(z)}{\dot{X}(z)} = \frac{-\lambda TH(z)}{1 - \lambda TH(z)} \tag{5--26}$$

From these transfer functions, we can understand how the error, and its rate, change with respect to the actual system. This is probably best done in the frequency domain, which leads us back to the stability region concept. These two transfer functions are of primary importance.

The first one gives the magnitude of the error with respect to the magnitude of the true output. Also notice that this is in the form of a sensitivity function for $\lambda TH(z)$. Clearly this should be less than 1 at all frequencies. Unfortunately, consideration of Figure 5.23 indicates that this is difficult to achieve. This figure shows the stability region for the AB-2 integrator plotted on the inverse Nyquist plane along with the lines of constant sensitivity.

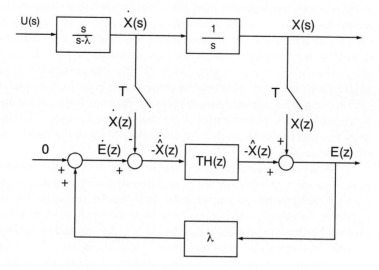

Figure 5.22   Configuration for simulation robustness analysis.

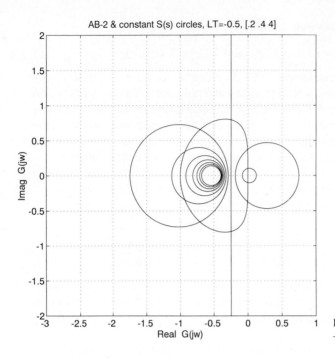

Figure 5.23 AB-2 and constant $S(s)$, LT = -0.5 [.2 .3 .5 .7 1 2 3 5 7 10].

Good error attenuation occurs at low frequencies; however, at higher frequencies, the stability region boundary breaks to the left to encircle λT, as it must for closed-loop stability. When this locus passes through $\lambda T/2$, error amplification is indicated. Fortunately, by choosing T small enough, this intersection occurs at high frequencies, and the amplification never becomes much greater than 1 because of the larger radius encirclement. That the sensitivity function gives the sensitivity to plant disturbances, or uncertainty, implies that the x-signal is the source of the plant uncertainty. Thus we want the sensitivity function to be small at those frequencies where the plant uncertainty is large; or alternatively, we allow the sensitivity function to be large where the disturbance signal is small. By choosing T small enough, the larger errors can be concentrated in the high frequencies, where the true system magnitudes are small, which keeps the error small also. Since the error signal is the difference between the true output, $x,$ and the simulated output, $\hat{x}$, it follows that this sensitivity function is also measuring the uncertainty between the discrete-time integrator and the true system $1/s$.

   The second transfer function gives the rate of error growth relative to the rate of true state. Also notice that this is in the form of a complementary sensitivity for $\lambda TH(z)$. Clearly we want the error rate to remain small. In particular, we want it to be less than 1 with respect to the rate of state changes; otherwise the error could grow faster than the state. Figure 5.24 shows the AB-2 stability region boundary plotted on the inverse Nyquist plane with the lines of complementary sensitivity. Here it is seen that the relative error rate can be made to be less than unity at all frequencies other than zero by choosing T small enough. That the complementary sensitivity function gives the sensitivity to measurement noise

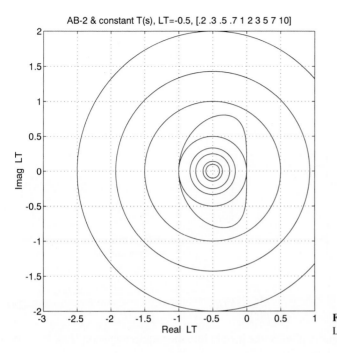

AB-2 & constant T(s), LT=-0.5, [.2 .3 .5 .7 1 2 3 5 7 10]

Figure 5.24 AB-2 and constant $T(s)$ circle, LT $= -0.5$ [.2:.2:4]

implies that the $\dot{x}$-signal is the source of the measurement noise with respect to the error rate in the simulation process. Thus we want the complementary sensitivity to be small at those frequencies where the measurement noise is large, that is, when the $\dot{x}$-signal is changing rapidly. The $\dot{x}$-signal will change at a rate near the natural frequencies of the particular system, at the λ's, and at the frequencies of the input signals. Fortunately, by choosing T so that the stability region boundary does not intersect the unity magnitude circle, we can ensure that the error rate will always have a signal-to-noise ratio greater than unity, except at DC, where it is unity. Notice also that this observation requires that the simulation timestep be chosen small enough to capture accurately the forcing function, u.

Another important transfer function is available when considering the integral relationship between x and $\dot{x}$, as in Figure 5.25. The resulting transfer function is

$$\frac{E(z)}{[\dot{x}(s)]} = \frac{\left[\dfrac{1}{s}\right]^* - TH(z)}{1 - \lambda TH(z)} \tag{5-27}$$

and is more complicated because of the sampled $1/s$ involved in the numerator. Rather than a frequency response analysis, a simple study of the numerator magnitude is considered. If the numerator can be made to be small, the error magnitude relative to the true state rate will be small. To accomplish this, we need to replace the sampled $1/s$ with its z-domain equivalent. Doing this and forcing the numerator to zero, we get

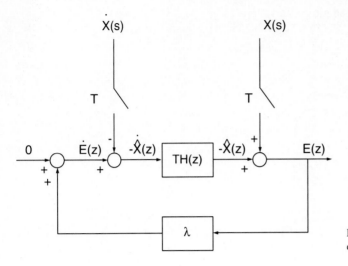

Figure 5.25   Configuration for simulation error robustness analysis.

$$\frac{T}{ln(z)} \approx TH(z) \tag{5-28}$$

This approximate identity gives some insight into the requirements on $H(z)$. Namely, $H(z)$ should approximate $1/ln(z)$ or

$$\frac{\rho(z)}{\sigma(z)} \approx 1n(z) \tag{5-29}$$

which is independent of the timestep (remember the numerator is still multiplied by T, which forces it to be small as T gets small). This is somewhat reminiscent of the Boxer-Thaler approach of Chapter 3. The better the rational transfer function for $H(z)$ can be made to approximate $ln(z)$, the better the resulting simulation. Unfortunately, the series expansion for $ln(z)$ is not very well behaved. Thus the usual control system approach of using dynamic error coefficient matching does not apply directly to this problem. However, matching up the two functions and their derivatives at a point (say $z = 1$), will work for solving the approximation problem. Doing this gives for the function matching

$$\frac{\rho(1)}{\sigma(1)} = ln(1) = 0 \tag{5-30}$$

which forces $\rho(1) = 0$, or that $(z - 1)$ is a factor of $\rho(z)$. Notice that this is the first of Lambert's equations. A match of derivatives with respect to z gives

$$\frac{\sigma(1)\rho'(1) - \rho(1)\sigma'(1)}{\sigma^2(1)} = \left[\frac{1}{z}\right]_{z=1} \tag{5-31}$$

Knowing that $\rho(1) = 0$, we reduce this to

$$\rho'(1) - \sigma(1) = 0 \tag{5-32}$$

which is the same as the second of Lambert's equations. Continuation of this process for higher derivatives gives the following general equation for the qth Lambert equation:

$$\left[\dfrac{d^q\left(\dfrac{\rho(z)}{\sigma(z)}\right)}{dz^q}\right]_{z=1} = \dfrac{d^q[ln(z)]}{dz^q}\Bigg|_{z=1} \tag{5–33}$$

It should also be recognized that this equation is simply the definition for the curvature of the stability region boundary at $z = 1$. Thus, the more accurate the simulation is from the Taylor series approach, the more of the derivatives in this equation will match, and thus the more nearly vertical the stability region will be at the λT-plane origin.

5.6 DISCUSSION OF SPECIFIC STABILITY REGIONS

Stability regions for several families of linear multistep methods are given in Appendix B. Specifically, the Adams-Bashforth methods are given for order equal to 1 to 4, the Adams-Moulton methods for order from 1 to 4, and the backward difference methods for order 1 to 4. It should be observed that in all cases, the stability regions shrink with increasing order. Also, it should be observed that the stability region of the implicit methods are always much larger than the stability regions of explicit methods having the same order. Note that the mappings are always k to 1.

5.7 PREDICTOR-CORRECTOR STABILITY REGIONS

Although some stability analysis was performed on predictor-corrector methods in the last chapter, stability regions were avoided. They are given in Appendix B for a variety of combinations of predictor and corrector, as well as number of iterations of the corrector. Some interesting features of these plots are "bubbles" and multiple branch cuts. The bubbles are caused by a 1-to-many mapping of the z-plane. The multiple branch cuts are caused by the use of multistep methods, yielding a k-to-1 mapping. Putting these together gives a mapping of k to $n + 1$, where k is the number of steps and n is the number of iterations of the corrector. As these stability regions become very complicated, many are plotted by using only points, rather than connecting them as was done in the previous section. It should also be observed that multiple corrections do not enlarge the size of the stability regions. Essentially they are limited, for multiple corrections, to a radius of approximately $1/\beta_k$. Thus correcting to converge will never recover the properties of the corrector completely.

5.8 PREDICTOR-CORRECTOR-MODIFIER STABILITY REGIONS

The effect of adding the computationally inexpensive and simple procedure of modifiers can be most easily determined by consideration of the corresponding stability regions. It should be noticed that the stability regions are somewhat more vertical than their unmodified versions near the λT-plane origin. This should be interpreted as improving the order of accuracy of the method by 1. Since the modifiers' derivation is based on removing the principal local truncation error, clearly the order of accuracy has gone up by 1. Furthermore, from Section 5.5, it follows that improving the order of accuracy implies having a more nearly vertical stability region near the λT-plane origin. The inclusion of modifiers clearly demonstrates this theory. It thus follows that modifiers can provide simulations of improved accuracy at a modest cost.

Clearly, multiple corrections can be combined with the modification procedure, but they should not provide any particular improvement in the stability region properties. Also, all the discussion of the properties of predictor-corrector stability regions equally applies to the modified versions.

5.9 HARDWARE-IN-THE-LOOP STABILITY

We now discuss the general problem of closed-loop real-time simulation. In real-time simulation, the timestep is the most important parameter. The simulation must finish in a required time frame. It is of great usefulness then to be able to predict the largest allowable timestep that still yields a stable and accurate simulation. Conversely, given the time frame and the corresponding simulation timestep, how accurate can the simulation method be made while still maintaining stability? It will be shown that the simulation must be designed not only for the open-loop system but also for the controller in the loop in order to maintain stability while still allowing for a maximal timestep. This situation often exists both in a real-time environment and in off-line control studies. The block diagram corresponding to these situations is shown in Figure 5.26.

In real-time simulation, the hardware to be tested is necessarily in the loop with the real-time simulation. In off-line control studies, some control strategy has been added into the feedback path, with the plant simulation in the forward path. In both cases, there is a closed-loop structure around the simulation that is usually designed from an open-loop point of view (the plant), if at all. Clearly, the feedback loop will change the system dynamics and can inadvertently destabilize the plant simulation or lead to an overly restrictive

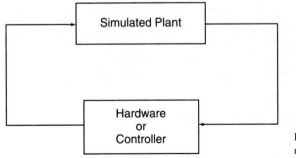

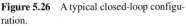

Figure 5.26   A typical closed-loop configuration.

timestep. When the feedback system is discrete by design, such as in the case of digital control of a continuous process, the timestep of the discrete simulation of the process must be faster than that of the controller or feedback loop. This is a specific case of the results of this section, the difference being that there are now two timesteps with $T_{system} < T_{feedback}$. We now present methods for obtaining guaranteed closed-loop stability by properly choosing the timestep.

The stability region contours developed in Chapter 4 can be used to determine stability of a system if all of the open-loop and closed-loop eigenvalues are known. We simply choose T to place all of the eigenvalues within the stability region. This can lead to a very conservative timestep since we are actually simulating only the feedforward loop. Alternatively, the resulting simulation will be guaranteed to be stable. What is necessary and sufficient for simulation stability is that all of the plant open-loop poles and all of the system closed-loop poles are contained within the stability region. A simple feedback case demonstrating this method is shown by replacing the "plant" and "controller" of Figure 5.26 with these two equations:

$$\text{Plant} = \frac{1}{s+1}$$

$$\text{Controller} = \frac{s-5}{s+10}$$

The open-loop poles of this system are $s_1 = -1$, $s_2 = -10$. The closed-loop poles are $s_3 \approx -0.4322$, $s_4 \approx -11.56$. Since all the poles are real, only the real-axis stability boundary need be considered. For Euler's explicit method, the stability requirement is $\lambda T > -2$. Since s_4 is the largest pole, $T < 0.1730$ is required for stability. The plant being simulated has its pole located at $s = -1$. Thus, open-loop design would yield erroneous results for the upper bound on T. Conversely, if the controller had fast poles, such as with a human in the loop, an unrealizably small timestep would be required if the controller open-loop poles were considered.

When it is not possible to determine the eigenvalues of a system, the stability region method can still be used to determine stability. The boundary of stability in the λT-plane becomes a boundary of stability in the s-plane directly or a boundary of stability in the z-plane through the mapping e^{sT}. A "frequency" response obtained about this boundary obeys the Nyquist criterion; that is, the system stability is based on the encirclements of the $-\lambda T$ point in the appropriate plane. This method works for static as well as dynamic feedback situations exactly like the normal Nyquist criterion. If we use the bilinear transform as a simulation method to convert from the s-plane to the z-plane, the resulting frequency response boundary is the jw-axis, and thus the Nyquist criterion does not change. This method is applied to the previous example. If Euler's explicit method is used to convert the system to the z-domain, the characteristic equation below results:

$$\Delta(z) = 1 + \left(\frac{T}{z-1+T}\right) + \left(\frac{z-1-5T}{z-1+10T}\right)$$

This equation is stable for $T < 0.1730$ (as before). (It is not actually necessary to obtain the characteristic equation before the method can be applied.) The Nyquist contours described above for two values of T are shown in Figure 5.27. From the figure we can discern stability of the $T = 0.15$ contour and instability of the $T = 0.4$ contour. In the above analysis, it was assumed that the transfer function of the feedback element was known. This is the case in off-line control studies. When there is actual hardware in the loop, often only the frequency response of the feedback is known. The above method can be applied in this case.

The conservativeness of the previous two methods can be observed by again considering the same system. Now the feedback system is $H(s)$, and it is discretized by replacing it with its zero-order-hold equivalent (ZOH) to obtain $H(z)$, which is appropriate in a real-time situation. The feedforward system is as before discretized by using Euler's explicit method. The characteristic equation of this system with T replaced by 1 in $H(z)$ also appears below.

$$H(s) = \frac{s+5}{s(s+10)}$$

$$H(z) = \frac{[\exp{(10T)}(10T+1)]z + [1 - 10T - \exp(10T)]}{20[\exp{(10T)}z^2 - (\exp{(10T)} + 1)z + 1]}$$

$$\Delta(z) = 1 + \left(\frac{T}{z-1+T}\right)\left(\frac{0.55(z - 9E - 2)}{(z-1)(z - 4.5E - 5)}\right)$$

The previous two methods dictate that $T < 0.1730$ for stability. The root locus of the characteristic equation (as a function of T) appears in Figure 5.28. Here it is assumed that the compensator timestep is fixed at $T = 1$, as it would be in a real situation. From this figure,

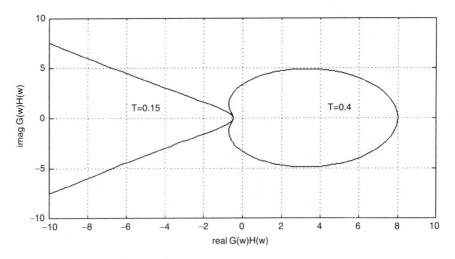

Figure 5.27   Nyquist contours for two values of T.

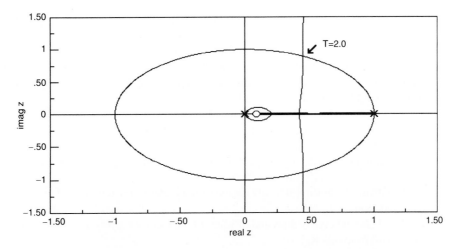

Figure 5.28   Root locus for hardware-in-the-loop simulation example.

the stable range of T is given as $0 < T < 2$; thus, the previous methods are over ten times as conservative as necessary.

In both cases, the stable range of T for the closed-loop system is controlled by the movement of the poles of the forward loop, not those of the feedback. Thus, in obtaining the root locus, we need only keep track of the poles of the simulated system. This is true, of course, only if the closed-loop feedback system is stable.

The method presented in this section also gives guaranteed bounds for closed-loop stability. Unfortunately, it requires complete knowledge of the entire closed-loop system and a complicated root locus as a function of the timestep.

Although the root locus approach works, it does not allow simple design in the familiar λT-plane. This next method will return to the λT-plane. Consider the general SISO characteristic equation of Figure 5.26, where $G(s)$ is the feedforward path and $H(s)$ is the feedback. The characteristic equation is $1 + G(s)H(s) = 0$. By partial fraction expansions, this can be written as $1 + [g_1 + g_2 + g_3 + \ldots + g_n]H(s) = 0$, $g_i = R_i/(s + \lambda_i)$, where for brevity, only simple eigenvalues (λ_i's) are considered. $H(s)$ can now be transformed directly to

$$H(s) = \frac{-1}{\displaystyle\sum_{i=1}^{n} \frac{R_i}{s + \lambda_i}}$$

where n is the order of $G(s)$. If s is now replaced by the right side of $s \approx \rho(z)/T\sigma(z)$, which provides the stability boundary of the discrete integrator, the following equation results after some algebra and application of the triangle inequality:

$$\lambda_i T + nTR_{max}\|H(s)\|_\infty \exp(j\theta) \leq \frac{\rho(z)}{\sigma(z)}$$

The term R_{max} denotes the maximum residue over all g_i's in the characteristic equation. The $\|G(s)\|_\infty$ −norm is defined as the maximum gain of the system $G(s)$ over all frequencies, or the least upper bound of the largest singular value over all frequencies. In the above equation, we have assumed that H is in continuous-time form as $H(s)$. If H is assumed to contain a ZOH, the results are the same since the ZOH has unity norm. The case for a general hold device is shown below:

$$\lambda_i T + n T R_{max} \|H(s)\|_\infty \exp(j\theta) \|\text{Hold}\|_\infty \leq \frac{\rho(z)}{\sigma(z)}$$

From the last two equations, we can determine the required T for stability as follows:

1. Obtain the normal stability region of the linear multistep method to be used for simulation.
2. Obtain the eigenvalues and residues of the feedforward system to be simulated, $G(s)$, and from this R_{max}.
3. Obtain the H_∞ norm of the feedback system $H(s)$; call it H_{max}.
4. Plot the λT products on the stability region (in the enclosed region, of course).
5. Draw a circle of radius $n*R_{max}*H_{max}$ centered at each eigenvalue determined in step 2.

For closed-loop stability, no part of any circle may extend beyond the bounds of the stability region.

This method is demonstrated with $G(s)$ and $H(s)$ given by

$$G(s) = \frac{1}{s+1}$$

$$H(s) = \frac{s-5}{s+10}$$

and Euler's explicit method as the linear multistep method. For this example, $n = 1$, $R_{max} = 1$, and $H_{max} = 1$. From Figure 5.28, for stability $T < 1$ is required. This is depicted in Figure 5.29 for $T = 0.8$.

A second example is given by

$$G(s) = \frac{3s+12}{(s+1)(s+10)}$$

$$H(s) = \frac{s+6}{(s+2)(s+100)}$$

and AB-2 is the linear multistep method. For this example, $n = 2$, $R_{max} = 2$, and $H_{max} = 0.03$. Again, for stability, $T < 0.1$. This example is depicted in Figure 5.30 for $T = 0.075$. From this last example, we note the extremely large gain margin on the norm of $H(s)$, as the radii are equal to 0.009.

Thus, this is a useful frequency domain method that can give improved bounds on T when there is a very fast system in the feedback loop. Some shortcomings of this method are the need for the feedback norm to be bounded, such as with an integrator in the control-

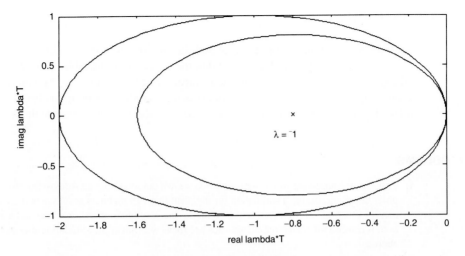

Figure 5.29 λT-product on the Euler stability region with robustness circle added.

ler, and the difficulty of obtaining R_{max} in the multi-input, multi-output (MIMO) case. Obviously, the conservatism increases with increasing plant order.

Several methods of guaranteeing closed-loop stability for a simulated system with hardware in the loop have been demonstrated. No one method is claimed to be the solution to the general problem of closed-loop simulation stability. Rather, they are intended to serve as additional tools for the simulation engineer. Several extensions of the above to the general MIMO case are needed. It is hoped that these results will encourage further research in this area.

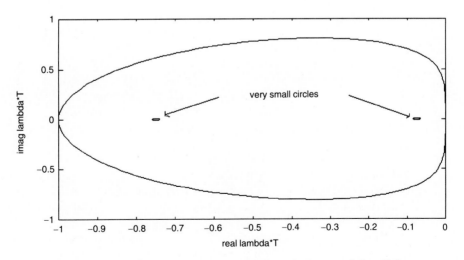

Figure 5.30 λT-products on the AB-2 stability region with robustness circles added.

5.10 CONCLUSIONS

The extreme importance and utility of stability regions in understanding the simulation process has been discussed. It is hoped that the inclusion of the tables of stability regions will aid users of given methods in their correct usage. Furthermore, by considering the closed-loop integration process in a robust control configuration, it has been possible to illuminate clearly most of the important concepts associated with linear multistep methods

PROBLEMS

1. Consider the hardware-in-the-loop situation shown below. Using an Adams-Bashforth second-order integrator, design a simulation for the plant that will maintain accuracy in the closed-loop control situation shown. Assume the computer has a sampler on the input (u) and a zero-order hold reconstructor on the output (x). Both of these will run at the simulation timestep that you choose.

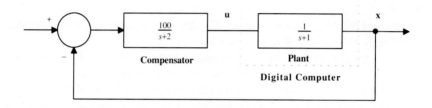

2. Analytically find the branch points in the stability region of the midpoint rule $H(z) = 2Tz/(z^2 - 1)$.

6

Runge-Kutta Methods

6.1 INTRODUCTION

At this point it is useful to have a general review. We started by discussing ZOH and operational substitution methods. These were particularly useful for transfer function representations and did not generate spurious poles. Unfortunately, a generally useful analysis for predicting stability and accuracy of simulations with these methods was not available, as they tended to be somewhat ad hoc.

We then developed linear multistep methods. These were most easily implemented in a state space representation; however, transfer function and block diagram implementations were still possible. These methods sacrifice the single-step nature to obtain improved accuracy, thereby yielding spurious poles. Fortunately, these methods allow powerful z-domain and complex variable tools to be used for the analysis of stability and accuracy.

Now we will develop Runge-Kutta methods. These methods require a state space representation for implementation. They effectively sacrifice linearity for accuracy in the sense that the open-loop integrator is not z-transformable. Thus Runge-Kutta methods are still single-step methods and generate no spurious poles. The analysis will be difficult; however, some tools will be borrowed from the linear multistep methods to simplify details.

The discussion can be motivated by considering Euler's method. Basically it is fast, explicit, requires no iterations, can be a substitution method, is a one-step method, requires no starting values, requires little memory of the past, and is linear for easy analysis. Its bad properties are that it is of low-order accuracy and has a limited stability region, particularly on the imaginary axis. This method is the basis for all other explicit methods. Linear multi-step methods overcome the disadvantages by sacrificing the one-step nature. Runge-Kutta methods will overcome the disadvantages by sacrificing the linear nature of the method.

From a control engineer's point of view, Runge-Kutta methods are not very attractive, as will be developed later. However they do possess some beautiful properties and should at least be understood. In all that follows, the Runge-Kutta methods will be assumed explicit unless otherwise indicated.

6.2 GRAPHICAL DERIVATION OF RUNGE-KUTTA METHODS

We have found that all Runge-Kutta methods are most easily explained by demonstrating their implementation graphically, which we now do for some simple Runge-Kutta methods. Consider the improved Euler method (a simple Runge-Kutta method):

$$x_{n+1} = x_n + \frac{T}{2} [f(t_n, x_n) + f(t_n + T, x_n + Tf(t_n, x_n))] \qquad (6\text{--}1)$$

Here the notation, $x_n = f(t_n, x_n)$, is used. This method can also be written in the form

$$x_{n+1} = x_n + \frac{T}{2} [K_1 + K_2], \text{ with}$$

$$K_1 = f(t_n, x_n) \qquad (6\text{--}2)$$

$$K_2 = f(t_n + T, x_n + TK_1)$$

where each K is called a stage in the method. This notation is perhaps better since it gives an indication of how the computation progresses. A similar notation is used for all Runge-Kutta methods in this chapter. The first stage is the derivative function evaluated at the left side of the interval. With that stage, an Euler step is taken to the right side of the simulation interval, as shown in Figure 6.1. There, the derivative function, f, is evaluated as the second stage. Following this, the two slopes, one at each end of the interval, are averaged and the average slope is used in one Euler step from the beginning of the interval to get the simulated value at the end of the interval. The resulting method is second-order accurate (as is shown later) and is an improvement over the usual forward Euler method, which is only first-order accurate.

All Runge-Kutta methods effectively work like this. By choosing time points with respect to the beginning of the interval, the idea is to predict where the output value will be at those points in time. Using those predictions, we can compute the derivative functions (or slopes). Finally an Euler step is taken from the beginning of the interval to the end of the interval by using a weighted average of all these derivatives. Each derivative evaluation is

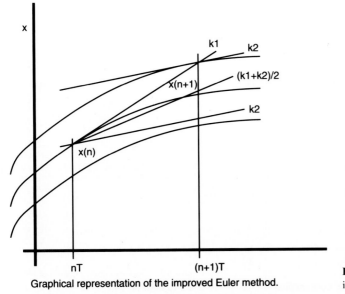

Graphical representation of the improved Euler method.

Figure 6.1   Graphical representation of the improved Euler method.

called a stage, and a particular way of characterizing Runge-Kutta methods is by the number of stages.

To illustrate the procedure further, we will consider another method. The modified Euler method is

$$x_{n+1} = x_n + TK_2, \text{ with}$$

$$K_1 = f(t_n, x_n) \qquad (6\text{--}3)$$

$$K_2 = f(t_n + T/2, x_n + (T/2)K_1)$$

Here the idea is to take a Euler step to the middle of the interval and find the slope there. This slope is then used in one Euler step from the beginning of the interval to the end of the interval to find the answer. It is demonstrated in Figure 6.2. This method is also second-order accurate. It will turn out that the improved Euler and the modified Euler methods are equivalent from a stability and accuracy viewpoint.

This graphical procedure can be implemented to understand the workings of any Runge-Kutta method. For methods with many stages, however, the graphical approach can become a little messy.

Several examples are now given to illustrate the use of Runge-Kutta methods.

Example 6.1

Here the modified Euler method is used to simulate the scalar first-order system considered in earlier chapters. The system state equation is

$$\dot{x} = -2x + 2u$$

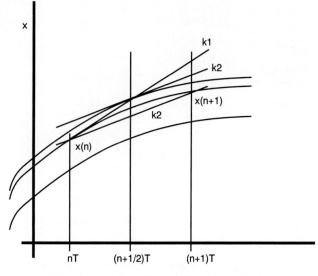

Figure 6.2   Graphical representation of the modified Euler method.

Graphical representation of the modified Euler method.

The modified Euler method used on this system is given below in a BASIC program. Notice that the input is now a sinusoid to demonstrate the required multirate input sampling.

```
T = 0.1
x = 0
for n = 0 to 99
        u = sin(n*T)                    Input at left end
        f1 = -2*x + 2*u                 Derivative at left end
        xh = x + 0.5*T*f1               Predicted state at middle
        uh = sin((n + 0.5)*T)           Input at middle
        f2 = -2*xh + 2*uh               Derivative at middle
        x = x + T*f2                    State at right end
next n
end
```

Example 6.2

We now consider the second-order oscillatory system,

$$\dot{x}_1 = x_2$$
$$\dot{x}_2 = -2x_1 - 2x_2 + 2u$$
$$y = x_1$$

The standard fourth-order Runge-Kutta method is chosen for the simulation. The BASIC program for this simulation is given below. The input is again a sinusoid to demonstrate the required multirate input sampling.

```
T = 0.1
x1 = 0
x2 = 0
for n = 1 to 250
        u = sin(nT)                              Input at left
        f11 = x2                                 Derivative at left
        f21 = -2*x1 - 2*x2 + 2*u                    (Stage 1)
        x1h = x1 + 0.5*T*f11                     State prediction in middle
        x2h = x2 + 0.5*T*f21
        uh = sin((n + 0.5)*T)                    Input in middle
        f12 = x2h                                Derivative in middle
        f22 = -2*x1h - 2*x2h + 2*uh                 (Stage 2)
        x1h = x1 + 0.5*T*f12                     State prediction in middle
        x2h = x2 + 0.5*T*f22
        uh = sin((n + 0.5)*T)                    Input in middle (have above)
        f13 = x2h                                Derivative in middle
        f23 = -2*x1h - 2*x2h + 2*uh                 (Stage 3)
        x1h = x1 + T*f13                         State prediction at right
        x2h = x2 + T*f23
        uh = sin((n + 1)*T)                      Input at right
        f14 = x2h                                Derivative at right
        f24 = -2*x1h - 2*x2h + 2*uh                 (Stage 4)
        x1 = x1 + (1/6)*T*(f11 + 2*f12 +               Euler step with
        2*f13 + f14)                                   averaged slopes
        x2 = x2 + (1/6)*T*(f21 + 2*f22 + 2*f23 + f24)
        pset(n,200 - 100*x1)
        pset(n,200 - 100*x2)
next n                                           Plot x1
end                                              Plot x2
```

6.3 THE GENERAL RUNGE-KUTTA METHOD

For general Runge-Kutta methods, we will not go into great detail concerning consistency and convergence. The interested reader should consult Lambert (1973), Butcher (1987), or Ralston and Rabinowitz (1978) for more information. These properties are important; however, it will be sufficient to remember that order of accuracy is determined by truncation of the usual Taylor series expansion for the exponential function.

The general expression for a Runge-Kutta method is

$$x_{n+1} = x_n + T \sum_{r=1}^{R} c_r K_r \text{ with} \tag{6-4}$$

$$K_1 = f(t_n, x_n) \text{ and} \tag{6-5}$$

$$K_r = f(t_n + Ta_r, x_n + T \sum_{s=1}^{r-1} b_{rs}K_s), r = 2, 3, \ldots R \qquad (6\text{–}6)$$

As is shown in the next section, the following are required for consistency:

$$a_r = \sum_{s=1}^{r-1} b_{rs} \text{ and } \sum_{r=1}^{R} c_r = 1$$

Here, the c's represent the weightings on the average slopes. Clearly, the sum of these weighting must be equal to 1. The a's represent the subinterval locations at which the derivatives are being evaluated. The b's represent the slope weightings for the intermediate stages and must clearly add up to the a of the particular intermediate stage. These relationships are fully derived in the next three sections. Finally, R is the total number of stages.

To derive methods, there is unfortunately no simple formula relating coefficients to truncation error coefficients, as there was for linear multistep methods with Lambert's equations. Useful relationships do exist that are considerably more complicated (Lambert, 1973; Ralston & Rabinowitz, 1978). Regrettably the resulting algebraic equations are nonlinear in the unknowns, and there are more unknowns than equations.

The next three sections discuss in detail the accuracy constraints on the coefficients. Section 6.4 considers accuracy for a linear system when the input is a constant during the sampling interval for simplicity. Section 6.5 extends these results to nonlinear systems when the input is constant during the sampling interval. Then Section 6.6 considers the more complicated situation of the input changing during the sampling interval. The sections following these consider stability and implementation issues.

6.4 LINEAR SYSTEM WITH PIECEWISE CONSTANT INPUT

We discuss here the development of the requirements on the coefficients for a three-stage, third-order accurate Runge-Kutta method when applied to a linear time-invariant system with an external input. The linear system with piecewise constant input is treated first because of its simplicity. The more general nonlinear case is considered in Section 6.5, and Section 6.6 treats the linear system when the input is time varying. The results obtained here can be used to derive coefficient constraints for first-, second-, and third-order accurate methods. This development can be extended to higher-order methods as well.

Here we assume that the input is piecewise constant over each simulation timestep of length T seconds. In many realistic situations the input varies continuously with time rather than being piecewise constant. If computation time is not a constraint, the simulation should sample the input every time a function evaluation is computed. We will show in Section 6.6 that failure to do this will result in a loss of accuracy. However, a piecewise constant assumption may be necessary if the simulation must be done in real time with serious computation time constraints. In some cases, for example, in a computer control system, the system input will really be piecewise constant. With the assumption of being piecewise constant, the input is

$$u(t) = u(nT) \equiv u_n \quad \forall t \, \varepsilon \, [nT, (n+1)T) \tag{6-7}$$

In general, the system to be simulated would be described by the differential equation

$$\dot{x}(t) = f[t, x(t), u(t)], \; x(0) \text{ given} \tag{6-8}$$

For the three-stage, third-order method, the Runge-Kutta equations are

$$\begin{aligned}
x_{n+1} &= x_n + T[c_1 K_1 + c_2 K_2 + c_3 K_3] \\
K_1 &= f(t_n, x_n, u_n) \\
K_2 &= f(t_n + a_2 T, x_n + Tb_{21}K_1, u_n) \\
K_3 &= f(t_n + a_3 T, x_n + Tb_{31}K_1 + Tb_{32}K_2, u_n)
\end{aligned} \tag{6-9}$$

Thus, the K_i terms are function evaluations computed at various points in time. The linear time-invariant differential equation with input included is

$$\dot{x}(t) = f[x(t), u(t)] = \lambda \, x(t) + u(t) \tag{6-10}$$

With this expression for the functional form of the system's differential equation, the three stages of the Runge-Kutta method can be written as

$$\begin{aligned}
K_1 &= \lambda \, x_n + u_n \\
K_2 &= \lambda[x_n + Tb_{21}(\lambda \, x_n + u_n)] + u_n \\
K_3 &= \lambda\{x_n + Tb_{31}(\lambda \, x_n + u_n) + Tb_{32}[\lambda(x_n + Tb_{21}(\lambda \, x_n + u_n)) + u_n]\} + u_n
\end{aligned} \tag{6-11}$$

Simplifying the last two expressions, we get

$$\begin{aligned}
K_2 &= \lambda x_n[1 + b_{21} \, \lambda T] + u_n[1 + b_{21} \, \lambda T] \\
&= (\lambda x_n + u_n) \, [1 + b_{21} \, \lambda T]
\end{aligned} \tag{6-12}$$

$$K_3 = (\lambda x_n + u_n) \, [1 + (b_{31} + b_{32}) \, (\lambda \, T) + b_{32}b_{21} \, (\lambda \, T)^2] \tag{6-13}$$

The expressions for K_1, K_2, and K_3 in the Runge-Kutta equation yield

$$\begin{aligned}
x_{n+1} &= x_n + Tc_1(\lambda x_n + u_n) \\
&+ Tc_2(\lambda x_n + u_n) \, (1 + b_{21}\lambda T) \\
&+ Tc_3(\lambda x_n + u_n) \, (1 + (b_{31} + b_{32}) \, \lambda \, T + b_{32}b_{21}(\lambda \, T)^2)
\end{aligned} \tag{6-14}$$

Combining terms by powers of λT, we produce the following result:

$$\begin{aligned}
x_{n+1} &= x_n[1 + (c_1 + c_2 + c_3)\lambda T + (c_2 b_{21} + c_3(b_{31} + b_{32}))(\lambda \, T)^2 + c_3 b_{32}b_{21}(\lambda T)^3] \\
&+ u_n[(c_1 + c_2 + c_3)T + (c_2 b_{21} + c_3(b_{31} + b_{32}))\lambda \, T^2 + c_3 b_{32}b_{21}\lambda^2 T^3]
\end{aligned} \tag{6-15}$$

The interstage consistency equations (mentioned in Section 6.3 and derived in the next section) can be used to give

$$\begin{aligned}
x_{n+1} &= x_n[1 + (c_1 + c_2 + c_3)\lambda \, T + (c_2 a_2 + c_3 a_3) \, (\lambda \, T)^2 + c_3 b_{32}a_2(\lambda \, T)^3] \\
&+ \left(\frac{1}{\lambda}\right)u_n[(c_1 + c_2 + c_3)\lambda \, T + (c_2 a_2 + c_3 a_3) \, (\lambda \, T)^2 + c_3 b_{32}a_2(\lambda \, T)^3]
\end{aligned} \tag{6-16}$$

where the factor $1/\lambda$ multiplying u_n is included so that the powers of λ and T will be the same in each term of the equation.

The constraints on the coefficients in equation 6–16 in terms of meeting accuracy requirements are developed as follows. Given a differential equation representing a linear, time-invariant system, as represented by equation 6–10, and given the assumption that the input $u(t)$ is constant over each length of time of T seconds, the exact solution to the differential equation at the sample points $(n + 1)T$ with $n \in \{0, 1, 2, 3, \dots \}$ is

$$x_{n+1} = e^{\lambda T}x_n + \left[\int_0^T e^{\lambda \tau}d\tau\right]u_n \tag{6–17}$$

or

$$x_{n+1} = [e^{\lambda T}]x_n + \left(\frac{1}{\lambda}\right)[e^{\lambda T} - 1]u_n \tag{6–18}$$

Using the power series expansion for $e^{\lambda T}$, we can write the exact solution

$$
\begin{aligned}
x_{n+1} = x_n &\left[1 + \lambda T + \frac{1}{2}(\lambda T)^2 + \frac{1}{6}(\lambda T)^3 + \dots\right] \\
&+ \left(\frac{1}{\lambda}\right)u_n\left[\lambda T + \frac{1}{2}(\lambda T)^2 + \frac{1}{6}(\lambda T)^3 + \dots\right]
\end{aligned}
\tag{6–19}
$$

Comparing the exact solution in equation 6–19 with the approximate solution in equation 6–16, we can easily see that the approximate solution will be most accurate if all of its coefficients of powers of λT match in value the corresponding coefficients in the exact solution. Given the form of equation 6–16, this can only be done up to the third power of λT, and this goal imposes the following constraints:

$$
\begin{aligned}
c_1 + c_2 + c_3 &= 1 \\
c_2a_2 + c_3a_3 &= \frac{1}{2} \\
c_3b_{32}a_2 &= \frac{1}{6}
\end{aligned}
\tag{6–20}
$$

Note that these constraints exactly equate the coefficients in the two equations through $(\lambda T)^3$, both in the terms involving x_n and in those with u_n. Therefore, the local truncation error introduced by the approximation will be $\mathcal{O}(T^4)$. A point to note in equation 6–20 is that the linear system model does not impose a constraint on the term $c_2a_2{}^2 + c_3a_3{}^2$. It will be seen in the next section that a more general nonlinear system description does cause a constraint on that term.

The expressions in equation 6–20 are used now to show the constraints on the coefficients for Runge-Kutta methods of first, second, and third order. For a one-stage, first-order accurate method, there is only one function evaluation, and that is evaluated at time t_n. Therefore, equation 6–20 gives

$$c_1 = 1 \tag{6–21}$$

and all the other coefficients are zero. Thus, the only one stage, first-order accurate Runge-Kutta method has the same equation as Euler's method, namely,

$$x_{n+1} = x_n + T\dot{x}_n \tag{6-22}$$

For a two stage, second-order accurate method, the constraints from equation 6–20 and the corresponding interstage consistency relation are

$$c_1 + c_2 = 1, \qquad c_2 a_2 = \frac{1}{2}, \qquad a_2 = b_{21} \tag{6-23}$$

Here, there are three equations and four unknowns. Hence there are an infinite number of two-stage, second-order accurate methods. For the improved Euler method, the choice of coefficients is

$$c_1 = c_2 = \frac{1}{2}, \qquad a_2 = b_{21} = 1 \tag{6-24}$$

and for the modified Euler method, the coefficients are

$$c_1 = 0, \qquad c_2 = 1, \qquad a_2 = b_{21} = \frac{1}{2} \tag{6-25}$$

For a three-stage, third-order accurate method there are a total of eight unknown coefficients: $c_1, c_2, c_3, a_2, a_3, b_{21}, b_{31},$ and b_{32}. To achieve third-order accuracy with the method, we have derived constraint relationships among the coefficients. Including the interstage consistency relationships, there are six such equations; the complete set of equations is explicitly shown in Section 6.5, equation 6–39. Therefore, there are two degrees of freedom in choosing values for the coefficients. Two of them may be chosen arbitrarily; the constraints will uniquely determine values for the remaining six coefficients. Thus, there is a doubly infinite family of Runge-Kutta methods that will satisfy these constraints. An example of one of these is Huen's third-order method. Even though coefficient $c_2 = 0$ for this method, we still need K_2 to compute the last stage:

$$x_{n+1} = x_n + \frac{T}{4}(K_1 + 3K_3), \text{ with}$$

$$K_2 = f(t_n, x_n)\left(t_n + \frac{T}{3}, x_n + \frac{T}{3}K_1\right) \tag{6-26}$$

$$K_3 = f\left(t_n + \frac{2T}{3}, x_n + \frac{2T}{3}K_2\right)$$

6.5 NONLINEAR SYSTEM WITH PIECEWISE CONSTANT INPUT

In this section, we derive the complete set of coefficient constraints needed to achieve third-order accuracy for the three-stage Runge-Kutta method. If the coefficients satisfy these constraints, the method will have third-order accuracy regardless of the form of the differential equation, as long as the input is constant over each sample period. The problem associated with the input changing during a sample period is discussed in the next section.

The system to be simulated is described by the nonlinear equation

$$\dot{x}(t) = f[t, x(t)], \qquad x(0) \text{ given} \tag{6-27}$$

To determine the form of the solution to equation 6–27, we will use a Taylor series expansion starting from the solution that exists at time $t = t_n$ to obtain the expression for x_{n+1}; this gives the following expression:

$$x_{n+1} = x_n + \frac{dx}{dt}\Big|_n T + \frac{1}{2}\frac{d^2x}{dt^2}\Big|_n T^2 + \frac{1}{6}\frac{d^3x}{dt^3}\Big|_n T^3 + \dots \tag{6-28}$$

Using equation 6–27 and the notation that

$$f(t,x) = f, \qquad \frac{\partial f(t,x)}{\partial t} = f_t, \qquad \frac{\partial f(t,x)}{\partial x} = f_x, \qquad \frac{\partial^2 f(t,x)}{\partial t \partial x} = f_{tx}, \qquad \text{etc.} \tag{6-29}$$

we get the following equations for the terms through T^3:

$$\frac{dx}{dt} = f$$

$$\frac{d^2x}{dt^2} = f_t + f_x f \tag{6-30}$$

$$\frac{d^3x}{dt^3} = f_{tt} + f_x f_t + (f_x^2 + 2f_{tx} + f_{xx}f)f$$

Therefore, the exact solution to the differential equation is

$$x_{n+1} = x_n + Tf_n + \frac{1}{2}T^2(f_t + f_x f)_n$$

$$+ \frac{1}{6}T^3[f_{tt} + f_x f_t + (f_x^2 + 2f_{tx} + f_{xx}f)f]_n + \text{higher-order terms} \tag{6-31}$$

For the three-stage, third-order accurate method, only the terms through T^3 will be written out explicitly since higher-order terms will not be used in the development of the constraints.

The accuracy constraints are developed as follows. First, each of the K_i terms in the Runge-Kutta approximate solution is expanded in a Taylor series about time t_n. Next, these expansions are substituted into the Runge-Kutta equation, and then terms in like powers of T between the expansions for the exact and approximate solutions are compared. Using the definitions for the K_i terms from equation 6–9, the series expansions are

$$K_1 = f \tag{6-32}$$

$$K_2 = f + f_t(a_2 T) + f_x(b_{21}Tf) + \frac{1}{2}[f_{tt}(a_2 T)^2 \tag{6-33}$$

$$+ 2f_{tx}(a_2 T)(b_{21}Tf) + f_{xx}(b_{21} Tf)^2] + \mathcal{O}(T^3)$$

$$K_3 = f + f_t(a_3 T) + f_x(b_{31} Tf + b_{32} TK_2) + \frac{1}{2}[f_{tt}(a_3 T)^2 + 2f_{tx}(a_3 T)(b_{31} Tf + b_{32} TK_2)$$
$$+ f_{xx}(b_{31} Tf + b_{32} TK_2)^2] + \mathcal{O}(T^3) \qquad (6\text{--}34)$$

Only terms through T^2 are shown in the above expansions since each of the K_i will be multiplied by T in the Runge-Kutta equation. When equations 6–32 through 6–34 are substituted into the Runge-Kutta equation and terms are collected by powers of T, the following approximate solution to the differential equation is obtained. To accomplish this, note that the expansion for K_2 in 6–33 must also be used in 6–34 wherever K_2 appears there. The algebra is tedious, keeping track of all the terms and powers of T, but the procedure is straightforward. Note that only the terms through T^3 are explicitly included in 6–35 since the method being considered is third-order accurate; there are not enough coefficients in the approximate solution to equate terms for higher powers of T.

$$x_{n+1} = x_n + T(c_1 + c_2 + c_3)f + T^2(c_2 a_2 f_t + c_2 b_{21} f_x f + c_3 a_3 f_t + c_3 b_{31} f_x f + c_3 b_{32} f_x f)$$
$$+ T^3 \left(\frac{1}{2} c_2 a_2^2 f_{tt} + c_2 a_2 b_{21} f_{tx} f + \frac{1}{2} c_2 b_{21}^2 f_{xx} f^2 \right)$$
$$+ T^3 \left(c_3 b_{32} a_2 f_x f_t + c_3 b_{32} b_{21} f_x^2 f + \frac{1}{2} c_3 a_3^2 f_{tt} \right) \qquad (6\text{--}35)$$
$$+ T^3 \left(c_3 a_3 (b_{31} + b_{32}) f_{tx} f + \frac{1}{2} c_3 b_{31}^2 f_{xx} f^2 + \frac{1}{2} c_3 b_{32}^2 f_{xx} f^2 + c_3 b_{31} b_{32} f_{xx} f^2 \right)$$
$$+ \mathcal{O}(T^4)$$

Equating like powers of T between equations 6–35 and 6–31 produces the following set of requirements on the coefficients:

$$T^1: (c_1 + c_2 + c_3)f = 1f \qquad (6\text{--}36)$$

$$T^2: (c_2 a_2 + c_3 a_3)f_t = \frac{1}{2} f_t$$
$$[c_2 b_{21} + c_3 (b_{31} + b_{32})] f_x f = \frac{1}{2} f_x f \qquad (6\text{--}37)$$

$$T^3: (c_3 b_{32} b_{21}) f_x^2 f = \frac{1}{6} f_x^2 f$$
$$(c_2 a_2^2 + c_3 a_3^2) f_{tt} = \frac{1}{3} f_{tt}$$
$$[c_2 a_2 b_{21} + c_3 a_3 (b_{31} + b_{32})] f_{tx} f = \frac{1}{3} f_{tx} f \qquad (6\text{--}38)$$
$$(c_3 b_{32} a_2) f_x f_t = \frac{1}{6} f_x f_t$$
$$[c_2 b_{21}^2 + c_3 (b_{31}^2 + 2 b_{31} b_{32} + b_{32}^2)] f_{xx} f^2 = \frac{1}{3} f_{xx} f^2$$

Realizing that the above expressions must be valid for arbitrary values of the function and its various partial derivatives, we easily see that equality in all those terms requires the following relationships to hold for the coefficients. These are the necessary constraints on the coefficients to provide third-order accuracy for the Runge-Kutta method.

$$c_1 + c_2 + c_3 = 1$$
$$c_2 a_2 + c_3 a_3 = \frac{1}{2}$$
$$a_2 = b_{21}$$
$$a_3 = b_{31} + b_{32} \tag{6-39}$$
$$c_3 b_{32} a_2 = \frac{1}{6}$$
$$c_2 a_2^2 + c_3 a_3^2 = \frac{1}{3}$$

Thus, constraints on the coefficients in the Runge-Kutta equation are obtained by matching terms in powers of T between the approximate and exact solutions. The matching is done up to a particular power of T to achieve a certain order of accuracy. In our example, third-order accuracy is achieved by matching terms through T^3. The interstage consistency constraints are seen to arise naturally from forcing equality between two equations that have different forms but that must have the same value even when multiplying different derivative terms. For example, it is easiest to see the requirement that $a_2 = b_{21}$ by noting the first and fourth expressions in equation 6–38. Once that relationship is established, the constraint that $a_3 = b_{31} + b_{32}$ follows directly from the two expressions in equation 6–37. Note from 6–38 that the constraint on the term $c_2 a_2^2 + c_3 a_3^2$ is not required unless the system's differential equation is of such a form that either f_{tt}, f_{tx}, or f_{xx} is nonzero. None of these conditions holds for the linear time-invariant system treated earlier; that is the reason the constraint did not appear in Section 6.4.

The nonlinear system considered in this section is modeled by the differential equation 6–27, in which an external input $u(t)$ does not appear. However, there is no loss of generality in 6–27 for the purposes of developing the accuracy constraints on the coefficients. The time dependence and nonlinear form of the model are sufficient to produce all the constraint equations that exist. Explicitly including an input term in 6–27 would not have changed the relationships shown in 6–39.

The above procedure can certainly be used to develop requirements on the coefficients for higher-order methods. Obviously, the tedium involved will increase dramatically with order of accuracy as additional K_i factors are involved and higher powers of T must be included in the expansions. The necessary constraint equations for four-stage, fourth-order accurate methods are found in Rosko (1972). There are 11 equations and 13 unknowns, so again there is a doubly infinite set of solutions. There are many popular four-stage, fourth-order accurate Runge-Kutta methods, and some of them are tabulated in Appendix B. Because of its widespread usage, the method given below is often referred to as the standard

Runge-Kutta method, or just the Runge-Kutta method. The coefficients in this method are due to Runge. Observe that there are two function evaluations at the center of the time interval. The graphical representation of these equations is rather messy and is not included.

$$x_{n+1} = x_n + \frac{T}{6}(K_1 + 2K_2 + 2K_3 + K_4), \text{ with}$$

$$K_1 = f(t_n, x_n)$$

$$K_2 = f\left(t_n + \frac{T}{2}, x_n + \frac{T}{2}K_1\right)$$

$$K_3 = f\left(t_n + \frac{T}{2}, x_n + \frac{T}{2}K_2\right)$$

$$K_4 = f(t_n + T, x_n + TK_3)$$

(6–40)

The methods just discussed all have the same number of stages as order of accuracy. In Chapter 7 we show how to use truncation error design equations to derive methods that have an order of accuracy less than the number of stages. This will allow more freedom in designing stability regions.

It turns out that there is a crude relationship between the number of stages and the achievable order of accuracy. The highest achievable order of accuracy for the given number of stages is given in Table 6.1 for Runge-Kutta methods with from 1 to 11 stages.

Unfortunately, only bounds on the number of stages are available for methods of higher than 8 (Butcher, 1987). It can be seen that after $R = 4$, the order of accuracy goes up more slowly than the number of stages. Thus, the law of diminishing returns applies. Adding more stages does not necessarily improve the order of accuracy proportionately, if at all. This explains, at least to some extent, the popularity of four-stage, fourth-order accurate methods. They have the highest order of accuracy of any method in which the number of function evaluations per time step does not exceed the order of accuracy.

TABLE 6.1

Number of Stages	Highest Achievable Order of Accuracy
1	1
2	2
3	3
4	4
5	4
6	5
7	6
8	6
9	7
10	7
11	8

6.6 LINEAR SYSTEM WITH MULTIRATE INPUT SAMPLING

This section considers the case in which the input is not piecewise constant over the sampling intervals. We will see how accuracy is lost if the assumption is wrongly made that the input is piecewise constant, and how that accuracy can be recovered by sampling the input at the times at which function evaluations are computed. For example, in the three-stage, third-order accurate Runge-Kutta method described here, function evaluations are computed at times

$$t_n, \qquad t_n + a_2 T, \qquad t_n + a_3 T \qquad\qquad (6\text{--}41)$$

Therefore, it will be assumed that the input signals

$$u_n, \qquad u_{n + a_2 T}, \qquad u_{n + a_3 T} \qquad\qquad (6\text{--}42)$$

are also available. In real-time simulation, this means that the input sampling device is operating at a higher frequency than the simulation is providing results. In batch simulation, (1) tabulated values for the input must be available at the finer time resolution, (2) a functional form of the input signal must be available so that the input can be computed at the appropriate times, or (3) some other part of the simulation must make those input values available at the proper times.

The procedure that will be followed is the same as in Section 6.5, namely, using Taylor series expansions of the exact and approximate solutions and comparing coefficients in like powers of T. The constraints presented in equation 6–39 will be used since they were developed for the general case. We will consider only linear systems in this section.

The system will be described by the linear differential equation with input, 6–10. The Taylor series for the exact solution is still given by equation 6–28. For our particular system, the necessary derivatives for the expansion are

$$\frac{dx}{dt} = f = \lambda x + u$$

$$\frac{d^2x}{dt^2} = f_t + f_x f + f_u \dot{u} = \lambda^2 x + \lambda u + \dot{u}$$

$$\frac{d^3x}{dt^3} = f_{tt} + f_x f_t + (f_x^2 + 2f_{tx} + f_{xx}f)f + (f_x f_u + 2f_{tu} + 2f_{xu}f + f_{uu}\dot{u})\dot{u} + f_u \ddot{u} \qquad (6\text{--}43)$$

$$= \lambda^3 x + \lambda^2 u + \lambda \dot{u} + \ddot{u}$$

Substituting these expressions into the Taylor series and collecting terms based on the variables evaluated at time $t = t_n$, we get the form for the exact solution we need, equation 6–44:

$$x_{n+1} = x_n\left[1 + \lambda T + \frac{1}{2}(\lambda T)^2 + \frac{1}{6}(\lambda T)^3 + \ldots\right]$$

$$+ \left(\frac{1}{\lambda}\right)u_n\left[\lambda T + \frac{1}{2}(\lambda T)^2 + \frac{1}{6}(\lambda T)^3 + \ldots\right]$$

$$+ \left(\frac{1}{\lambda^2}\right)\dot{u}_n\left[\frac{1}{2}(\lambda T)^2 + \frac{1}{6}(\lambda T)^3 + \ldots\right]$$

$$+ \left(\frac{1}{\lambda^3}\right)\ddot{u}_n\left[\frac{1}{6}(\lambda T)^3 + \ldots\right] + \text{higher−order terms}$$

(6–44)

Note that if the input is piecewise constant over sample intervals, the time derivatives of $u(t)$ evaluated at t_n would be zero, and equation 6–44 then reduces to equation 6–19. If the input is assumed to be piecewise constant in the simulation, and assuming that the coefficient constraints are satisfied, the approximate solution is

$$x_{n+1} = x_n\left[1 + \lambda T + \frac{1}{2}(\lambda T)^2 + \frac{1}{6}(\lambda T)^3\right]$$

$$+ \left(\frac{1}{\lambda}\right)u_n\left[\lambda T + \frac{1}{2}(\lambda T)^2 + \frac{1}{6}(\lambda T)^3\right]$$

(6–45)

Comparing equations 6–44 and 6–45, we can see that if the input actually is piecewise constant, the first term different between the two equations is proportional to T^4 since the derivative terms of u_n are zero; therefore, the method really is third-order accurate in that case. However, if the input is assumed to be piecewise constant but actually is not, the first term different between the approximate and exact solutions is proportional to T^2, being due to the $du(t)/dt$ term. Therefore, if the input actually changes during a timestep, but that information is not included in the simulation, two orders of accuracy are lost. If the input changes rapidly relative to the value of T, significant simulation error will result.

To circumvent this loss in order of accuracy, it is necessary to sample the input signal at each time argument where a function evaluation is computed. This is referred to as multirate input sampling (MIS). The approximate solution is obtained in a manner similar to the development of equation 6–15, now taking into account the form of the input signal given by equation 6–42. The corresponding equation is

$$x_{n+1} = x_n[1 + (c_1 + c_2 + c_3)\lambda T + (c_2 b_{21} + c_3(b_{31} + b_{32}))(\lambda T)^2 + c_3 b_{32} b_{21}(\lambda T)^3]$$

$$+ \left(\frac{1}{\lambda}\right)u_n[c_1\lambda T + (c_2 a_2 + c_3 b_{31})(\lambda T)^2 + c_3 b_{32} a_2(\lambda T)^3]$$

$$+ \left(\frac{1}{\lambda}\right)u_{n + a_2 T}[c_2\lambda T + c_3 b_{32}(\lambda T)^2]$$

(6–46)

$$+ \left(\frac{1}{\lambda}\right)u_{n + a_3 T}[c_3\lambda T]$$

To compare the approximate and exact solutions, all the input signals must be evaluated at time t_n. To accomplish this, the two signals at the intermediate sample points can be expanded in a Taylor series about u_n, and these expressions can be substituted into equation 6–46. The perturbations from t_n for the two signals are a_2T and a_3T, respectively. The expansions are

$$u_{n+a_2T} = u_n + \dot{u}_n(a_2T) + \frac{1}{2}\ddot{u}_n(a_2T)^2 + \ldots$$

$$u_{n+a_3T} = u_n + \dot{u}_n(a_3T) + \frac{1}{2}\ddot{u}_n(a_3T)^2 + \ldots$$

(6–47)

Using these expansions and the coefficient constraints from equation 6–39, we get the form of the approximate solution with MIS, which can be used for comparison with the actual solution:

$$
\begin{aligned}
x_{n+1} = x_n &\left[1 + \lambda T + \frac{1}{2}(\lambda T)^2 + \frac{1}{6}(\lambda T)^3 \right] \\
&+ \left(\frac{1}{\lambda} \right) u_n \left[\lambda T + \frac{1}{2}(\lambda T)^2 + \frac{1}{6}(\lambda T)^3 \right] \\
&+ \left(\frac{1}{\lambda^2} \right) \dot{u}_n \left[\frac{1}{2}(\lambda T)^2 + \frac{1}{6}(\lambda T)^3 \right] \\
&+ \left(\frac{1}{\lambda^3} \right) \ddot{u}_n \left[\frac{1}{6}(\lambda T)^3 + \frac{1}{12}a_2(\lambda T)^4 \right]
\end{aligned}
$$

(6–48)

$$+ \text{ higher-order derivative terms in } u_n$$

Comparison with the exact solution, equation 6–44, shows that the MIS has restored the order of accuracy to its correct value of three, since all terms match through T^3. It should be pointed out that MIS does recover the loss in order of accuracy, which should lead to more accurate simulation results. However, the actual error in the simulation is a strong function of the particular value of T used in the simulation; rapidly varying input signals will need a smaller value of T than slowing varying signals to achieve the same actual accuracy.

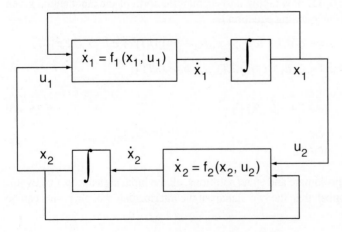

Figure 6.3 A multi-block simulation problem.

One disadvantage to MIS is the difficulty of applying the Runge-Kutta method on a block-by-block basis in a multiblock system. Consider Figure 6.3, which shows two subsystems connected together in a feedback configuration. The same comments will also apply to a series connection of elements. From the figure, it is clear that the output x_1 of the first subsystem is the input u_2 of the second subsystem, and vice versa. With MIS, the value of the input to a block must be available at intermediate times relative to the basic sample period T. For example, with the modified Euler method, the integration for each subsystem would be performed by the following algorithm:

$$x_{n+1} = x_n + TK_2, \text{ with}$$
$$K_1 = f(x_n, u_n)$$
$$K_2 = f\left(x_n + \frac{T}{2}f(x_n, u_n), u_{n+\frac{1}{2}}\right) = f\left(x_n + \frac{T}{2}K_1, u_{n+\frac{1}{2}}\right)$$

(6–49)

The exact value of the input to the block at the intermediate time point, $u_{n+1/2}$, is assumed to be available. The output of each block, however, is produced at the end of the time interval, that is, x_{n+1}. Therefore, MIS cannot be applied directly in a block-by-block fashion. To avoid the loss in order of accuracy that would occur if the inputs were assumed to be piecewise constant, the normal approach would be to combine the two subsystems into a single system. The Runge-Kutta integration method could then be applied in the usual manner to the complete system, with multirate sampling on the external input. An alternative would be to make the intermediate state computation $x_n + TK_1$ available at the output of the block for use as the input to the next block. The accuracy of those input values would be less, since they are being computed from the first-order accurate Euler method, but they would prevent the necessity of combining the blocks into a single system and would also provide better accuracy than assuming that all of the inputs were piecewise constant.

6.7 TRANSFER FUNCTIONS AND WEAK STABILITY

It was mentioned in Section 6.1 that Runge-Kutta methods obtain their accuracy by giving up the linear nature of linear multistep methods. It turns out that we can still develop transfer functions and evaluate the frequency response of the methods. Consider the linear test equation 6–10 and the improved Euler method, given by

$$x_{n+1} = x_n + \frac{T}{2}(K_1 + K_2)$$
$$K_1 = \lambda x_n + u_n$$
$$K_2 = \lambda(x_n + TK_1) + u_{n+1}$$

(6–50)

If the expression for K_1 and K_2 are substituted into the Runge-Kutta equation, the following result is obtained:

$$x_{n+1} = x_n\left(1 + \lambda T + \frac{(\lambda T)^2}{2}\right) + \frac{T}{2}[u_{n+1} + u_n(1 + \lambda T)]$$

(6–51)

If we take the z-transform of equation 6–51 in the normal manner, the transfer function becomes

$$\frac{X(z)}{U(z)} = \frac{\dfrac{T}{2}\,[z + (1 + \lambda T)]}{z - \left(1 + \lambda T + \dfrac{(\lambda T)^2}{2}\right)} \tag{6–52}$$

Since this is the closed-loop transfer function for this Runge-Kutta method when the linear test equation is being integrated, the closed-loop pole for the simulation is seen to be

$$\text{Closed-loop pole at } z = 1 + \lambda T + \frac{(\lambda T)^2}{2} \tag{6–53}$$

For comparison, we will now form the transfer function for the modified Euler method, which is given by

$$\begin{aligned}
x_{n+1} &= x_n + TK_2 \\
K_1 &= \lambda x_n + u_n \\
K_2 &= \lambda\left(x_n + \frac{T}{2}K_1\right) + u_{n+\frac{1}{2}}
\end{aligned} \tag{6–54}$$

Expanding the expression for K_2 and substituting that into the Runge-Kutta equation yields

$$x_{n+1} = x_n\left[1 + \lambda T + \frac{(\lambda T)^2}{2}\right] + T\left(u_{n+\frac{1}{2}} + \frac{\lambda T}{2}u_n\right) \tag{6–55}$$

From this, the z-transform for the modified Euler method is easily obtained:

$$\frac{X(z)}{U(z)} = \frac{T\left(z^{1/2} + \dfrac{\lambda T}{2}\right)}{z - \left(1 + \lambda T + \dfrac{(\lambda T)^2}{2}\right)} \tag{6–56}$$

A comparison of equations 6–56 and 6–52 produces several interesting facts. First, the denominators of the two transfer functions are identical, so that the closed-loop pole for each method is given by equation 6–53. Thus, the stability properties of the two methods are identical. It turns out that all two-stage, second-order accurate Runge-Kutta methods have the same closed-loop pole. A further point concerning the closed-loop pole comes from equations 6–17 and 6–18. The exact solution to the linear test equation at the sample points involves the complex exponential $\exp(\lambda T)$, and the series expansion for that is

$$e^{\lambda T} = 1 + \lambda T + \frac{(\lambda T)^2}{2} + \frac{(\lambda T)^3}{6} + \dots \tag{6–57}$$

Thus, the closed-loop pole for all two-stage, second-order accurate methods is given by a truncated form of the series expansion for $\exp(\lambda T)$, and this observation can be generalized to higher-order methods. The fact that we can equate the Taylor series expansion for the exponential function with the Runge-Kutta z-plane pole location is extremely significant. This means that all we need to do to determine the closed-loop pole for a Runge-Kutta method is to truncate the Taylor series for the exponential at the appropriate order of T. The result is the pole location as a function of λT. Therefore, the following fact is presented.

FACT

For any explicit R-stage, Rth-order accurate Runge-Kutta method, the closed-loop pole is located at

$$z = 1 + \lambda T + \frac{(\lambda T)^2}{2} + \frac{(\lambda T)^3}{6} + \ldots + \frac{(\lambda T)^R}{R!} \qquad (6\text{–}58)$$

This result is true only for R-*stage, R*th-order methods. Any method that is fifth or higher order has a slightly modified pole location.

Analysis of the closed-loop pole suggests the use of the root locus technique. Although this can be done, the usual construction rules for the root locus only work for a first-order Runge-Kutta method, which is necessarily Euler's method. The root loci for negative real λT are given in Figures 6.4 to 6.7 for first- through fourth-order Runge-Kutta methods. It should be noticed that the principal root leaves the unit circle through the left side for odd orders and through the right side for even orders. This observation can yield some insight into instabilities that might arise in the simulation of a given system with a given order method.

Looking again at the transfer functions in equations 6–56 and 6–52, we see that the numerators are different, so that the two methods will not have the same frequency re-

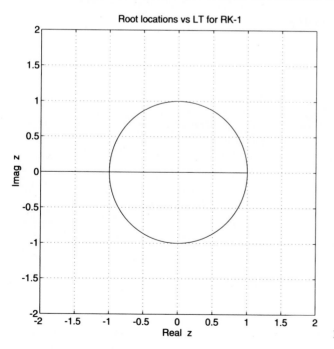

Figure 6.4   Root locations vs. LT for RK-1.

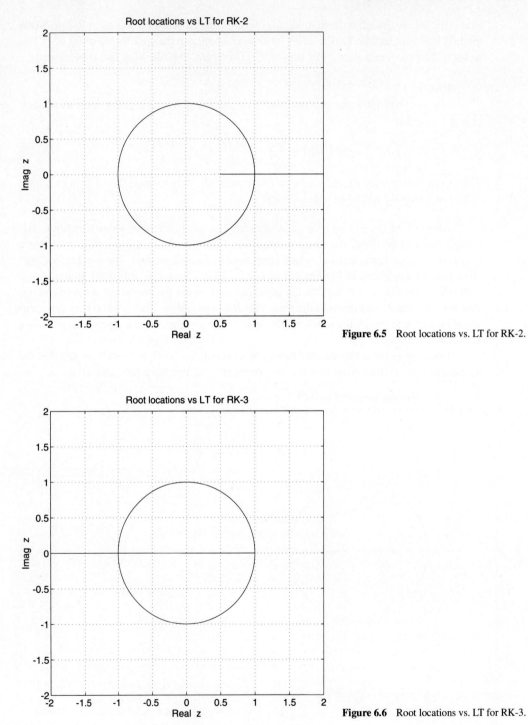

Figure 6.5 Root locations vs. LT for RK-2.

Figure 6.6 Root locations vs. LT for RK-3.

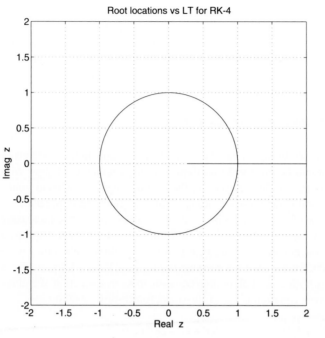

Figure 6.7   Root locations vs. LT for RK-4.

sponse. Specifically, note that the transfer function for the modified Euler method has a fractional power of z. This is due to the fact that the method evaluates the function in the interior of the time interval between t_n and t_{n+1}. Since this often happens with Runge-Kutta methods, fractional powers of z in the numerator are a common occurrence. Although this is generally not seen in other transfer functions, the definitions of zeros and frequency response are made in the usual way. Since the frequency response of a discrete-time transfer function is obtained by letting

$$z^n = e^{jn\omega T}, \qquad j = \sqrt{-1} \tag{6–59}$$

the same substitution as usual is made; in the case of fractional powers, $n < 1$.

Two further important observations should be made. Since all R-stage, Rth-order Runge-Kutta methods have the same characteristic equation, it then follows that they must be equivalent from a stability viewpoint. They must also have the same stability region. This bears repeating.

FACT

All explicit R-stage, Rth-order Runge-Kutta methods have the same principal root, the same stability region, the same closed-loop root locus, and the same stability properties. Unlike the root locus for a normal transfer function, the root locus for an explicit Runge-Kutta method is not affected by the zeros of the transfer function since they do not appear in the closed-loop characteristic equation.

the same closed-loop pole for any value of λT. They also have the same region of absolute stability, as do all of the infinite number of other two-stage, second-order accurate methods. The differences in the methods only become apparent when time-dependent inputs are considered. The different numerators produce different frequency responses. This yields different time responses when the input is not piecewise constant and multirate input sampling is used.

6.8 STABILITY REGIONS FOR RUNGE-KUTTA METHODS

The stability regions for the first- through the fourth-order Runge-Kutta methods are given in Figures 6.8 to 6.11. It should be observed that these stability regions get bigger with increasing order, which is contrary to the behavior of the stability regions for linear multistep methods. This is one reason why high-order Runge-Kutta methods are so popular.

Unfortunately, increasing the order also increases the number of stages. Increasing the number of stages linearly increases the computation time since each stage requires a derivative evaluation. This was not the case for linear multistep methods unless they were used in a correct to convergence mode. To return to stability regions, Runge-Kutta methods require the solution to a fairly high-order polynomial equation in λT as a function of z-plane radius and angle. The following equation is for a four-stage, fourth-order method:

$$\frac{(\lambda T)^4}{24} + \frac{(\lambda T)^3}{6} + \frac{(\lambda T)^2}{2} + \lambda T + (1 - re^{j\theta}) = 0 \tag{6-60}$$

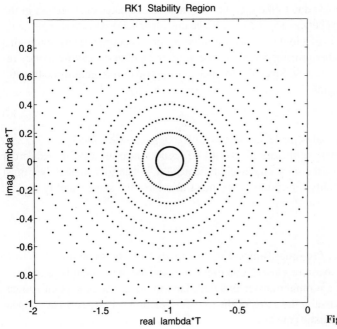

Figure 6.8 RK-1 stability region.

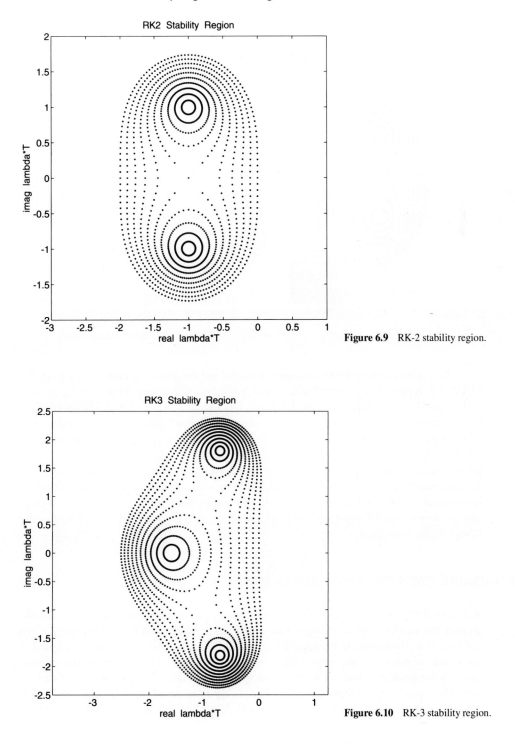

Figure 6.9   RK-2 stability region.

Figure 6.10   RK-3 stability region.

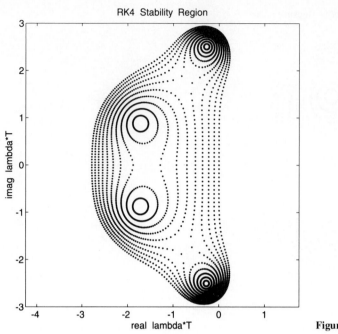

Figure 6.11   RK-4 stability region.

This would be very difficult to solve for λT if not for control system software such as MATLAB. It should be observed that the result is a 1-to-R mapping. That is, every point in the z-plane maps into R points in the λT-plane, R being the number of stages of the method. It is interesting to examine the real z-plane axis. Setting $\theta = 0$ gives merely a standard root locus to plot as a function of r. The z-plane origin, $r = 0$, maps into the R-images in the λT-plane. There are no finite λT-plane zeros; hence, there is a λT-plane pole excess of R. A given trajectory on the positive root locus as a function of r leaves the images of $z = 0$, moves through the stability region boundary when $r = 1$, and then goes to infinity along the appropriate asymptote. Each trajectory is the boundary between one image of the z-plane and another in the λT-plane. An example is shown in Figure 6.12 for the fourth-order Runge-Kutta methods.

6.9 VARIABLE TIMESTEP RUNGE-KUTTA METHODS

Although Runge-Kutta methods suffer from several disadvantages, as discussed in the last section, the variable timestep Runge-Kutta methods are extremely powerful when a given system can be formulated in the appropriate manner. Specifically, if it is appropriate and/or acceptable to put the given system into one large state space representation, the variable timestep Runge-Kutta methods would be an appropriate choice for performing the simulation. Their power arises from the availability of an estimate of the local truncation error directly from the method at essentially no cost. With this estimate, it is then possible to

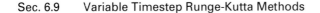

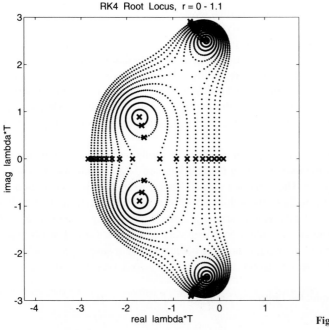

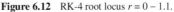

Figure 6.12   RK-4 root locus $r = 0 - 1.1$.

change the timestep freely, with no particular past data shuffling or filling in since Runge-Kutta is a one-step method. Usually, if the local error at a particular time is too large, the timestep is appropriately reduced and the step is taken over again until the error is acceptably small. Thus, all that is lost is a little computation time during this process. Alternatively, if the error is sufficiently small, the timestep can be increased, which allows for a faster simulation.

With these properties, the variable timestep Runge-Kutta method has been a very popular explicit method choice by users. It is particularly useful for nonlinear systems, in which the powerful stability region idea can only give an approximate choice of simulation timestep via the system Jacobian matrix. For example, if a given nonlinearity is discontinuous, as in many engineering problems because of relays, limiters, and so on, the derivative can significantly change as the simulation method steps over the discontinuity. Before the end of the current simulation interval, but after crossing the discontinuity, significant errors in the simulated system states can arise. A variable timestep Runge-Kutta method is able to detect this error and reduce its timestep, so that an acceptably small error occurs at the discontinuity. Another specific application of this method is to systems whose dynamics change significantly during the simulation. A system may start out with all of its states changing very slowly, only to have some of them occasionally become fast. Systems of this type are called stiff systems and will be discussed at length in Chapter 7. A fixed timestep method would require its simulation timestep to be based on the fastest behavior and thus have a small timestep for the entire simulation. The variable timestep methods, however, would be able to choose a small timestep when necessary and use a large one when possible. The overall effect, then, is a sufficiently accurate simulation, accomplished in as little time as possible for an explicit method of given order.

Users are often faced with unknown nonlinear systems that can readily be put into state space form. Thus, the variable timestep Runge-Kutta methods have become popular because they require almost no preliminary knowledge of the system beyond its describing equations, and therefore little analysis by the user. Alternatively, these methods usually cannot be used for simulations that require a fixed simulation timestep, or frametime; this includes real-time applications.

Unfortunately, variable timestep Runge-Kutta methods are often overdone because of lack of knowledge by the user. Often, a method is taken from a textbook or from the available simulation software. The variable timestep Runge-Kutta methods most often presented, unfortunately, are usually of the Fehlberg, Verner, or Merson type and are at least fifth-order accurate. Consequently, at least six stages are required along with the associated interstage multiples. The resulting simulation is then very complicated and slow because of the complexity of the algorithm. What follows is intended to be a simple introductory example to variable timestep Runge-Kutta methods and is also a very powerful, and fast, method by itself.

Consider, for example, the improved Euler method:

$$
\begin{aligned}
x_{n+1} &= x_n + \frac{T}{2} [K_1 + K_2] \\
K_1 &= f(t_n, x_n) \\
K_2 &= f(t_n + T, x_n + TK_1)
\end{aligned}
\tag{6--61}
$$

When this method is implemented, the first stage is a normal Euler step and by itself is a first-order accurate approximation of the next value of x. Correcting this Euler step with the trapezoidal rule gives the improved Euler method, which gives a second-order accurate approximation of the next value of x. The difference between the output of the first stage and the second stage gives an estimate of the error. Rewriting this method as follows will clarify the intended implementation:

$$
\begin{aligned}
K_1 &= f(t_n, x_n) \\
\hat{x}_{n+1} &= x_n + TK_1 \\
K_2 &= f(t_n + T, x_n + TK_1) \\
x_{n+1} &= x_n + \frac{T}{2} [K_1 + K_2] \\
e_n &= x_{n+1} - \hat{x}_{n+1}
\end{aligned}
\tag{6--62}
$$

With this formulation, it is clear that the cost of the error estimate is a simple and fast vector subtract. The size of this error is easily monitored by using some vector norm. When error norm exceeds the user-defined maximum allowable error, the timestep can be reduced. After timestep reduction, the last step should be repeated so that large errors are not allowed to propagate, a particularly important issue for nonlinear systems. Finally, it should be noted that the error estimate results directly from within an otherwise useful Runge-

Kutta method; that is, nothing special is done to obtain this estimate. Error estimates of this type are said to be imbedded in the method. All of the variable timestep methods discussed here are based on a similar type of embedding.

As mentioned above, the error estimate is based on some norm of the error vector. Clearly many norms are possible, but the most often used are the 1, 2, and infinity norms. For an m-vector e, these are

$$|e|_1 = |e_1| + |e_2| + \ldots + |e_m|$$ (6–63)

$$|e|_2 = \sqrt{|e_1|^2 + |e_2|^2 + \ldots + |e_m|^2}$$ (6–64)

$$|e|_\infty = \max |e_j| \qquad \text{for } j = 1 \text{ to } m$$ (6–65)

Here, the 1- or ∞-norms are suggested since they require little calculation.

Remember, as long as the particular norm of the error estimate stays below the user-defined maximum, the simulation timestep can stay the same or increase. Alternatively, if the error norm is too large, the timestep should be decreased. The remaining question is, then, how does one change the timestep? Many methods are available in the literature for doing so; however, the best discussion is probably found in Thomas (1986). The main problem in choosing the timestep is keeping the timestep from swinging rapidly from small to large and back. If the timestep must be reduced frequently, there is much recalculation at each point in time, and thus the simulation will be slower than necessary. Alternatively, it is desirable to make the timestep as large as possible for the given tolerance; otherwise the simulation will again be slower than necessary. Consequently, the scheme for varying the timestep must possess some probing, to increase the timestep, along with caution, to keep the timestep from getting too large too often (*probing* and *caution* are terms borrowed from the adaptive control area, Goodwin & Sin, 1984).

The following method is suggested for anyone making a first attempt at implementing his or her own variable timestep Runge-Kutta method.

1. Choose E_{max} and E_{min} at least one order of magnitude apart and choose an initial T; start the simulation.
2. If $\text{norm}(e) > E_{max}$, $T = T/2$; repeat step.
3. If $\text{norm}(e) < 0.8\,E_{max}$, $T = 1.05T$ for the next step.
4. If $E_{max} > \text{norm}(e) > 0.8\,E_{max}$, keep the current T for the next step.
5. If $\text{norm}(e) < E_{min}$, $T = 1.25T$ for the next step.

If more complexity is allowable or better timestep control is required, the reader should consult Thomas (1986). Those methods make the timestep an explicit function of the current error. Clearly, many approaches to this procedure are available, and users should be encouraged to experiment with their own methods for changing T.

Appendix B contains several variable timestep Runge-Kutta methods and the stability regions associated with the predictor stages and the corrector stages, which are used in obtaining the error estimate. Also there is some information concerning the utility of the particular methods. The reader is further warned against the use of the Merson methods and the Fehlberg methods of order greater than 4, as they give erroneous error predictions (Butcher, 1987).

6.10 CONCLUSIONS

The Runge-Kutta methods are considerably more complicated than linear multistep methods since they sacrifice linearity to improve accuracy. We have shown that they are generally well suited to nonlinear systems given in state space representations, particularly the variable timestep Runge-Kutta methods. For systems given in block diagram form, however, Runge-Kutta methods can lose significant accuracy unless the entire system can be transformed into one large state space representation.

PROBLEMS

1. Consider the fourth-order, fourth-stage Runge-Kutta method applied to x'_1 $x_2 x'_2 = w^2 x_1$. Determine the system's closed-loop poles for various values of w and T, and discuss.

2. The system

$$u(s)/y(s) = 8.77(s^2 + 3.51s + 12442.5)/(s^3 + 3.51s^2 + 43548.6s + 109144)$$

represents the dynamics of a supersonic inlet. The input is the bypass door position and the output is the shock position.

   a. Using the standard fourth-order Runge-Kutta integrator, obtain a step response and an impulse response for this system.

   b. Plot the closed-loop poles of the integration process in the z-plane as a function of T for the system.

   c. Geometrically explain what the integrator does.

3. Simulate the system shown below by using the standard fourth-order Runge-Kutta integrator. This block diagram represents a linearized model of a magnetic levitation control system. G_c is given as $G_c(s) = K(s + 40)/(s + 400)$, $K = 50$. The other parameters of the system are $x_0 = 0.008m$, $m = 0.068kg$, $R = 28\Omega$, $L = 0.438H$, $i_0 = 0.76A$, $C = 7.39E - 5\ Nm^2/A^2$, $k = 1.756N/A$, $B = 1.143E3$ V/m.

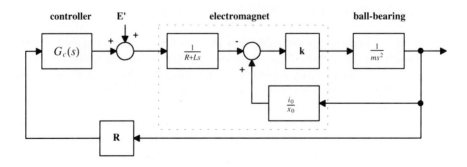

4. Sketch the graphical interpretation of the following second-order, two-stage Runge-Kutta method, for one timestep.

$$\dot{x}_n = f(t_n, x_n)$$

$$x_{n+1} = x_n + \frac{T}{4}(3K_1 + K_2)$$

$$K_1 = f(t_n, x_n)$$

$$K_2 = f(t_n + 2T, x_n + 2TK_1)$$

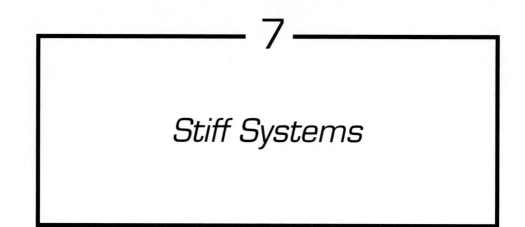

7

Stiff Systems

7.1 INTRODUCTION

In chapters 4, 5, and 6 we studied the numerical integration of ordinary differential equations. We saw how the stability of the simulation and the accuracy of its results depended on the value of the step size T. For linear multistep methods, the size of the stability region decreases with increasing order of accuracy; for Runge-Kutta methods, the reverse is true. With either class of method, however, there is an upper limit of the value of T for acceptable results. From a practical point of view, the simulation engineer would like to use as large a value of T as possible, consistent with the stability and accuracy constraints of the problem. For larger values of T, fewer timesteps have to be taken to simulate the system over a given period of time. This makes the simulation run faster and decreases the amount of round-off error incurred.

Some systems present special problems for the engineer. These systems require a small value of T to maintain simulation stability, and they also have to be simulated over a long (relative to T) period of time to get a complete picture of the time response of the system. This type of system is called stiff, and stiff systems have provided many interesting problems for the simulation engineer.

This chapter examines the simulation problem of stiff systems and presents several stability definitions that have been developed specifically for them. Several integration methods, developed by us and others, are explained and examined in terms of their ability to deal with stiff systems.

7.2 DEFINITION AND EXAMPLES

A stiff system is one that has dynamics operating on two different time scales. There is a fast mode, which decays very quickly in response to initial conditions or a changing input, and a slow mode, which takes much longer for its transient response to decay. In a stiff system, the rates of decay of the fast and slow modes will be very different, often by more than three orders of magnitude. The stiffness arises from the fact that the system has some eigenvalues with large magnitude negative real parts, corresponding to the fast modes, and other eigenvalues with small magnitude negative real parts, the slow modes. Thus, a stiff system has the property that

$$\text{Stiffness ratio} = \frac{\max | Re(\lambda) |}{\min | Re(\lambda) |} \gg 1 \tag{7–1}$$

In real systems that are stiff, the ratio of largest to smallest eigenvalues are commonly in the range of 10^3 to 10^6. This large ratio of eigenvalues is where the problem associated with stiffness arises. The large negative eigenvalues are the fastest components of the system, and they have an effect on the response only for a short period of time since they decay to zero quickly. Unfortunately, the simulation timestep is limited by these large eigenvalues since the product of the timestep with each eigenvalue must lie inside the stability region for whatever simulation method is being used.

The small negative eigenvalues are the dominant components in the system because their transient response takes the longest time to decay. These are usually the most important dynamics of the system, but their evolution in the simulation takes a great deal of time because of the small timestep dictated by the fast states. This is the problem of stiffness.

DEFINITION
A system is stiff if the timestep required to produce a stable simulation of the system is significantly smaller than the timestep required to produce an accurate simulation of the system assuming that stability is not an issue.

Example 7.1
The system to be simulated is defined as

$$\begin{bmatrix} \dot{x}_1 \\ \dot{x}_2 \end{bmatrix} = \begin{bmatrix} 0 & -10 \\ 10 & -101 \end{bmatrix} \begin{bmatrix} x_1 \\ x_2 \end{bmatrix} \tag{7–2}$$

Its characteristic equation is

$$| \lambda I - A | = \begin{vmatrix} \lambda & 10 \\ -10 & \lambda + 101 \end{vmatrix} = \lambda^2 + 101\lambda + 100 = 0 \tag{7–3}$$

and the system eigenvalues are

$$\lambda = -1, -100 \qquad\qquad (7\text{--}4)$$

so the stiffness ratio is 100, making the system only moderately stiff. The response due to the pole at -1 takes much longer to decay than the response due to the pole at -100. Unfortunately, the timestep must be chosen to maintain simulation stability for both poles. The largest possible choice for most explicit linear multistep methods would be about $T = 0.01$ seconds. The transient response due to the dominant pole at -1 would take approximately five seconds to decay to 1 percent of its initial value. Thus, a simulation showing an initial condition response would take at least 500 iterations to provide an accurate simulation while maintaining stability. Figure 7.1 shows the response of the system described by equation 7–2 to initial conditions $x_1 = x_2 = 10$. The timestep used for these plots is $T = 0.0025$ seconds. The figure illustrates that the fast component decays to 10 percent of its initial value over 20 times more rapidly than the slow component. The demonstration of stiffness would be even more dramatic if the two states were decoupled, so that the system matrix would be $A = \text{diag}(-1\ -100)$. In that case, state x_2 would continue to decay rapidly, reaching a value of 0.1 in approximately 0.5 seconds.

This example demonstrates the problem of stiffness; you must integrate slowly to produce values for a component that is not important, whereas it takes a long time to get the important results. It should be pointed out that this definition is based on starting with a system model. Improved modeling techniques, or model reduction methods, could be applied to the given physical system to obtain a nonstiff mathematical model that ignores the fast dynamics. The problems associated with stiffness assume that the model cannot be changed, and all that can be done is to create specialized numerical methods to deal with the problems.

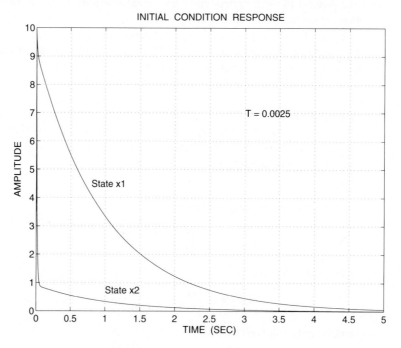

Figure 7.1   Exact initial condition response.

It should be emphasized that the simulation timestep cannot be changed to a larger value after the fast dynamics decay to sufficiently small values unless the system model is altered. The fast dynamics are still present in the system, and even though they do not contribute much to the total response, their simulation must still be stable.

A second example (Gear, 1968) is presented to further illustrate the stiffness issue.

Example 7.2

The system is described by a third-order, nonlinear model given by

$$
\begin{aligned}
\dot{x}_1 &= -0.013x_1 - 1000x_1x_3, & x_1(0) &= 1 \\
\dot{x}_2 &= -2500x_2x_3, & x_2(0) &= 1 \\
\dot{x}_3 &= -0.013x_1 - 1000x_1x_3 - 2500x_2x_3, & x_3(0) &= 0
\end{aligned}
\tag{7-5}
$$

When these equations are linearized, the eigenvalues of the Jacobian are in the following ranges:

$$
\begin{aligned}
\lambda_1 &= -9.3 \times 10^{-3} \rightarrow -4.0 \times 10^{-3} \rightarrow -6.3 \times 10^{-3} \\
\lambda_2 &= 0 \\
\lambda_3 &= -3.5 \times 10^{3} \rightarrow -3.8 \times 10^{3}
\end{aligned}
\tag{7-6}
$$

The stiffness ratio for this system is on the order of 10^6. The simulation will be done with a second-order Adams-Bashforth-Moulton (ABM-2) predictor-corrector with a timestep $T = 0.00025$ seconds. This provides simulation accuracy for the fastest state. The first step is taken by using Euler's method to provide the second starting value needed by the AB-2 method. Figure 7.2 shows the response of the three states. As can be seen, the first two states have changed by less than 1 percent of their initial value in one-half second, whereas x_3 reaches a nearly constant value in two milliseconds.

A similar but slightly different problem to the above occurs when the system has eigenvalues with large imaginary parts. These correspond to components with high-frequency oscillations. Even if there are no eigenvalues with large real parts, a small value of T is required to produce an accurate simulation of the oscillations as well as to produce a stable simulation. This is the difference between a stiff system and a system with high-frequency, slowly decaying oscillations. In the former, a small T is required only for stability, not for accuracy; in the latter, a small T is required both for stability and accuracy. For the stiff system, numerical methods can be developed to allow the use of a larger T to provide an accurate simulation of the important dynamics without losing stability. For the system with large imaginary components in its eigenvalues, a small T is needed if an accurate simulation is desired. See the following example.

Example 7.3

For a system with state equations

$$
\begin{bmatrix} \dot{x}_1 \\ \dot{x}_2 \\ \dot{x}_3 \end{bmatrix} =
\begin{bmatrix} -102 & -10201 & -1000100 \\ 1 & 0 & 0 \\ 0 & 1 & 0 \end{bmatrix}
\begin{bmatrix} x_1 \\ x_2 \\ x_3 \end{bmatrix}, \quad
x(0) = \begin{bmatrix} 10 \\ 10 \\ 10 \end{bmatrix}
\tag{7-7}
$$

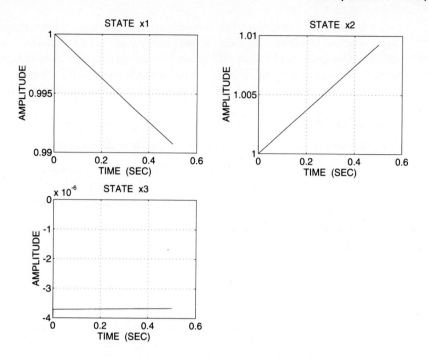

Figure 7.2   Simulation response for Example 7.2.

the eigenvalues are

$$\lambda = -100, -1 \pm j100 \qquad (7-8)$$

Figure 7.3 shows the response of this system to the initial conditions given in equation 7–7. Each of the states exhibits high-frequency oscillations whose amplitudes have an exponential decay. The rate of the exponential decay is controlled by the real part of the complex eigenvalues, and the frequency of oscillation is controlled by the imaginary part. Assuming that we wanted an accurate simulation that showed the oscillations, the value of T would have to be based on their frequency, not on the exponential decay. Thus, both from an accuracy standpoint and for stability, a small value of T would be needed. This situation separates this type of problem from that of stiff systems.

7.3 STABILITY DEFINITIONS

Several definitions pertaining to simulation stability when dealing with stiff systems are presented here. These definitions have been motivated at least in part by the special need of stiff systems, which require a much smaller timestep for stability than for accuracy. Their purpose is to define characteristics of integration algorithms when simulation stability is independent of the choice of T.

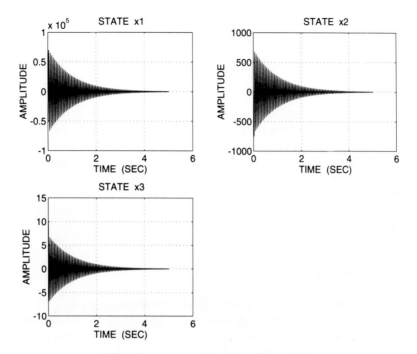

Figure 7.3   Simulation response for Example 7.3.

DEFINITION

An integration method is *A*-stable if and only if its region of absolute stability is the entire left half of the complex λT plane (Dahlquist, 1963; Lambert, 1973).

With an *A*-stable method, the value of the timestep *T* can be chosen without regard to the system eigenvalues since any λT product will be inside the stability region, assuming that the $Re(\lambda) < 0$. This seems like a very desirable situation. With an *A*-stable method, *T* can be based on accuracy, and simulation stability is guaranteed. However, there are two problems associated with *A*-stability.

The first problem is one of existence. Are there any linear multistep integration methods that in fact are *A*-stable? Most of the methods examined in chapters 4 and 5 had finite stability regions. First, we will look at explicit methods, which can be dealt with by the following fact.

FACT

No explicit linear multistep or Runge-Kutta method, regardless of step number or order of accuracy, can be *A*-stable.

The reason for this can easily be seen from the discussion of linear multistep methods in Chapter 4. Viewing the integration algorithm as a transfer function, we see that an explicit method always has a higher-degree denominator than numerator. For a general *k*th-order method, the transfer function is

$$H(z) = \frac{T \sum\limits_{i=0}^{k-1} b_i z^i}{z^k + \sum\limits_{i=0}^{k-1} a_i z^i} = \frac{T\sigma(z)}{\rho(z)} \tag{7–9}$$

Since linear multistep methods have more poles (roots of $\rho(z) = 0$) than zeros (roots of $\sigma(z) = 0$), at least one branch of the root locus, plotted in the z-plane as a function of λT, will go to infinity as λT increases without bound. Since the region of stability in the z-plane is the unit circle, instability is guaranteed for any explicit method for sufficiently large λT. Although the analysis is more difficult for Runge-Kutta methods, the same conclusions are reached.

Implicit methods can be A-stable, but it is not a common occurrence. Only low-order methods can have that property.

FACT

The highest order of accuracy for an implicit linear multistep method that is A-stable is 2.

A-stable implicit methods that have been discussed are the first-order backward difference and the trapezoidal algorithms. The trapezoidal method is second-order accurate, and it is the second-order A-stable algorithm with the smallest error constant ($C_3 = -1/12$). The root locus plots for these methods are shown in Chapter 4 to lie entirely within the unit circle. Therefore, for any value of λT, with $Re(\lambda) < 0$, those methods are absolutely stable. Another A-stable implicit method is the second-order backward difference algorithm, which is

$$y_{n+2} = \frac{4}{3} y_{n+1} - \frac{1}{3} y_n + \frac{2T}{3} \dot{y}_{n+2} \tag{7–10}$$

and has an error constant $C_3 = -2/9$. The two facts about A-stability presented here show that it is an extremely stringent requirement for a linear multistep method.

The second problem with A-stability is one of convergence. Since only implicit methods can have that property, and implicit methods generally have to be iterated at each timestep to obtain a solution, stability must be examined in this context also. If an accurate solution cannot be obtained because the iterations do not converge, the properties of the stability region for the integration method are of no concern. From Chapter 4, recall that for an implicit method, convergence of the iterations requires that

$$T < \frac{1}{b_k |\lambda|_{max}} \tag{7–11}$$

Thus, there is a restriction on the maximum value of T when an implicit method is used, even if the method is A-stable. If T exceeds this maximum value, the iterations of the solution will diverge. Therefore, A-stability of a method is not a panacea for the problem of dealing with stiff systems.

Several additional definitions related to A-stability follow.

DEFINITION

An integration method is $A(\theta)$-stable if and only if its region of absolute stability contains an infinite slice of the λT-plane defined by $\pm\theta$, as measured from the negative real axis. If $\theta = 0$, the method is $A(0)$-stable (Lambert, 1973).

DEFINITION

An integration method is $A(a)$-stable if and only if its region of absolute stability includes the entire half of the complex λT-plane to the left of $-a$.

Figure 7.4 illustrates these two definitions of stability. Clearly, no explicit method can satisfy either of these requirements since they force the integration method to be stable for increasingly large values of λT. The only implicit A-stable method whose order of accuracy p exceeds its step number k is the trapezoidal rule (Lambert, 1973).

DEFINITION

Consider Figure 7.5. A method is stiffly stable if its region of absolute stability contains both regions R and S and the method is accurate in producing simulation results when λT lies within region S and $Re(\lambda) < 0$ (Lambert, 1973).

An integration method that is stiffly stable and is applied to a stiff system accommodates the fast system eigenvalues by ensuring simulation stability; λT lies within the stable region R, even for the fastest dynamics. The slow and important dynamics are simulated both stably and accurately since the λT product would fall within region S for those dynamics. Thus, the value of T could be chosen to place the important λT within region S without being concerned about loss of stability because of the fast eigenvalues. Note that for those

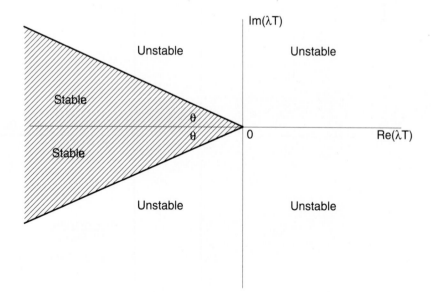

Figure 7.4a $A(\theta)$ stability.

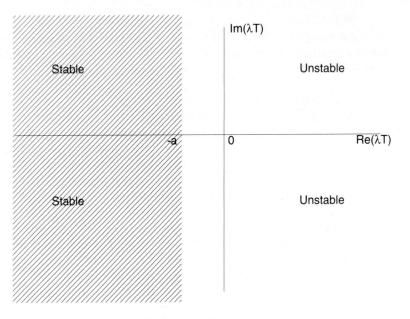

Figure 7.4b   *A(a)* stability.

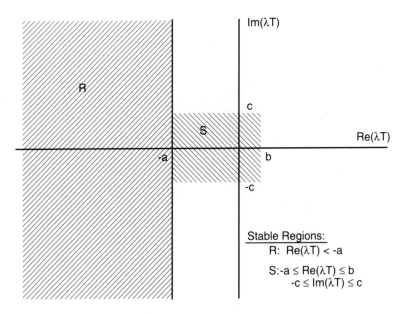

Figure 7.5   Stiff stability.

values of $\lambda T < -a$, the stiffly stable method is effectively $A(a)$-stable. The method is also $A(\theta)$-stable for

$$\theta = \tan^{-1}\left(\frac{c}{a}\right) \tag{7-12}$$

One last pair of definitions of stability will be presented, motivated by the following situation. Consider the first-order linear test equation

$$\dot{y}(t) = \lambda y(t), \qquad y(0) = y_0 \tag{7-13}$$

We know that the exact solution to this homogeneous differential equation is

$$y(t) = y_0 e^{\lambda t} \tag{7-14}$$

If $\lambda < 0$, the solution $y(t)$ decays with increasing time, asymptotically approaching zero. The more negative λ is, the faster the response decays. Specifically, as $\lambda \rightarrow -\infty$, $y(t)$ decays to zero instantaneously. In this case, with a one-step numerical integration method, y_{n+1} should be equal to zero regardless of the value of y_n. Let us look at the trapezoidal method, which we know is A-stable, and see how it deals with this situation. Note that iteration of the solution is not a concern here since the system is linear. If we substitute the linear test equation into the trapezoidal rule and constrain λ to be negative, we get

$$y_{n+1} = y_n \left(\frac{1 - \dfrac{|\lambda T|}{2}}{1 + \dfrac{|\lambda T|}{2}} \right) \tag{7-15}$$

As the magnitude of λ increases without bound, the simulation will produce the following result:

$$y_{n+1} = -y_n \tag{7-16}$$

This solution is obviously wrong. It not only fails to decay to zero in one timestep, it never decays at all, continually oscillating between $\pm y_0$. Thus, although the integration method is A-stable, having no problem with stability, the solution produced by the method for fast system dynamics is completely erroneous. The accuracy for that scenario is missing.

Now let us consider the first-order backward difference method, which is also A-stable. Making the same substitution of the test equation into the integration algorithm, we get for $\lambda < 0$

$$y_{n+1} = y_n \left(\frac{1}{1 + |\lambda T|} \right) \tag{7-17}$$

We see from equation 7–17 that the numerical solution will decay monotonically with time, as it should to be an accurate representation of the true solution. Also, as $\lambda \rightarrow -\infty$, the solution will decay to zero in one timestep. Therefore, this method maintains accuracy as well as stability. The following definition is based on this observation.

DEFINITION

A one-step method is L-stable if and only if it satisfies the following two conditions (Lambert, 1973):

1. It is A-stable.

2. The following relationship holds:

$$\left| \frac{y_{n+1}}{y_n} \right| \to 0 \text{ as } \lambda T \to -\infty \tag{7-18}$$

Thus, only implicit methods can be L-stable since the first requirement for that property is A-stability. The definition of L-stability is very stringent, however.

THEOREM

A linear one-step numerical integration method is L-stable if and only if it has the difference equation form (Beale & Hartley, 1987)

$$y_{n+1} = -a_0 y_n + T b_1 \dot{y}_{n+1} \tag{7-19}$$

or equivalently has the transfer function

$$H(z) = \frac{T b_1 z}{z + a_0} \tag{7-20}$$

with the coefficient constraints

$$|a_0| \leq 1 \tag{7-21}$$

$$b_1 > 0 \tag{7-22}$$

PROOF

"If " (sufficiency)

Assume that the integration method has the form and coefficient constraints stated in the theorem. We will show that these assumptions guarantee that both parts of the definition of L-stability are satisfied by the integration method. First, A-stability will be shown. To be A-stable, the root locus of the integration method with the linear test equation must lie entirely inside the unit circle for all values of $\lambda T < 0$. With the test equation, the closed-loop transfer function of the algorithm, assuming an inherently stable system, is

$$H_{cl}(z) = \frac{T b_1 z}{z(1 + |\lambda T| b_1) + a_0} \tag{7-23}$$

The closed-loop pole, which governs simulation stability, is

$$z = \frac{-a_0}{1 + |\lambda T| b_1} \tag{7-24}$$

From equation 7-20, it is seen that the open-loop zero and pole of the integration method are located at the origin of the z-plane and at $-a_0$, respectively. Since the root locus begins at the open-loop pole and ends at the open-loop zero, and $b_1 > 0$ corresponds to a root locus plot for positive gain, the closed-loop pole will lie on the real axis between the open-loop pole and open-loop zero for all $\lambda T < 0$. Thus, the closed-loop pole is always inside the unit circle, and the condition for A-stability is established. To finish the proof of sufficiency, substitute the linear test equation into equation 7–19 and rearrange to get

$$\frac{y_{n+1}}{y_n} = \frac{-a_0}{(1 + |\lambda T| b_1)} \tag{7–25}$$

Since $b_1 > 0$ and a_0 has finite magnitude, this ratio goes to zero as the λT product increases without bound. This proves that the proposed form and coefficient constraints will always produce an L-stable integration method.

   "Only if" (necessity)
   For this part of the proof, we show that violating the form or coefficient constraints yields an integration method that either is not A-stable or has a ratio for y_{n+1}/y_n, which does not go to zero as λT increases. First consider $|a_0| > 1$. This places the open-loop pole outside the unit circle. Therefore, for sufficiently small $|\lambda T|$ the closed-loop pole is also outside the unit circle, so the method is not A-stable. Second, consider the case when $b_1 < 0$. This produces a root locus plot corresponding to negative gain, so the closed-loop poles lie on that part of the real axis not between the open-loop pole and zero. Therefore, for some $\lambda T < 0$, the closed-loop pole will be outside the unit circle. Again, the method is not A-stable. Next, we consider the situation when $b_1 = 0$. Looking again at the ratio in equation 7–25, we see that $y_{n+1}/y_n = -a_0$, regardless of the value of λT. This violates the second part of the definition for L-stability. Lastly, assume that the one-step integration method has a more general form than given in equation 7–19. Specifically, assume the form has a b_0 coefficient, so that it has the y_{n+1}/y_n ratio shown below:

$$\frac{y_{n+1}}{y_n} = \frac{-(a_0 + |\lambda T| b_0)}{(1 + |\lambda T| b_1)} \tag{7–26}$$

As $\lambda \to -\infty$, this ratio goes to $-b_0/b_1$, so L-stability is lost unless $b_0 = 0$. This is the most general form for a linear one-step integration method. Therefore, if either of the coefficient constraints are violated or if the algorithm form is different from that which was stated in the theorem, the method cannot be L-stable. This completes the proof.

   In Chapter 4, consistency of an integration method was introduced and shown to be a requirement for the numerical solution generated by the method to converge to the true solution. For a linear multistep method described by equation 7–19, consistency requires that $a_0 = -1$ and $b_1 = 1$. Thus, the only consistent L-stable linear multistep method is the first-order backward difference algorithm. There are no R-stage, Rth-order Runge-Kutta methods, with $R > 1$, which are L-stable.

   The concept of L-stability can be extended to methods of higher step numbers. The following definition concludes this section.

   DEFINITION
   A k-step linear multistep method is $L(k)$-stable if and only if (Beale & Hartley, 1987)

   1. It is A-stable.
   2. $y_{n+k} \to 0$ as $\lambda T \to -\infty$.

With the stable linear test equation, the k-step method in closed-loop is

$$(1 + |\lambda T| b_k) y_{n+k} = - \sum_{i=0}^{k-1} (a_i + |\lambda T| b_i) y_{n+i} \tag{7-27}$$

Following the logic given in the proof of the above theorem, it can be shown that to be $L(k)$-stable, a method must be implicit, all the coefficients b_0 through b_{k-1} must be zero, and the coefficients b_k and a_0 through a_{k-1} must be selected so that the method is A-stable. With these conditions, y_{n+k} is

$$y_{n+k} = \frac{-\sum_{i=0}^{k-1} a_i y_{n+i}}{(1 + |\lambda T| b_k)} \tag{7-28}$$

Three $L(2)$ methods are given below; the first two are first-order accurate; the third is second-order accurate:

$$\begin{aligned} y_{n+2} &= y_n + 2T\dot{y}_{n+2} \\ y_{n+2} &= 1.5 y_{n+1} - 0.5 y_n + 0.5 T \dot{y}_{n+2} \\ y_{n+2} &= \frac{4}{3} y_{n+1} - \frac{1}{3} y_n + \frac{2T}{3} \dot{y}_{n+2} \end{aligned} \tag{7-29}$$

7.4 STABILITY REGION PLACEMENT METHODS

We have seen that the simulation of stiff systems poses a serious problem for the engineer. Stability requires a small timestep for most methods, and yet a larger T would be satisfactory from an accuracy consideration. The stability definitions in the last section define integration methods that allow the choice of T to be based on accuracy rather than stability. However, all explicit methods are excluded by these definitions, and convergence requirements limit the applicability of the definitions even for implicit methods. What is needed is an explicit integration method that will be stable for a particular system, but with the value of the timestep T chosen to meet constraints of accuracy and computation time.

We developed the stability region placement (SRP) family of integration methods to accomplish this (Hartley & Beale, 1985). The philosophy is the following:

1. Choose the value of T that will satisfy the accuracy and computation time constraints imposed on the problem.
2. Design an integration algorithm that will have a stability region that encloses the λT product for each of the system eigenvalues, using T from step 1.

This section describes the design procedure in terms of a first-order system with a fast eigenvalue. The procedure shows that the stability region of a consistent, zero-stable explicit method can be much larger than we have seen before. For a higher-order system, the slower eigenvalues will often be accommodated automatically. The next section extends this procedure directly to systems with more than one variable.

The SRP method uses a pole placement technique to determine the values of the a_i and b_i coefficients. For the specific value of λ in the system being simulated, and for the value of T chosen to produce an accurate simulation, the coefficients are selected so that all of the closed-loop poles of the simulation are at prescribed locations inside the unit circle. A set of simultaneous linear equations is developed to permit the computation of these coefficients. To determine what these locations should be, consider the stable linear system

$$\dot{y}(t) = \lambda y(t) + u(t), \qquad \lambda < 0 \tag{7–30}$$

whose solution is

$$y(t) = y(0)e^{\lambda t} + \int_0^t e^{\lambda(t-\tau)}u(\tau)d\tau \tag{7–31}$$

If the input is piecewise constant, the discretized version of equation 7–31 is

$$y_{n+1} = e^{\lambda T}y_n + \left(\int_0^T e^{\lambda \tau}d\tau\right)u_n \tag{7–32}$$

This shows that the natural response of the real system is controlled by $\exp(\lambda T)$, which is the exact z-plane mapping of a continuous-time pole at λ with a sampling period T. To produce accurate results, the simulation should model this response as well as possible. For a k-step explicit method, the closed-loop simulation equation is

$$y_{n+k} = -\sum_{i=0}^{k-1}(a_i - \lambda Tb_i)y_{n+i} + T\sum_{i=0}^{k-1}b_iu_{n+i} \tag{7–33}$$

For fixed values of λ and T, proper choice of the a_i and b_i coefficients allows arbitrary placement of the closed-loop poles. By comparison with equation 7–32, it can be seen that the following choice should be made:

$$a_{k-1} - \lambda Tb_{k-1} = -e^{\lambda T} \tag{7–34}$$

since these coefficients are associated with the signal 1 timestep before the time at which the solution is being generated. This choice takes care of one of the closed-loop poles. In a multistep method, a decision must be made about the other k-1 poles. They have no counterpart in the analytic solution to the differential equation of the continuous-time system. These spurious poles can be placed anywhere inside the unit circle. The choice of location that we use for the spurious poles is the origin of the z-plane. This puts those poles farthest away from the unit circle, making that the "most stable" choice for the spurious poles. That decision gives the following constraints on the coefficients:

$$\begin{align} a_i - \lambda Tb_i &= 0, \qquad \forall\ i\ \varepsilon\ [0, (k-2)] \\ a_{k-1} - \lambda Tb_{k-1} &= -e^{\lambda T} \end{align} \tag{7–35}$$

Equation 7–35 represents a set of k equations in $2k$ unknowns, which cannot be solved uniquely, and nothing has been said yet about accuracy. An additional set of equations is needed, and this is supplied by Lambert's equations, presented in Chapter 4. For a method to be pth-order accurate, error constants C_0 through C_p must be zero and C_{p+1} must be nonzero. Thus, a k-step, pth-order explicit linear multistep method will have $p+1$ constraints on the $2k$ algorithm coefficients. Lambert's equations are repeated below, where $a_k = 1$ is assumed:

$$C_0 = \sum_{i=0}^{k} a_i$$

$$C_1 = \sum_{i=0}^{k} (ia_i - b_i) \qquad (7\text{-}36)$$

$$C_g = \frac{1}{g!} \sum_{i=1}^{k} i^g a_i - \frac{1}{(g-1)!} \sum_{i=1}^{k} i^{g-1} b_i, \quad g \geq 2$$

Since k additional equations are needed to compute uniquely the method's coefficients, the following relationship must hold between the order of accuracy and the step number of the algorithm:

$$p = k - 1 \qquad (7\text{-}37)$$

In summary, the $2k$ a_i and b_i coefficients for an explicit SRP method are determined for a particular λT by simultaneously solving k pole placement equations, arising from stability considerations, and the first k of Lambert's equations, establishing the order of accuracy. The explicit methods are identified by the label SRP$N_1 N_2$, where N_1 is the order of accuracy and N_2 is the magnitude of λT for which the method is designed. With this labeling, SRP110 represents an explicit two-step, first-order accurate method designed for $\lambda T = -10$. SRP21000 represents an explicit three-step, second-order accurate method that was designed to simulate a system with $\lambda T = -1000$. Examples illustrate this procedure.

Example 7.4
We derive the coefficients for an explicit two-step, first-order method. The accuracy constraints are

$$C_0 = a_0 + a_1 + 1 = 0$$
$$C_1 = a_1 + 2 - b_0 - b_1 = 0 \qquad (7\text{-}38)$$

The other two equations come from placing the closed-loop poles. From equation 7–35, these equations are

$$a_0 - \lambda T b_0 = 0$$
$$a_1 - \lambda T b_1 = -e^{\lambda T} \qquad (7\text{-}39)$$

Collecting these four equations together and putting them in matrix form, we get

$$\begin{bmatrix} 1 & 1 & 0 & 0 \\ 0 & 1 & -1 & -1 \\ 1 & 0 & -\lambda T & 0 \\ 0 & 1 & 0 & -\lambda T \end{bmatrix} \begin{bmatrix} a_0 \\ a_1 \\ b_0 \\ b_1 \end{bmatrix} = \begin{bmatrix} -1 \\ -2 \\ 0 \\ -e^{\lambda T} \end{bmatrix} \qquad (7\text{-}40)$$

The values of the coefficients for various values of λT are shown in Table 7.1. As λT gets more and more negative, the pole at $\exp(\lambda T)$ rapidly approaches zero.

TABLE 7.1 SRP COEFFICIENTS, $k = 2$, $p = 1$

λT	a_0	a_1	b_0	b_1
-10	0.9000045	-1.9000045	-0.09000045	0.1899959
-100	0.9900000	-1.9900000	-0.00990000	0.0199000
-1000	0.9990000	-1.9990000	-0.00099900	0.0019990

Figures 7.6a through 7.6c show the regions of absolute stability for these three methods. The shapes are similar, but the sizes of the regions have been scaled to accommodate the specific values of λT.

Example 7.5

Here we derive the coefficients for a three-step, second-order accurate explicit method for a system whose eigenvalue is $\lambda = -40$. The value of the timestep T is chosen to be 0.5 seconds, so the λT product is -20, and the method is labeled SRP220. The three-step method has six coefficients, so there are three stability equations and three accuracy equations. Lambert's equations for this example are

$$C_0 = a_0 + a_1 + a_2 + 1 = 0$$
$$C_1 = a_1 + 2a_2 + 3 - b_0 - b_1 - b_2 = 0 \qquad (7\text{--}41)$$
$$C_2 = \frac{1}{2}(a_1 + 4a_2 + 9) - b_1 - 2b_2 = 0$$

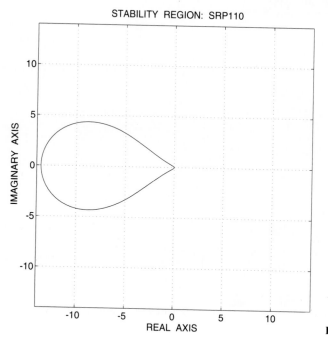

Figure 7.6a   Stability region: SRP110.

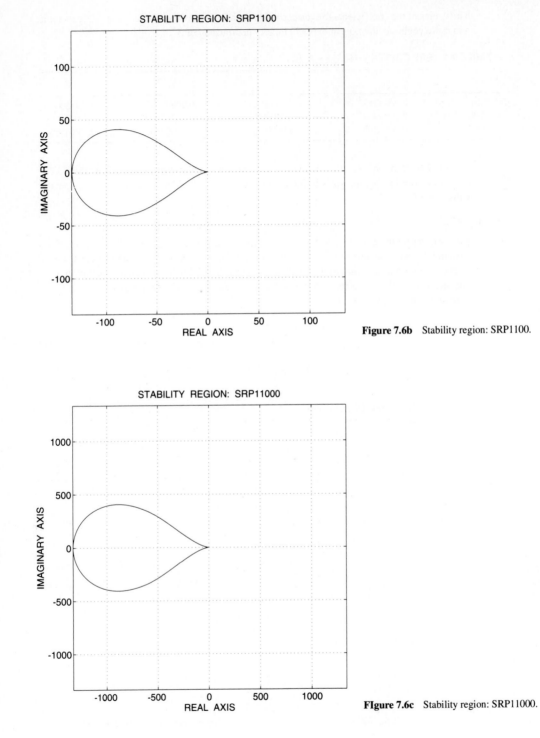

Figure 7.6b   Stability region: SRP1100.

FIgure 7.6c   Stability region: SRP11000.

and the pole placement equations for stability are

$$
\begin{aligned}
a_0 + 20b_0 &= 0 \\
a_1 + 20b_1 &= 0 \\
a_2 + 20b_2 &= -e^{-20}
\end{aligned}
\tag{7-42}
$$

From these we get the following set of simultaneous equations to solve for the coefficients:

$$
\begin{bmatrix}
1 & 1 & 1 & 0 & 0 & 0 \\
0 & 1 & 2 & -1 & -1 & -1 \\
0 & 0.5 & 2 & 0 & -1 & -2 \\
1 & 0 & 0 & 20 & 0 & 0 \\
0 & 1 & 0 & 0 & 20 & 0 \\
0 & 0 & 1 & 0 & 0 & 20
\end{bmatrix}
\begin{bmatrix}
a_0 \\ a_1 \\ a_2 \\ b_0 \\ b_1 \\ b_2
\end{bmatrix}
=
\begin{bmatrix}
-1 \\ -3 \\ -4.5 \\ 0 \\ 0 \\ -2.06 \times 10^{-9}
\end{bmatrix}
\tag{7-43}
$$

The values are given in Table 7.2, and the stability region is shown in Figure 7.7. The major region of absolute stability for this method is the interior of the leftmost closed loop in the figure. Not visible in the figure because of the scale used is a small region of stability very close to the origin. The remainder of the λT-plane, including the right loop in Figure 7.7, are unstable regions for this method. Thus, simulation stability would be lost if this method were used with a system having a smaller magnitude λT (unless it was extremely small), but it is stable for the specified values.

Implicit SRP methods can also be derived. This adds one additional unknown coefficient, so an additional equation is needed. Since implicit methods generally require iteration at each timestep to obtain a solution, and this iteration constrains the relationship between b_k and T, the following expression will be used to determine b_k:

$$
b_k = \frac{1}{2|\lambda T|}
\tag{7-44}
$$

This choice is not the only feasible one, but it provides a value for b_k that allows convergence in a reasonable amount of time without making b_k so small that it is negligible. Since one unknown and one equation have been added to the set of simultaneous equations, there is no change in the relationship between step number and order of accuracy given in equation 7–37. The labeling procedure for implicit SRP methods is the same as for explicit methods except that SRPI replaces SRP. An example will illustrate the procedure.

Example 7.6

We use the same value for timestep and eigenvalue as in Example 7.5, namely, $\lambda T = -20$. We derive the coefficients for a second-order implicit method. The accuracy equations are the

TABLE 7.2 SRP220 COEFFICIENTS

Term	$i = 3$	$i = 2$	$i = 1$	$i = 0$
a_i	1	−2.8775	2.805	−0.9275
b_i	0	0.143875	−0.14025	0.046375

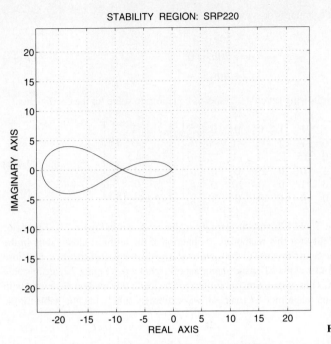

STABILITY REGION: SRP220

Figure 7.7 Stability region: SRP220.

same as in the previous example, except that b_3 is no longer zero; they are given in equation 7–45.

$$C_0 = a_0 + a_1 + a_2 + 1 = 0$$
$$C_1 = a_1 + 2a_2 + 3 - b_0 - b_1 - b_2 - b_3 = 0 \qquad (7\text{–}45)$$
$$C_2 = \frac{1}{2}(a_1 + 4a_2 + 9) - b_1 - 2b_2 - 3b_3 = 0$$

The pole placement equations for this example are still given by equation 7–42. The convergence constraint for b_3 is given in equation 7–46.

$$b_3 = \frac{1}{40} = 0.025 \qquad (7\text{–}46)$$

and the set of equations to be solved for the coefficients is given in equation 7–47. Table 7.3 gives the coefficient values, and Figure 7.8 shows the stability region for the SRPI220 method. The region of absolute stability is the interior of the leftmost loop in the figure. Like the explicit

TABLE 7.3 SRPI220 COEFFICIENTS

Term	$i = 3$	$i = 2$	$i = 1$	$i = 0$
a_i	1	−2.81625	2.7075	−0.89125
b_i	0.025	0.1408125	−0.135375	0.0445625

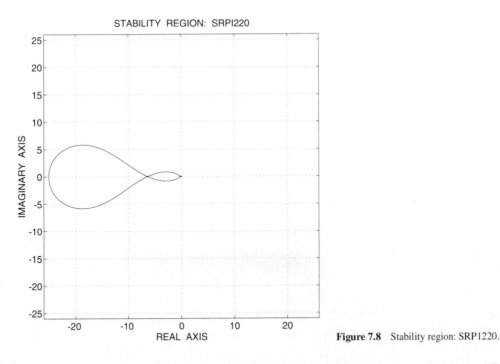

Figure 7.8   Stability region: SRP1220.

method, the SRPI220 has a small region of stability close to the origin, which is not visible in the figure.

$$
\begin{bmatrix}
1 & 1 & 1 & 0 & 0 & 0 & 0 \\
0 & 1 & 2 & -1 & -1 & -1 & -1 \\
0 & 0.5 & 2 & 0 & -1 & -2 & -3 \\
1 & 0 & 0 & 20 & 0 & 0 & 0 \\
0 & 1 & 0 & 0 & 20 & 0 & 0 \\
0 & 0 & 1 & 0 & 0 & 20 & 0 \\
0 & 0 & 0 & 0 & 0 & 0 & 1
\end{bmatrix}
\begin{bmatrix}
a_0 \\ a_1 \\ a_2 \\ b_0 \\ b_1 \\ b_2 \\ b_3
\end{bmatrix}
=
\begin{bmatrix}
-1 \\
-3 \\
-4.5 \\
0 \\
0 \\
-2.06 \times 10^{-9} \\
0.025
\end{bmatrix}
\tag{7-47}
$$

Figure 7.9 shows the stability regions of the explicit SRP220 and implicit SRPI220 methods. The region for the implicit method is seen to be larger than that for the explicit method in all directions. Thus, the implicit method can tolerate larger variations in the value of λT than can the explicit method. The small regions of stability near the origin are barely visible in the figure. The small circle in Figure 7.9 marks the design value of λT used for the development of each of the methods.

7.5 MATRIX STABILITY REGION PLACEMENT METHODS

The SRP design procedure in Section 7.4 allows the engineer to develop an integration algorithm that will be absolutely stable for any specified λT product. This allows the stable simulation of fast systems without using the excessively small value of T that would be

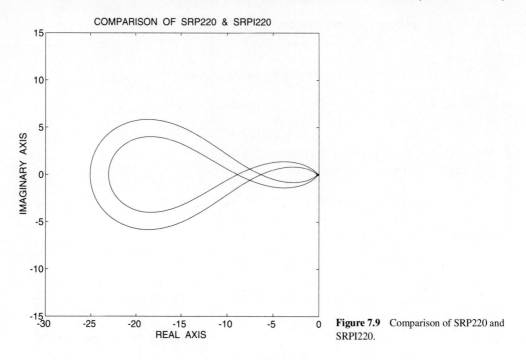

COMPARISON OF SRP220 & SRPI220

Figure 7.9 Comparison of SRP220 and SRPI220.

imposed by many classical methods. Only a single value of λT appears in the design equations, so the procedure can only be applied directly to one eigenvalue of a system.

For a system with one fast mode and one or more slow modes, the value of T could be chosen to provide an accurate and not overly time-consuming simulation of the important (slow) modes, and the integration algorithm could be designed to place its stability region around the λT product corresponding to the fast mode. It is hoped that the algorithm would provide a stable simulation of the slow modes as well. However, as we have seen for the second-order accurate methods, the stability region does not necessarily include all the real axis from zero to the design value of λT. Therefore, stability would not be guaranteed for the slow modes even though it is for the fast mode.

What we would like to have is a design procedure that will guarantee simulation stability for all of the system eigenvalues while still allowing the timestep to be based on the important dynamics. The matrix stability region placement (MSRP) procedure accomplishes this (De Abreu-Garcia & Hartley, 1990; Hartley & Beale, 1987). The approach taken parallels that presented in the last section for the scalar case; with the MSRP, the concepts are generalized to the vector case. We show only the derivation for two-step algorithms here; the derivation is extended to higher step numbers in De Abreu-Garcia and Hartley.

The explicit two-step MSRP integration algorithm is defined to have the form

$$y_{n+2} = -A_1 y_{n+1} - A_0 y_n + T(B_1 \dot{y}_{n+1} + B_0 \dot{y}_n) \tag{7–48}$$

where y is an N-dimensional vector, and the A_i and B_i are N-by-N matrices. With the MSRP integration algorithm, the value of a particular element of the y vector at time $(n + 2)T$ is a function of all the state variables and their derivatives because the matrices in equation 7–48 are not diagonal. This is in contrast to the classical use of integration algorithms for multistate systems, where the integration of a state variable is not explicitly a function of the other variables. The coupling between states would appear only in the differential equation itself.

The unknowns in equation 7–48 are the $4N^2$ elements in the A_i and B_i matrices. A total of $4N^2$ equations, imposing accuracy and stability constraints, will be needed to solve uniquely for those unknowns. A pole placement procedure will be used to place the $2N$ eigenvalues of the closed-loop simulation equation, and a generalization of Lambert's equations will be used to obtain the specified order of accuracy.

The linear differential equation, representing the system to be simulated or its linearization, is

$$\dot{y}(t) = Jy(t) + Gu(t) \qquad (7\text{–}49)$$

where J is the N-by-N system matrix, or the Jacobian if the system is nonlinear, and $u(t)$ is an external input signal. The solution to equation 7–49 has the same form as the scalar results in equation 7–31 and is given by

$$y(t) = e^{Jt}y(0) + \int_0^t e^{J(t-\tau)}Gu(\tau)d\tau \qquad (7\text{–}50)$$

Under the assumption that the input is piecewise constant, this becomes

$$y_{n+1} = e^{JT}y_n + \left[\left(\int_0^T e^{J\tau}d\tau\right)G\right]u_n \qquad (7\text{–}51)$$

When we substitute the system's differential equation 7–49 into the MSRP method shown in equation 7–48, the closed-loop simulation response is

$$y_{n+2} = -\sum_{i=0}^{1}(A_i - TB_iJ)y_{n+i} + T\sum_{i=0}^{1}B_iGu_{n+i} \qquad (7\text{–}52)$$

By comparing equations 7–52 and 7–51 and by analogy with the scalar case, we see that the following relationship between A_1 and B_1 should be established so that the primary closed-loop poles of the simulation will correspond to the exact z-plane mapping of the continuous-time system:

$$A_1 - TB_1J = -e^{JT} \qquad (7\text{–}53)$$

The spurious closed-loop poles will be placed at the origin of the z-plane, as before:

$$A_0 - TB_0J = O_{N \times N} \qquad (7\text{–}54)$$

Equations 7–53 and 7–54 impose a total of $2N^2$ constraints on the $4N^2$ coefficients. The remaining constraints will come from accuracy considerations. As with the scalar case, we will get the same number of equations as unknowns if the order of accuracy is 1 less than the step number, equation 7–37. Fortunately, Lambert's equations can be applied directly to the vector case with the appropriate change in notation. For the two-step, first-order accurate MSRP method, the equations are

$$C_0 = A_0 + A_1 + I_N = O_{N \times N}$$
$$C_1 = A_1 + 2I_N - B_0 - B_1 = 0_{N \times N}$$

(7–55)

where the N-by-N identity matrix corresponds to $a_k = 1$ in the scalar case. Equation 7–55 adds the necessary additional number of constraints, so unique solutions can be obtained for the A_i and B_i matrices, assuming the system matrix J is invertible. If all of the system eigenvalues have negative real parts, the inverse of J will exist. Assuming J^{-1} exists, we can eliminate the B_i from equations 7–53 through 7–55, and compute the A_i matrices from

$$A_1 = (e^{JT} - I_N)(JT)^{-1} - 2I_N$$
$$A_0 = (I_N - e^{JT})(JT)^{-1} + I_N = -(A_1 + I_N)$$

(7–56)

The B_i matrices can then be computed from these results:

$$B_1 = (A_1 + e^{JT})(JT)^{-1}$$
$$B_0 = A_0(JT)^{-1}$$

(7–57)

As long as the system matrix J is not too ill conditioned, the calculations in equations 7–56 and 7–57 can be conveniently and accurately carried out with available software packages, for example MATLAB. The following example illustrates the procedure and shows simulation results obtainable by the method.

Example 7.7

A third-order nonlinear model of a turbojet engine can be written as a set of two algebraic and three differential equations, as shown below and in Example 4.4 (Brennan & Leake, 1975; De Abreu-Garcia & Hartley, 1990; Hartley & Beale, 1987).

$$T_3 = 0.64212 + 0.35788x_3^2$$

(7–58)

$$W_{3d} = 1.3009x_3 - 0.13982\left[x_1 + \sqrt{x_1^2 + 0.41688x_3^2 - 0.0899x_1x_3}\right]$$

(7–59)

$$\dot{x}_1 = W_{fd}\left[\frac{0.93586x_1}{x_2} + 31.486\right] + 21.435W_{3d}T_3 - \frac{53.86x_1^2}{x_2}$$

(7–60)

$$\dot{x}_2 = 37.78W_{3d} - 38.448x_1 + 0.66849W_{fd}$$

(7–61)

$$\dot{x}_3 = \frac{1.258}{x_3}\left(\frac{x_1^2}{x_2} - W_{3d}x_3^2\right)$$

(7–62)

The states represent combustor pressure, combustor density, and rotor speed, respectively. The variables T_3 and W_{3d} are the compressor discharge temperature and compressor discharge mass flow, respectively, and W_{fd} is the input fuel mass rate. Linearization about $W_{fd} = 1$ gives the following expression for the system Jacobian:

$$J = \begin{bmatrix} -112.5 & 53.3 & 42.60 \\ -48.11 & 0.205 & 47.92 \\ 2.841 & -1.269 & -4.117 \end{bmatrix} \qquad (7\text{–}63)$$

The eigenvalues of the Jacobian are at $\lambda = -2.85$, -32.19, and -81.37. The stiffness ratio is 28.5, so the system is not terribly stiff, but there is a moderately wide spread in the dynamics of the system. Since the fastest eigenvalue puts the stability limit on timestep, let us determine the largest value of T that might be reasonable for the second-order Adams-Bashforth (AB-2) method. Since all the eigenvalues are real, we only need to consider the stability region along the negative real axis. As we saw in Chapter 4, the boundary of the absolute stability region for the AB-2 method intersects the real axis at $\lambda T = -1$. Therefore, the largest value of T to get a stable simulation would be

$$T_{\max - AB\text{-}2} = \frac{1}{81.37} = 0.0123 \qquad (7\text{–}64)$$

However, this value of T does not provide any margin for a change in the eigenvalue, and it does not provide relative stability even with no change. A more reasonable selection for T would be one-half the value given in equation 7–64. This would provide much more accuracy and stability robustness to the simulation. With this logic, $T = 0.006$ seconds will be the timestep chosen by the AB-2 method. If Euler's method were going to be used instead, the timestep given in equation 7–64 might be acceptable from an accuracy viewpoint since its region of absolute stability extends to -2 along the real axis.

If we consider that part of the transient response controlled by the eigenvalue at -2.85 to be the most important, it will take approximately 2 seconds for the system to reach a steady-state solution. With a timestep of 0.006 seconds, over 300 timesteps will be needed to model this response. If the simulation had to be done in real time, all of the computations at each timestep would have to be performed in 6 milliseconds.

The dominant time constant for this system is $1/2.85 = 0.351$ seconds. If we assume that the transient response due to that eigenvalue can be accurately modeled if we compute five simulation steps per time constant, the value of T could be

$$T = \frac{1}{5 \times 2.85} = 0.07 \text{ seconds} \qquad (7\text{–}65)$$

This is over ten times the value needed to get a stable and accurate simulation if the AB-2 method is used.

With the MSRP procedure, we can design an integration algorithm that will produce a stable simulation with this larger value of T, or any other value. First, we design the simulation with the value of T given by equation 7–65. We also compute the MSRP matrices by using a value of $T = 0.02$ seconds for comparison. We perform the simulation of the nonlinear dynamics with these two MSRP algorithms and compare those results with an Euler integration obtained with a timestep of 0.002 seconds.

With the Jacobian from equation 7–63 and the timestep from equation 7–65, the MSRP matrices are

T = 0.07 SECONDS

$$A_0 = \begin{bmatrix} 0.9564 & -0.2430 & -0.5889 \\ 0.2050 & 0.4536 & -0.7407 \\ -0.0214 & 0.0037 & 0.1114 \end{bmatrix}, \quad A_1 = \begin{bmatrix} -1.9564 & 0.2430 & 0.5889 \\ -0.2050 & -1.4536 & 0.7407 \\ 0.0214 & -0.0037 & -1.1114 \end{bmatrix} \quad (7\text{--}66)$$

$$B_0 = \begin{bmatrix} -0.0703 & -0.1337 & -0.2405 \\ 0.1158 & -0.3493 & -0.2971 \\ -0.0100 & 0.0025 & -0.4610 \end{bmatrix}, \quad B_1 = \begin{bmatrix} 0.1140 & 0.3767 & 0.8295 \\ -0.3209 & 0.8957 & 1.0378 \\ 0.0314 & -0.0062 & 1.3496 \end{bmatrix} \quad (7\text{--}67)$$

With the same Jacobian, and a timestep of $T = 0.02$ seconds, the matrices are

T = 0.02 SECONDS

$$A_0 = \begin{bmatrix} 0.6582 & -0.2645 & -0.3050 \\ 0.2354 & 0.1015 & -0.3617 \\ -0.0161 & 0.0058 & 0.0371 \end{bmatrix}, \quad A_1 = \begin{bmatrix} -1.6582 & 0.2645 & 0.3050 \\ -0.2354 & -1.1015 & 0.3617 \\ 0.0161 & -0.0058 & -1.0371 \end{bmatrix} \quad (7\text{--}68)$$

$$B_0 = \begin{bmatrix} -0.2503 & -0.1052 & -0.1101 \\ 0.0940 & -0.4720 & -0.1290 \\ -0.0062 & 0.0024 & -0.4873 \end{bmatrix}, \quad B_1 = \begin{bmatrix} 0.5921 & 0.3697 & 0.4151 \\ -0.3293 & 1.3705 & 0.4907 \\ 0.0223 & -0.0082 & 1.4502 \end{bmatrix} \quad (7\text{--}69)$$

Figure 7.10 shows the response of the system to the initial conditions $x_1 = 0.5383$, $x_2 = 1.775$, $x_3 = 0.5459$ for the two MSRP methods and for Euler's method. The timestep for the Euler

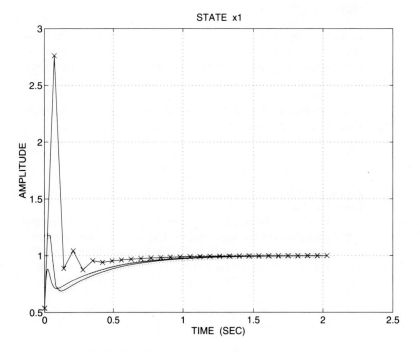

Figure 7.10a Simulation response for Example 7.7, state x_1.

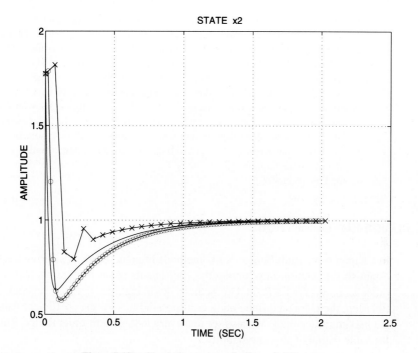

Figure 7.10b   Simulation response for Example 7.7, state x_2.

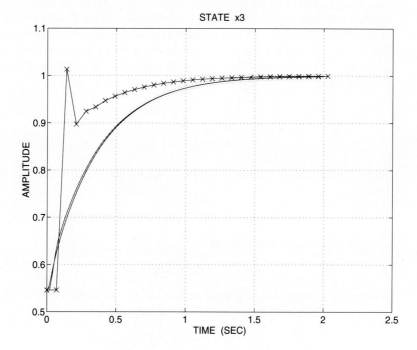

Figure 7.10c   Simulation response for Example 7.7, state x_3.

integration is 0.002 seconds, and that simulation represents the exact response for this example. For each of the state variables, the solid curve is the Euler response, the line with circles is the MSRP method with $T = 0.02$ seconds, and the line with Xs is the MSRP method with $T = 0.07$ seconds. As the plots show, all the simulations are stable, and the steady-state values for the three simulations are equal. Therefore, even with fast eigenvalues and a large timestep, the MSRP design procedure produces an integration algorithm that gives a stable simulation and is accurate for the slower dynamics. It is evident in the figure that with $T = 0.07$ seconds, there is considerable error for the first few timesteps because of the faster modes being ignored and the nonlinearity. After three or four timesteps, however, the solution converges quickly to the exact solution. When T is reduced to 0.02 seconds, an accurate simulation is obtained over most of the time duration. This value of T is still larger than the value proposed for the AB-2 method by a factor of 3.

7.6 RUNGE-KUTTA METHODS

In Chapter 6 we study Runge-Kutta methods in great detail. We derive the equations that establish the relationships among the coefficients of an R-stage method to achieve Rth-order accuracy, for $1 \leq R \leq 3$. For methods involving more than one function evaluation per timestep, there are an infinite number of possibilities for the coefficient values in order to get maximum accuracy. We also see that every R-stage, Rth-order Runge-Kutta method has the same region of absolute stability.

In this section, we look at enlarging the region of stability by modifying the coefficient relationships. Making the stability region larger allows the method to be applied to systems with faster eigenvalues without reducing the value of T, an important benefit when dealing with stiff systems. Of course, there must be a price paid for the increased size of the stability region. The price is a reduction in the order of accuracy. For convenience, we focus on two-stage methods. The maximum order of accuracy for an explicit method is 2. From Chapter 6, the two-stage Runge-Kutta method is defined by

$$x_{n+1} = x_n + T(c_1 K_1 + c_2 K_2), \text{ with}$$
$$K_1 = f(t_n, x_n, u_n) \tag{7-70}$$
$$K_2 = f(t_n + a_2 T, x_n + b_{21} T K_1, u_{n+a_2 T})$$

and the coefficient constraints for the two-stage method to achieve second-order accuracy are

$$c_1 + c_2 = 1, \qquad a_2 = b_{21}, \qquad c_2 a_2 = \frac{1}{2} \tag{7-71}$$

When the linear test equation

$$\dot{x}(t) = \lambda x(t) + u(t) \tag{7-72}$$

is substituted into the Runge-Kutta equations and the closed-loop transfer function is determined (as in Section 6.7), the closed-loop simulation pole is located at

$$z = 1 + (c_1 + c_2)\lambda T + (c_2 a_2)(\lambda T)^2 \tag{7-73}$$

Recall that the s-plane pole at λ maps into the z-plane pole $\exp(\lambda T)$, and that when the Runge-Kutta coefficients are chosen to achieve second-order accuracy the closed-loop pole is at the truncated approximation of $\exp(\lambda T)$. To determine the stability region for the method, it is necessary to map the z-plane stability boundary (the unit circle) into the λT plane via the pole equation equation 7–73. This is done through

$$(c_2 a_2)(\lambda T)^2 + (c_1 + c_2)\lambda T + 1 - [z]_{z=e^{j\theta}} = 0, \qquad 0 \le \theta \le 2\pi \qquad (7\text{–}74)$$

This equation must be solved for λT for all values of θ to get the stability boundary. All two-stage, second-order accurate methods have the same boundary (see Section 6.7). We will see that if the order of accuracy is reduced to 1, the stability boundary can be increased in size.

For second-order accuracy, the constraint $c_2 a_2 = 1/2$ must be satisfied. First-order accuracy is maintained as long as $c_1 + c_2 = 1$. If we require only first-order accuracy and define $c_2 a_2 = 1/g$, the equation for the stability boundary becomes

$$(\lambda T)^2 + g\lambda T + g(1 - [z]_{z=e^{j\theta}}) = 0, \quad 0 \le \theta \le 2\pi \qquad (7\text{–}75)$$

For a fixed value of g, the values of a_2 and c_2 can be chosen arbitrarily, as long as $c_1 + c_2 = 1$. The effect of g on the size of the stability region can be determined by considering a root locus perspective and letting $z = +1$. Since the stability region equation is a 1-to-2 mapping from the z-plane to the λT-plane, the point $z = +1$ fortunately maps into the left and right real-axis crossings of the stability region. From equation 7–75, with $z = +1$, the real-axis crossings are seen to be at

$$\lambda T = 0 \quad \text{and} \quad \lambda T = -g \qquad (7\text{–}76)$$

Thus, it would appear that choosing g to be large and positive would increase the size of the stability region. This is indeed the case. However, the stability region cannot be made arbitrarily large by letting g become unbounded. At approximately $g = 8$, the stability region splits apart, giving two stable regions separated from each other by a region of instability. Thus, for a method with a single, connected region of stability, the maximum value for the product λT is -8. Nonetheless, this is a tremendous improvement over the two-stage method with second-order accuracy, which has $g = 2$. Hence a speedup of 4 is possible when using $g = 8$.

Figures 7.11 to 7.13 show the stability regions for several two-stage methods. The stability region for $g = 2$ in Figure 7.11 is second-order accurate; all the other curves correspond to methods that are first-order accurate. It is easily seen that the only advantage to the curve for $g = 1$ is directly up the imaginary axis. For $g = 4$ and $g = 6$, the increase in the size of the stability region along the real axis is readily apparent in Figure 7.12. In Figure 7.13, we see the splitting of the stability region for $g \ge 8$.

The accuracy of these methods can be analyzed by looking at the principal error constant for the local truncation error. For $g = 2$, the local truncation error is proportional to T^3; for other values of g, it is proportional to T^2. In those cases, the principal error constant is the difference between the actual value for $c_2 a_2$ and its value required for second-order accuracy. That is,

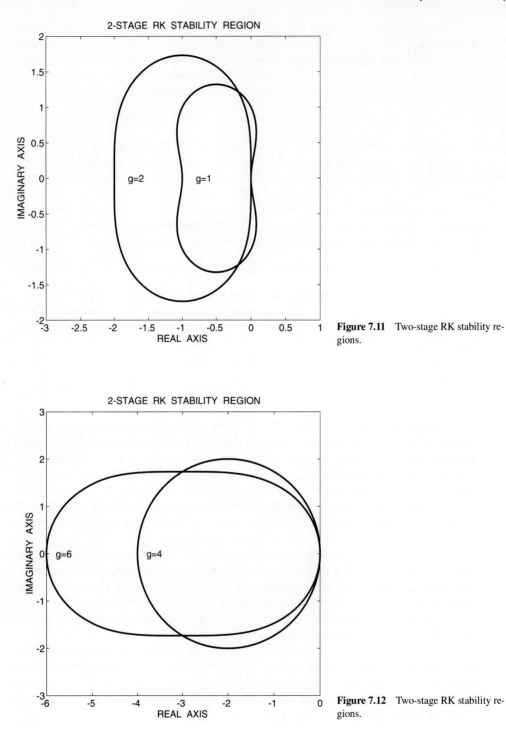

Figure 7.11 Two-stage RK stability regions.

Figure 7.12 Two-stage RK stability regions.

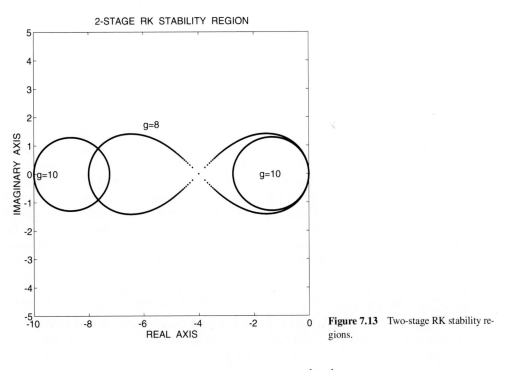

Figure 7.13 Two-stage RK stability regions.

$$\text{Principal error constant} = \frac{1}{2} - \frac{1}{g}, \qquad g \neq 2 \qquad (7\text{-}77)$$

Since we have assumed that g is positive, the error constant for these methods is always less than 0.5. Thus, they have a smaller error constant than Euler's method. Therefore, the extra stage of computation does provide some increase in accuracy as well as a larger stability region than the one-step, first-order accurate Euler method.

A choice for the coefficients in the method that allows easy multirate input sampling is

$$a_2 = \frac{1}{2}, \qquad c_2 = \frac{1}{a_2 g} = \frac{2}{g}, \qquad c_1 = 1 - c_2 \qquad (7\text{-}78)$$

For $g = 6$, these values are $a_2 = 1/2$, $c_2 = 1/3$, and $c_1 = 2/3$. With these coefficients, the principal error constant is 1/3. It should be noted that making the choice $a_2 = 1/2$ allows an easy computation of a second-order accurate solution by using the modified Euler method. Once K_1 and K_2 are computed, the second-order solution would be obtained by letting $c_1 = 0$ and $c_2 = 1$ in the Runge-Kutta equation. By comparing the first- and second-order solutions, an estimate of the local truncation error is possible at every timestep, as in the Runge-Kutta-Fehlberg method (Lambert, 1973), at almost no additional cost in computation. This information could then be used in the simulation in any way desired by the user, but only if the second-order Runge-Kutta method ($g = 2$) were stable at the chosen value of T.

Which value of g to use can be based on considerations of accuracy. In particular, the location of the λT-plane mappings of the point $z = 0$ should be investigated. The farther to the left in the plane that the rightmost of these points is, the larger the accurate (tuned) portion of the stability region. For the first-order, two-stage Runge-Kutta methods, a simple root locus of the equation

$$(\lambda T)^2 + g(\lambda T + 1) = 0 \tag{7–79}$$

with gain g shows that the farthest to the left this point can be occurs when $g = 4$; then both λT roots equal -2. When the gain is smaller, say, $g = 2$, as in the second-order methods, the $z = 0$ mappings are complex and farther to the right of $\lambda T = -2$. This reduces the accurate portion of the stability region. When the gain is larger than 4, the two $z = 0$ mappings are real and to the left and right of $\lambda T = -2$. As g goes to infinity, there is only one finite mapping, and it is at $\lambda T = -1$, as this is Euler's method. From this, we can conclude that $g = 4$ provides the largest region of accuracy, although $g = 6$ possesses a much larger region of absolute stability. To use this method with $g = 6$, it is necessary to choose T to place the product of T with the largest negative real eigenvalue near the leftmost $z = 0$ mapping, which is approximately $\lambda T = -4.7$. This is true for any value of g greater than 4. Unfortunately, the methods with $g > 4$ will usually require a good separation between the system eigenvalues so that many of the λT products will not be placed between the two $z = 0$ mappings in the λT-plane.

Another observation is that the $g = 4$ method is really Euler's method applied twice with the given time interval, and thus has the size of its stability region doubled. This is seen in Figure 7.12, where the stability region for $g = 4$ is a circle of radius 2 centered at -2. This realization yields a good test for the Runge-Kutta stability region. Compare the stability region for a particular Runge-Kutta method with the stability region of Euler's method multiplied by the number of stages in the Runge-Kutta method. From this it can be concluded that most standard Runge-Kutta methods provide improvements only in Euler's stability region along the imaginary axis; the great potential improvement along the real axis is generally unrealized.

Example 7.8

To demonstrate the utility of this method, it is applied to the third-order Brennan and Leake jet engine model discussed in Example 7.7. The same linearized version is used for this example. The system matrices are

$$J = \begin{bmatrix} -112.5 & 53.3 & 42.60 \\ -48.11 & 0.205 & 47.92 \\ 2.841 & -1.269 & -4.117 \end{bmatrix}, \qquad G = \begin{bmatrix} 0.2733 \\ 0.4667 \\ 0.2667 \end{bmatrix} \tag{7–80}$$

corresponding to the linear equation

$$\dot{x}(t) = Jx(t) + Gu(t), \qquad x(0) = \begin{bmatrix} -0.5 \\ 0.7 \\ -0.5 \end{bmatrix} \tag{7–81}$$

Two sets of responses are obtained for each of the three Runge-Kutta methods tried. The first is the response to the initial conditions shown in equation 7–81 with no external input. The second is the step response with zero initial conditions. The first Runge-Kutta method is the modified Euler algorithm, which has $g = 2, a_2 = 1/2, c_1 = 0, c_2 = 1$, and $T = 0.01$. This method is second-order accurate and is used as the basis of comparison for the other two methods. One of these has $g = 6, a_2 = 1/2, c_1 = 2/3, c_2 = 1/3$, and $T = 0.05$. The last method has $g = 4, a_2 = 1/2, c_1 = 1/2, c_2 = 1/2$, and $T = 0.03$. The stability regions for these methods appear in Figures 7.11 and 7.12.

The initial condition responses of the three methods are shown in Figure 7.14. With $T = 0.05$, corresponding to $g = 6$, there are obviously some errors in states x_1 and x_2 for the first 0.1 second. After that, there is little difference between the three curves. State x_3 has only small differences in the curves over the complete time interval. This slight degradation in the simulation accuracy is often acceptable to get the speedup in computations. The step responses for the example are shown in Figure 7.15. The responses are almost identical in all three states. It should be noted that the $g = 4$ method provides a speedup of 3 over the second-order Runge-Kutta method, and the $g = 6$ method provides a speedup of 5. It should also be noted that the second-order method would not be stable for either of the larger timesteps. If $T = 0.01$ did not meet computation time constraints, a two-stage, first-order method could provide the required speedup in computation without severely degrading accuracy. Note that these methods are derived through simple adjustments in the coefficients in the usual two-stage, second-order accurate Runge-Kutta methods.

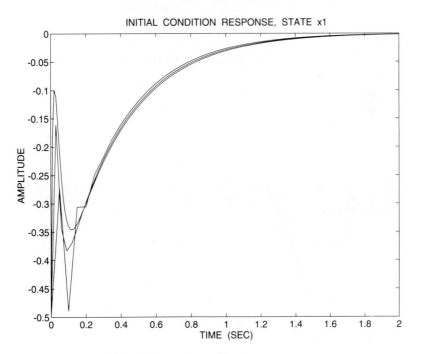

Figure 7.14a   Initial condition response, state x_1.

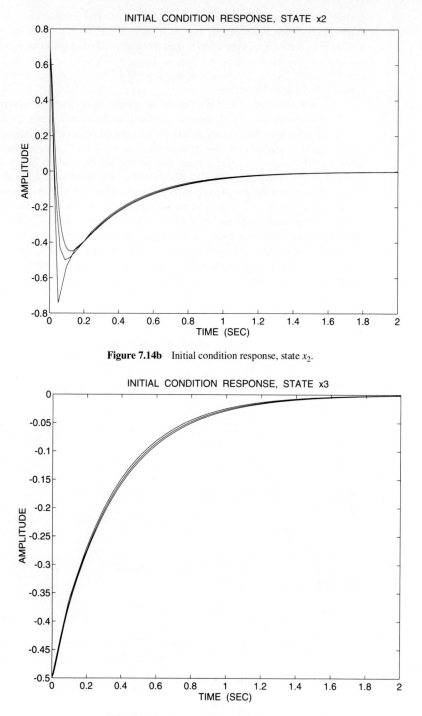

Figure 7.14b   Initial condition response, state x_2.

Figure 7.14c   Initial condition response, state x_3.

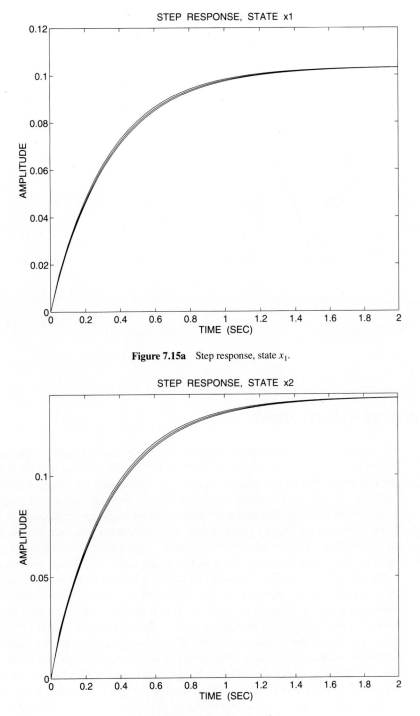

Figure 7.15a   Step response, state x_1.

Figure 7.15b   Step response, state x_2.

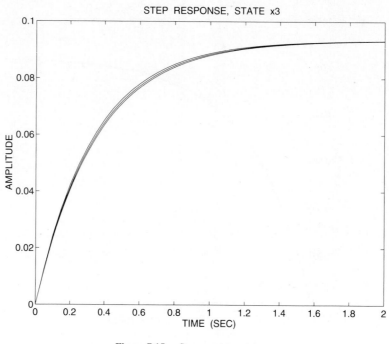

Figure 7.15c   Step response, state x_3.

7.7 INVERSE NYQUIST ARRAY

One approach for simulating a stiff system of higher than first order is the matrix stability region placement method, discussed in Section 7.5. The drawbacks with this approach is the low order of accuracy relative to the step number and the fact that there is coupling between states in the integration process. This latter problem increases the number of computations per timestep. The major advantage, of course, is that the simulation is guaranteed to be stable as long as the system matrix does not change. Another, subtle, benefit of the MSRP method is that there is no need to identify eigenvalues with individual state variables or to worry about the amount of coupling between states. The design process generates the correct integration matrices for the given system matrix regardless of the coupling.

An alternative approach to the simulation of systems with coupled dynamics is through the inverse Nyquist array (INA). This is discussed in the context of multivariable control system design by a number of authors, including Maciejowski (1989) and Van de Vegte (1990). Under certain conditions, this technique allows the simulation engineer to choose an integration operator for a particular system state independently of the operators used for the other states. The INA technique makes use of Gershgorin bands. A theorem based on Gershgorin (Kreyzig, 1988) states that for an n-by-n matrix J, the eigenvalues of the matrix are contained in the union of a set of circles. Each circle has the form

$$\text{Center} = J_{ii}, \qquad \text{Radius} = \sum_{j=1}^{j=n} |J_{ij}|, \quad j \neq i \tag{7-82}$$

or

$$\text{Center} = J_{ii}, \qquad \text{Radius} = \sum_{j=1}^{j=n} |J_{ji}|, \quad j \neq i \tag{7-83}$$

If the elements of the matrix are functions of some variable, such as frequency, there is a corresponding circle for each value of that variable. The set of these circles for a particular diagonal element of the matrix is called a Gershgorin band.

The motivation of the INA technique and the use of the Gershgorin bands is to determine whether or not a given multivariable coupled system can be treated as a set of uncoupled scalar systems for the purpose of analyzing the closed-loop system's stability. We know that for a scalar system, simulation stability requires that the product λT must lie within the region of stability for the particular integration method being used. In the multivariable case, the same is true; all of the λT products must be in the stability region. A problem arises when we want to use different integrators for different state variables. For example, we may want to use the AB-4 method for the important variables to achieve high accuracy, but for some rapidly decaying variables, where only stability is a concern, Euler's method might be entirely suitable. However, in a multivariable system with a nondiagonal system matrix, the eigenvalues are not associated with individual state variables, and the stability analysis problem becomes more difficult.

The INA method is introduced from the perspective of control system design; the connection to simulation is then shown. It is assumed that the reader has some knowledge of polar plots of a system's frequency response and the concept of the Nyquist stability criterion. Only a brief review of that will be given here. Standard texts on the subject are Phillips and Harbor (1991) and D'Azzo and Houpis (1981) for continuous-time systems and Phillips and Nagle (1990) and Kuo (1992) for discrete-time systems.

Consider a scalar closed-loop system, whose open-loop transfer function is

$$L(z) = kG(z)H(z) \tag{7-84}$$

where k is a constant gain value. Then the closed-loop characteristic equation for the system is

$$F(z) = 1 + kG(z)H(z) = 0 = \frac{1}{k} + G(z)H(z) \tag{7-85}$$

The stability of the closed-loop system can be evaluated by making a polar plot of the frequency response of the open-loop transfer functions $G(z)H(z)$ and counting the number of encirclements this plot makes of the point $z = -1/k$. As discussed in Chapter 5, the Nyquist stability criterion states that the number of these encirclements is related to the number of unstable closed-loop poles through the following equation:

$$N = Z_o - P_o \tag{7-86}$$

where N is the number of clockwise encirclements of the $-1/k$ point by the polar plot, Z_o is the number of closed-loop poles outside the unit circle, and P_o is the number of open-loop poles outside the unit circle. For a stable closed-loop system, Z_o is zero; therefore, the requirement for stability is $N = -P_o$, that is, one counterclockwise encirclement of the $-1/k$ point for each unstable open-loop pole.

Instead of the direct polar plot described above, the inverse polar plot can also be made. This comes from rewriting equation 7–85 as

$$F(z) = 1 + kG(z)H(z) = 0 = \frac{1}{G(z)H(z)} + k \qquad (7\text{–}87)$$

Now the polar plot of $1/G(z)H(z)$ can be made, and the number of encirclements of the point $z = -k$ can be determined. The Nyquist stability criterion is then stated as

$$N = Z_o - P_z \qquad (7\text{–}88)$$

where N is now the number of clockwise encirclements of the $-k$ point by the inverse polar plot, Z_o is still the number of closed-loop poles outside the unit circle, and P_z is the number of open-loop zeros outside the unit circle. For a stable closed-loop system, Z_o is zero; therefore, the requirement for stability is $N = -P_z$, that is, one counterclockwise encirclement of the $-k$ point for each unstable open-loop zero.

When we move to a multi-input, multioutput system, the situation becomes more complicated. The open-loop transfer function is replaced by a matrix of transfer functions. In general, closed-loop stability cannot be determined by inspection of any or all of the polar plots of the individual transfer functions. Except in certain situations, the Nyquist stability criterion has to be applied to the plots of the eigenvalues of the transfer matrix (the characteristic loci). Although this can be done, and it always gives the correct answer, it requires additional computation to determine the characteristic loci, and their polar plots do not necessarily look like the "normal" polar plot we get from an individual transfer function. Therefore, some of the intuitive feel of the plots is lost. The multivariable closed-loop characteristic equations for the direct and inverse forms of the plot are

$$\left|F(z)\right| = \left|I + KG(z)H(z)\right| \qquad (7\text{–}89)$$

and

$$\left|F(z)\right| = \left|(G(z)H(z))^{-1} + K\right| \qquad (7\text{–}90)$$

where K is a diagonal matrix of constant gains.

We consider only the inverse Nyquist array, which comes from equation 7–90; the direct Nyquist array from equation 7–89 is essentially the same. The INA technique is based on the concept of the transfer matrix $(G(z)H(z))^{-1}$ being diagonally dominant. To be diagonally dominant, the frequency response magnitude of the diagonal element in the ith row, $(G_{ii}(z)H_{ii}(z))^{-1}$, must be greater than the sum of the magnitudes of the off-diagonal elements in the same row; this must be true for every row in the transfer matrix. Thus, to be diagonally dominant, we must have

$$\left| \frac{1}{G_{ii}(z)H_{ii}(z)} \right| > \sum_{j=1}^{n} \left| \frac{1}{G_{ij}(z)H_{ij}(z)} \right| \tag{7-91}$$

$$z = e^{j\omega T}, \qquad 1 \le i \le n, \qquad j \ne i$$

The importance of the diagonal dominance property is that if a transfer matrix has that property, closed-loop stability can be determined from applying the Nyquist stability criterion to the polar plots of only the diagonal elements of the matrix. From a stability standpoint, those diagonal elements sufficiently characterize the entire matrix.

Diagonal dominance is checked in the following manner. Given an open-loop n-by-n transfer matrix $KG(z)H(z)$ and closed-loop characteristic equation as in equation 7–90,

1. For each i, make a polar plot of $1/[G_{ii}(z)H_{ii}(z)]$, with $z = \exp(j\omega T)$.
2. At each value of ωT, add a circle to the plot whose center is $1/|G_{ii}(z)H_{ii}(z)|$ and whose radius is $\Sigma 1/|G_{ij}(z)H_{ij}(z)|(i \ne j)$ at the same frequency (the set of circles indexed over ωT is the Gershgorin band for that diagonal element).
3. If each Gershgorin band excludes the corresponding $-K_{ii}$ point (the $-K_{ii}$ is not inside the band), the transfer matrix is diagonally dominant.
4. If step 3 is true, count the number of encirclements of the $-K_{ii}$ point made by the polar plot of $1/[G_{ii}(z)H_{ii}(z)]$ and apply the Nyquist stability criterion based on that diagonal element.
5. The total number of encirclements for the complete system is the sum of the encirclements from the individual diagonal elements.

Therefore, if the matrix is diagonally dominant, the system stability can be analyzed as a set of n scalar uncoupled systems. This is a very convenient property for the matrix to have; unfortunately, not all matrices have it.

One important fact should be noted at this time. If one of the Gershgorin bands does include the corresponding $-K_{ii}$ point, the matrix is not diagonally dominant and more complex procedures must be used to get a guaranteed correct verdict on stability. However, a Gershgorin band including the $-K_{ii}$ point does not mean that the system is unstable, only that a correct stability decision cannot be guaranteed by examining the diagonal elements. Thus, diagonal dominance is a sufficient, but not a necessary, condition for the diagonal elements to hold the answer to closed-loop stability.

After that introduction into Nyquist arrays and Gershgorin bands, it is time now to relate those concepts to the simulation problem. Assume that the system to be simulated has the linear differential equation

$$\dot{x} = Jx + u \tag{7-92}$$

and that the integration operator for the ith state is a linear multistep method with transfer function

$$H(z) = \frac{T\sigma_i(z)}{\rho_i(z)} \tag{7-93}$$

Then, the closed-loop characteristic equation is given by

$$F(z) = I_n - T \operatorname{diag}\left[\frac{\sigma_i(z)}{\rho_i(z)}\right] J = 0 = \operatorname{diag}\left[\frac{\rho_i(z)}{\sigma_i(z)}\right] - TJ \qquad (7-94)$$

The zeros of equation 7–94 are the closed-loop simulation poles. The term $\rho_i(z)/\sigma_i(z)$ is the inverse transfer function of the ith integration operator, which we know is the expression used for plotting the boundary of absolute stability. Thus, the stability boundary of a linear multistep integration method is the inverse polar plot of its transfer function.

What we would like to be able to do is to determine the overall simulation stability, using particular integration operators for particular state variables with the chosen value of T, by plotting the operators' stability regions and applying the concept of Gershgorin bands and the inverse Nyquist stability criterion to the individual plots. If the diagonal dominance property is present, the simulation stability can be evaluated from these plots. To use this approach, we rewrite equation 7–94 in the following way:

$$\left\{ diag\left[\frac{\rho_i(z)}{\sigma_i(z)}\right] - J_2 T \right\} - J_1 T = 0 = M(z) - J_1 T \qquad (7-95)$$

where J_1 is an n-by-n diagonal matrix whose entries are the diagonal elements of the Jacobian matrix J, and J_2 contains all the off-diagonal elements of J and has zeros on its main diagonal. Thus

$$J_1 = diag[J_{ii}], \qquad J_2 = J - J_1 \qquad (7-96)$$

If the matrix $M(z)$ in equation 7–95 is diagonally dominant with respect to the $J_1 T$ points, the overall simulation stability can be determined by counting encirclements made by each stability region around the corresponding $J_{ii} T$ point. Note that varying T varies both the $J_{ii} T$ point as well as the radius of the Gershgorin band associated with the particular integration operator. If $M(z)$ is diagonally dominant, then for the integration of the ith state variable, J_{ii} can be looked at as the ith eigenvalue; therefore, $J_{ii} T$ has to be inside the region of absolute stability of the integration operator for that state variable.

One procedure to check for diagonal dominance is the same as previously discussed in the context of control system design. Given an n-dimensional system matrix J and a set of n linear multistep methods,

1. Form the matrix $M(z)$, which has as its diagonal elements the reciprocals of the transfer functions of the integration operators and which has as its off-diagonal terms the off-diagonal terms of the Jacobian matrix J multiplied by T.
2. For each i, plot the boundary of the integration operator's stability region, which is the polar plot of the ith diagonal element of $M(z)$.
3. At each value of ωT, add a circle to the plot whose center is $|\rho_i(z)/\sigma_i(z)|$ and whose radius is the sum of $|J_{ij} T|$ (the set of circles indexed over ωT is the Gershgorin band for that diagonal element, with all the circles for a particular diagonal element having the same radius in this application).
4. If each Gershgorin band excludes the corresponding $J_{ii} T$ point ($J_{ii} T$ is not inside the band), the transfer matrix is diagonally dominant.

5. If step 4 is true, the stability of the entire integration process is guaranteed if each $J_{ii}T$ is inside the stability region of the ith integration operator.

Because the radius of the Gershgorin band for a particular diagonal element is constant in this application of simulation stability, being the sum of the absolute values of the off-diagonal elements in J, an equivalent and simpler procedure is the following:

1. For each i, plot the boundary of the ith integration operator's stability region.
2. To each plot add a single circle whose center is $J_{ii}T$ and whose radius is the sum of the absolute values $|J_{ij}T|$ (the Gershgorin circle for the ith row of J);
3. If each Gershgorin circle excludes the corresponding stability region boundary plot (does not intersect the curve), the system has the diagonal dominance property for the chosen value of T and the selected integration operators.
4. If step 3 is true, the stability of the entire integration process is guaranteed if each $J_{ii}T$ is inside the stability region of the ith integration operator.

Two examples illustrate this approach.

Example 7.9

The system to be simulated is linear and second order, with the continuous-time model

$$\begin{bmatrix} \dot{x}_1(t) \\ \dot{x}_2(t) \end{bmatrix} = \begin{bmatrix} -6 & 2 \\ -3 & -10 \end{bmatrix} \begin{bmatrix} x_1(t) \\ x_2(t) \end{bmatrix}, \qquad \begin{bmatrix} x_1(0) \\ x_2(0) \end{bmatrix} = \begin{bmatrix} 10 \\ 10 \end{bmatrix} \qquad (7\text{–}97)$$

The open-loop eigenvalues of the system matrix J are $\lambda = -8 \pm j1.414$. The magnitude of the eigenvalues are 8.124, and the phase angles are $\pm170°$. To illustrate, we will assume that the AB-2 method is going to be used to integrate the first state variable and that Euler's method is going to be used for the second state variable. The initial choice for the timestep is $T = 0.1$. With these choices, the closed-loop characteristic equation, in the form of equation 7–95, is

$$\left\{ \begin{bmatrix} \dfrac{z^2 - z}{1.5z - 0.5} & 0 \\ 0 & z - 1 \end{bmatrix} - \begin{bmatrix} 0 & 0.2 \\ -0.3 & 0 \end{bmatrix} \right\} - \begin{bmatrix} -0.6 & 0 \\ 0 & -1 \end{bmatrix} = 0 \qquad (7\text{–}98)$$

Thus, the two $J_{ii}T$ points are at -0.6 and -1 for the first and second state variables, respectively, and the radii of the Gershgorin circles are 0.2 and 0.3. Figure 7.16 shows the stability region plots with the Gershgorin bands superimposed. The small x in each plot is the corresponding $J_{ii}T$ point. The plots come from applying the first procedure for determining diagonal dominance. Two major points can be determined by inspection of the plots in Figure 7.16. First, the Gershgorin band for each state variable–integration operator combination excludes the respective $J_{ii}T$ point. Therefore, the system is diagonally dominant, and simulation stability can be determined by applying the usual technique to the two plots independently, using the $J_{ii}T$ point as the normal λT value. The second observation is that each $J_{ii}T$ point is inside the region of stability for the corresponding integration operator; therefore, the complete simulation will be stable with the two methods chosen and $T = 0.1$. Thus, for this example, stability of the simulation is guaranteed only by ensuring that each $J_{ii}T$ point is in the region of stability for the appropriate integration method.

To illustrate the second procedure for determining diagonal dominance, consider the

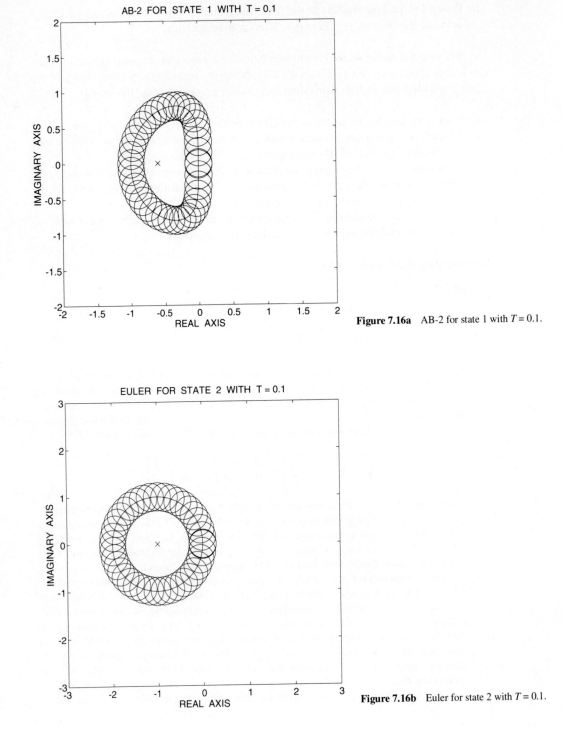

Figure 7.16a   AB-2 for state 1 with $T = 0.1$.

Figure 7.16b   Euler for state 2 with $T = 0.1$.

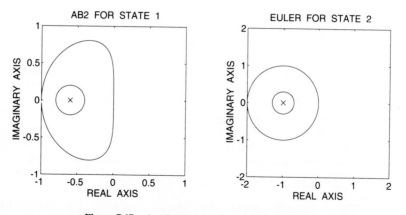

Figure 7.17   a) AB-2 for state 1.   **b)** Euler for state 2.

plots in Figure 7.17. The x's represent the $J_{ii}T$ points as before, and the circles surrounding them are the corresponding Gershgorin circles. Note that each of them is contained entirely within the respective stability region. The small o's in the plots are the actual λT locations. Obviously, all the λT products are inside both stability regions, so the simulation should be stable.

Some further insight into the INA technique can be gained from Figure 7.18. Here both stability regions, both Gershgorin circles, both $J_{ii}T$ points, and the actual λT are plotted on the

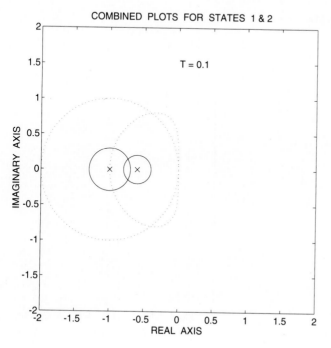

Figure 7.18   Combined plots for states 1 and 2.

same graph. One thing to note is that although the λT are inside the AB-2 stability region, they are fairly close to the boundary. Although the simulation will be stable, the transient response may not be accurate. Another point is that the Gershgorin circle for state x_2 intersects the stability boundary for the AB-2 method. Therefore, if the AB-2 method had been chosen for the second state variable rather than the first, the system would not have the diagonal dominance property. Also, the $J_{ii}T$ point would not be inside the AB-2 stability region. However, the simulation would still be stable, since the λT products are inside both stability regions. This shows that diagonal dominance is not a necessary condition for simulation stability. Figure 7.19 shows the same information for $T = 0.05$. The Gershgorin circles have smaller radii, the $J_{ii}T$ values are closer to the origin, and the simulation should be more accurate. This is borne out in Figure 7.20, which shows the simulated time response for the two states with $T = 0.1$ and $T = 0.05$.

The next example applies the INA technique to the Brennan and Leake jet engine model. The fact that diagonal dominance is only a sufficient condition for simulation stability to be inferred from the $J_{ii}T$ points is again illustrated.

Example 7.10

The Jacobian matrix J is the same as in equation 7–80, with eigenvalues $\lambda = -2.85, -32.19$, and -81.37. Euler's method is used for the first and second states, and the AB-2 method is used for the third state. The timestep is chosen to be $T = 0.0061$, which places the fastest eigenvalue in the region of relative stability of the AB-2 method, as mentioned in Example 7.7. With this value for T, the $J_{ii}T$ points are at $-0.6863, 0.0013$, and -0.0251; the radii of the corresponding Gershgorin circles are $0.5850, 0.5858$, and 0.0251. Figure 7.21 shows the

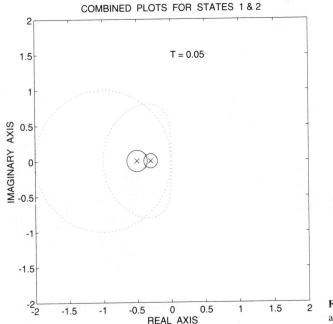

COMBINED PLOTS FOR STATES 1 & 2

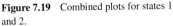

Figure 7.19 Combined plots for states 1 and 2.

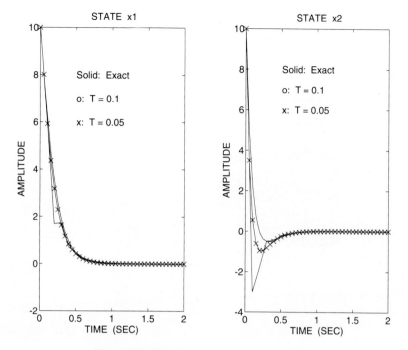

Figure 7.20   Simulation responses for Example 7.9.

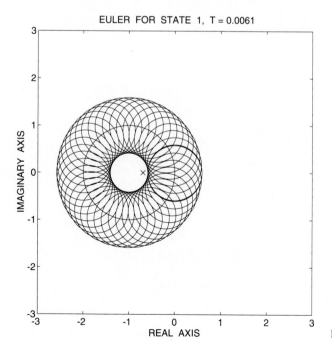

Figure 7.21a   Euler for state 1, $T = 0.0061$.

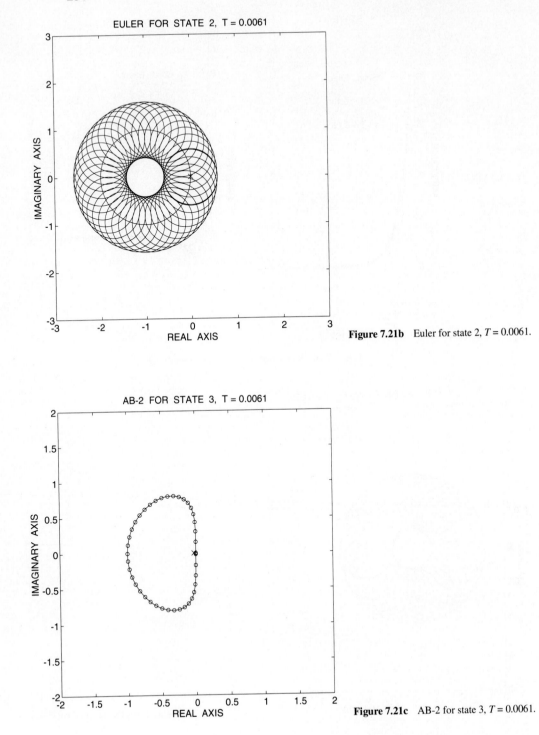

Figure 7.21b   Euler for state 2, $T = 0.0061$.

Figure 7.21c   AB-2 for state 3, $T = 0.0061$.

Gershgorin bands on the appropriate stability region boundary; the $J_{ii}T$ points are indicated with the small x's. For state 1, it is seen that the Gershgorin band does exclude the $J_{ii}T$ point and that the point is well away from the left side of the stability boundary. By itself, this looks like good news, but we must check the other two states also. For state 2, the $J_{ii}T$ point is clearly inside the Gershgorin band since the point is essentially at the origin. Also, the $J_{ii}T$ point is outside the stability region since it is positive. This situation will not change by varying the value of T. Thus, diagonal dominance is not present for this system. For state 3, the $J_{ii}T$ point and radius of the Gershgorin band have the same value, regardless of T. Figure 7.22 shows the combined plots of all of the stability regions and Gershgorin bands. The x's represent the $J_{ii}T$ points, and the o's represent the actual λT products. One of the o's and two of the x's are essentially overlapping near the origin. As the figure shows, all of the eigenvalues are inside all of the stability regions, so the simulation should be stable. This is verified in Figure 7.23, which shows the initial condition response of the system. The dashed lines in the figure are the exact responses, and the solid lines are for the simulation with the Euler and AB-2 methods. It is clear from the figure that the simulation is stable and that high accuracy is obtained.

This example has brought out some interesting points. As mentioned before, closed-loop simulation stability is still possible when diagonal dominance is absent, as it is in this example. In addition, when the system is not diagonally dominant, the simulation can still be stable even if the $J_{ii}T$ point is outside the appropriate stability region. So once again we have the conclusion that having diagonal dominance is a sufficient but not necessary condition for using the diagonal elements of the J matrix as the actual eigenvalues to assess simulation stability.

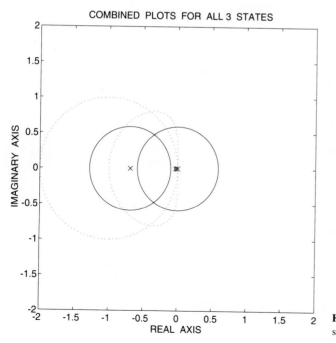

Figure 7.22 Combined plots for all three states.

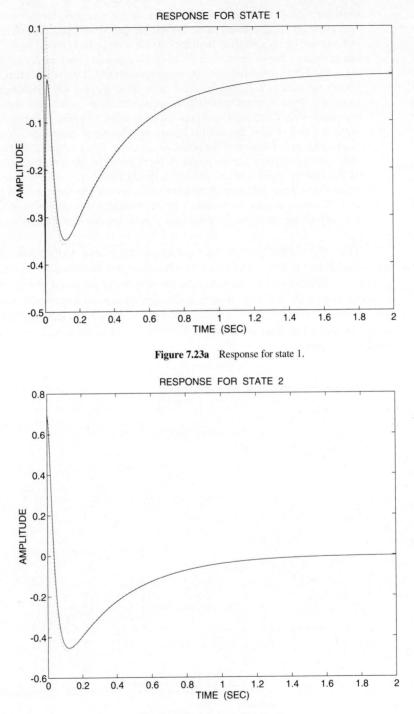

Figure 7.23a   Response for state 1.

Figure 7.23b   Response for state 2.

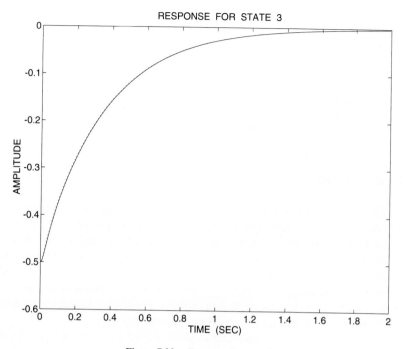

Figure 7.23c   Response for state 3.

When diagonal dominance is not present for a particular system–integration operator combination, it is instructive to check on which side of the stability region the $J_{ii}T$ point is included by the Gershgorin band. If it is near the left side of the stability region, the designer must be very careful. In this case, the $J_{ii}T$ point is negative, and the actual λT value may be even more negative. Therefore, there is a good chance that the simulation will be either unstable, or at least very inaccurate. This will not always be the case, but a complete analysis should be made before doing the simulation. If the intersection of the Gershgorin band with the $J_{ii}T$ point occurs on the right side of the stability region, things are a little more relaxed. First, if the real λT is negative and close to the origin, it generally indicates not only a stable but also a very accurate simulation. This is obviously what we want. Second, if the real λT is positive, it indicates a system that itself is unstable; therefore, the simulation should also be unstable, so that it provides an accurate representation of the actual response. Thus, we conclude that the $J_{ii}T$ point being included by the Gershgorin band on the left side of the stability boundary is generally a much more serious condition than being included on the right side of the boundary.

The inverse Nyquist array can be used to study simulation stability when different timesteps as well as different integration operators are used for each state variable. In this case, equation 7–93 becomes

$$H(z) = \frac{T_i \sigma_i(z)}{\rho_i(z)} \qquad (7\text{--}99)$$

and equation 7–94 becomes

$$diag\left[\frac{\rho_i(z)}{\sigma_i(z)}\right] - diag[T_i]J = 0 \tag{7–100}$$

The rest of the procedure is the same as before. Each row of the Jacobian is multiplied by the value of T_i corresponding to the integration operator being used for that state variable. If diagonal dominance is achieved, the overall simulation stability can be assessed by comparing the $J_{ii}T_i$ point with the appropriate algorithm's stability region.

7.8 CONCLUSIONS

This chapter studies some of the problems facing the simulation engineer when the system being simulated is stiff. Stiffness is defined in terms of two time scales being present in the problem, a long time scale for the important dynamics and a much shorter time scale for the fast transients in the system. Unfortunately, for most integration methods, the value of the simulation timestep must be based on those fast dynamics. This results in a large number of computations to simulate accurately the slower, important dynamics. A second problem arises when the simulation must be done in real time. In that case, a minimum value for the timestep will often be imposed by the hardware.

Several definitions relating to stiffness and several techniques for dealing with the simulation of stiff systems have been presented. Most of these attack the problem by designing an integration operator with an enlarged region of absolute stability. The SRP, MSRP, and reduced order of accuracy Runge-Kutta methods are examples of this approach. The inverse Nyquist array gives sufficient conditions under which the simulation stability of a multivariable system can be assessed by considering only the diagonal elements of the Jacobian, with each state having its own integration operator. For stiff systems, when the Gershgorin circles for the fast and slow eigenvalues do not overlap, the value of T can be chosen for accuracy and computation time, and stability of the fast eigenvalues can be obtained by choosing the right integration method for them.

There are other methods for simulating stiff systems. One well-known technique is known as Gear's method (Lapidus & Seinfeld, 1971). This is an implicit method that uses a measure of the local truncation error to adaptively vary the timestep during the simulation. Many examples of its application have been reported in the literature. However, it is a very complex algorithm, generally not suited for real-time simulation; therefore, it will not be discussed further in this book. Implicit Runge-Kutta methods can also be used for stiff systems. The interested reader is referred to Butcher (1987) for a discussion of this topic.

The engineer involved in the simulation of dynamic systems should be aware that many real systems are in fact stiff, having stiffness ratios of one thousand or larger. The engineer should also realize that instabilities observed in a simulation may be due to the use of a timestep that was based on the important dynamics, and it may be too large to maintain stability in the fast dynamics. The techniques discussed in this chapter can be utilized to solve the trade-off between simulation stability and computational burden while maintaining an accurate simulation of the important dynamics in the system.

PROBLEMS

1. A client gives you the system in the block diagram below. You are asked to design a real-time simulation, using linear multistep methods with $T = 0.1$ seconds. Furthermore, you must consider each block as a separate unit, as it may be necessary to redesign the continuous-time units after their performances are evaluated. Give some indication of the accuracy, robustness, and stability of your simulation to convince the client of its utility. Find the system response to a unit step.

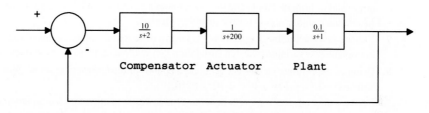

2. The system $x' = -75x + u$ is to be simulated in real time, where the hardware constrains $T = 0.1$. Derive an explicit two-step method that is first-order accurate, has its principal closed-loop pole at the direct mapping of the s-plane pole, and has its spurious closed-loop pole at the origin.

3. Following is an intermediate design model for the first all-electric aircraft brake. You are to design a real-time simulation of this system, using $T = 0.001$ seconds. Using your design, simulate the system step response; that is, get the applied braking torque as a function of braking signal.

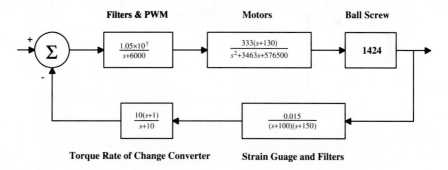

4. Consider the block diagram below. Suggest a choice of simulation timestep when Euler's method is used to simulate this system.

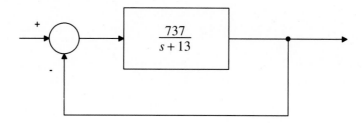

5. Consider the SRP12 integrator

$$H(z) = (0.7163z - 0.2838)/(z^2 - 1.5675 + 0.5675).$$

    a. What is the order of accuracy?
    b. Sketch the root locus. Indicate the principal and spurious loci.
    c. What is the largest real value of λT for stability?
    d. Where are the spurious and principle roots when $\lambda T = -2$ Is this good?

6. Design an integrator that is at least first-order accurate and has all of its spurious roots at the origin and its principal roots and the exact mapping of λT when $\lambda T = -2$. Plot a root locus for real λT and the stability region for the resulting method.

7. You are given a nonlinear system that is known to be stiff in some regions of space. Discuss how to implement a variable timestep simulation that will maintain accuracy and speed.

8. Using an appropriate technique, generate a bound for the largest pole (in magnitude) of the system $G(s) = 1/(s^4 + 17s^3 + 81s^2 + 105s + 50)$.

9. Derive a three-stage, first-order accurate Runge-Kutta method that has an enlarged stability region with respect to the usual Runge-Kutta methods. Include the stability region in your answer. Is this method useful for on-line real-time simulation of a single plant, and why?

10. Obtain a step response of the following system by using the AB-2 method with $T = 0.1$ seconds. Explain what you observe.

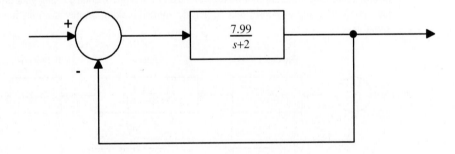

8

Nonlinear Systems

8.1 INTRODUCTION

Life is really nonlinear (Ortega & Poole, 1981). Almost everything around us is associated with some fundamentally nonlinear physics. Unfortunately, the mathematics associated with nonlinear dynamic systems is not very accessible or useful. The mathematics for linear systems, however, is relatively simple. Thus, engineers initially consider most systems to be linear just so they can begin to understand their behavior, realizing all the while that this belief is incorrect. Fortunately, this works much of the time. However, it does not always work, and we must always be aware of what nonlinear systems can do. From the point of view of control theory, if we do not know what the system is going to do, we cannot control it. From a simulation point of view, if we do not know what the system is capable of doing, we will not know if our simulation is reasonable. Thus, it is important to be aware of the types of behavior demonstrable by nonlinear systems so that the strange responses that sometimes occur are not totally unexpected.

We categorize nonlinearities in dynamic systems into continuous and discontinuous nonlinearities. Continuous nonlinearities usually arise out of the physics of the systems involved. These are more mathematical in nature. A good example is the gravitational force acting on a pendulum, which is proportional to $\sin(x)$. Discontinuous nonlinearities are more often engineered into a system. Good examples of these are relays, saturation, hysteresis, and gear backlash. Neither of these nonlinearity types are particularly easy to deal with; however, they each have developed specific tools.

The major question to be addressed here is this: What do nonlinear systems do? A bottom-up hierarchy will be developed to answer this question.

8.2 FIRST-ORDER CONTINUOUS-TIME SYSTEMS

First-order autonomous time invariant systems can all be written as

$$\dot{x} = f(x) \tag{8-1}$$

Several important things can be determined from this equation. The values of x where $f(x) = 0$ are the steady-state points of the system. These are also called fixed points. Contrary to linear thinking, more than one steady state can exist simultaneously. The slope of $f(x)$ at these points, the system Jacobian, which is df/dx, determines the local stability of the point. If df/dx is negative at a fixed point, the point is stable and is called an attractor. If df/dx is positive at a fixed point, the point is unstable and is called a repeller. If $df/dx = 0$, the point is a saddle, which is attracting from one side and repelling from the other. The geometry of this equation indicates that the attractors and repellers alternate on the x-axis (see Figure 8.1). It should be noted that no oscillations or undamped responses are possible. Basically, first-order systems are damped.

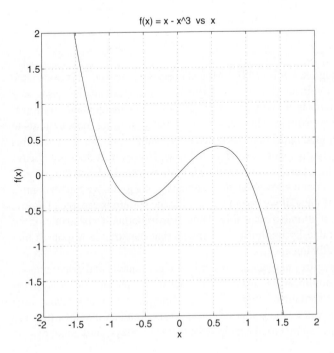

Figure 8.1 $f(x) = x - x^3$ vs. x.

Consider the system

$$\dot{x} = x - x^3 + u$$

with the input $u = 0$. The plot of the derivative versus x is given in Figure 8.1. The fixed points can be found either from the intersections with the x-axis, or from

$$\dot{x} = 0 = x - x^3$$

which gives x_{ss} = 0, +1, and −1. The local stability of these points can be determined graphically from the slopes at these points. Mathematically, this is

$$J = \left[\frac{\partial \dot{x}}{\partial x}\right]_{x_{ss}} = 1 - 3x_{ss}^2$$

Thus, the slope at $x = 0$ is +1, and the slope at $x = +1$ or $x = -1$ is −2. The fixed point at the origin is unstable, therefore, and is called a repeller. The fixed points at $x = +1$ or $x = -1$ are stable and are called attractors. It should be observed that these slopes near the fixed points are effectively equivalent to the system poles of a linearized version of the system.

Another thing to observe is that each attractor possesses a basin of attraction, that is, a region in space where any initial condition in the space is attracted to that particular attractor. For this example, the basin of attraction for the $x = +1$ point is all $x > 0$. The basin of attraction for the $x = -1$ point is all $x < 0$.

8.3 LINEARIZATION

Before going on to second-order systems, it is important to talk about the linearization procedure as it is used by almost all control engineers at some time. The general vector form is given first, and then it is applied to the scalar system of the last section.

Consider the vector system

$$\dot{x} = f(x,u) \tag{8–2}$$

Assume that u reaches some fixed value u_{ss}. It is then possible to find the steady-state value of x where the derivative vector is identically zero by solving

$$f(x_{ss}, u_{ss}) = 0 \text{ for } x_{ss} \tag{8–3}$$

The purpose of linearization is to find the behavior of the system near one of the fixed points. Thus we set

$$x = x_{ss} + \delta x \text{ and } u = u_{ss} + \delta u \tag{8–4}$$

Substituting this back into the original vector system, we get

$$\dot{x}_{ss} + \delta \dot{x} = f(x_{ss} + \delta x, u_{ss} + \delta u) \tag{8–5}$$

Notice that the derivative of x_{ss} is identically zero. Now expanding the right side in a Taylor series and keeping the first two terms, we obtain

$$\delta \dot{x} = f(x_{ss}, u_{ss}) + \left[\frac{\partial f}{\partial x} \right]_{x_{ss}, u_{ss}} \delta x + \left[\frac{\partial f}{\partial u} \right]_{x_{ss}, u_{ss}} \delta u \qquad (8\text{--}6)$$

Notice that the first term in this series is zero by definition. We then have

$$\delta \dot{x} = J \delta x + G \delta u \qquad (8\text{--}7)$$

where

$$J = \text{system Jacobian matrix} = \left[\frac{\partial f}{\partial x} \right]_{x_{ss}, u_{ss}} \qquad (8\text{--}8)$$

and

$$G = \text{input Jacobian matrix} = \left[\frac{\partial f}{\partial u} \right]_{x_{ss}, u_{ss}} \qquad (8\text{--}9)$$

This last equation should be recognized as the usual state space representation used by most control engineers. Thus the eigenvalues of the system Jacobian matrix are the poles of the linearized system and determine the stability of the resulting system. Improved techniques for computing the input Jacobian, G, are given in Danielli and Krosel (1979).

For the example in the last section, we have

$$\delta \dot{x} = \delta x + \delta u$$

near the origin, and

$$\delta \dot{x} = -2\, \delta x + \delta u$$

near $x = +1$ or $x = -1$. Thus using the small perturbation variables, we can determine a control law for behavior near one of the fixed points.

The procedure for discrete-time nonlinear systems is very similar to that presented here, except that the fixed points are determined by setting $x_{n+1} = x_n = x_\infty$ and solving the resulting equation. Also, the eigenvalues of the resulting Jacobian must be inside the unit circle for stability.

8.4 SECOND-ORDER SYSTEMS

Second-order systems are considerably more complicated than first-order systems. Although multiple attractors and repellers are still possible, oscillations are now possible in addition to the damped responses of first-order systems. Furthermore, a new form of attractor-repeller is available. This is usually called a limit cycle. In the phase plane ($\dot{x}$ vs x), this is a closed curve that shows continuous oscillation. Whereas a fixed point is a zero-dimensional object, a limit cycle, being a closed curve, is a one-dimensional object. These, too, can be attracting or repelling and can coexist with other types of attractors and repellers. It should be noted that limit cycles should not be confused with conservative oscillations. A conservative system, such as

$$\ddot{x} + x = 0 \quad \text{or} \quad \ddot{x} + \sin(x) = 0 \tag{8-10}$$

will oscillate forever with an amplitude determined by its initial conditions. Any slight perturbation will cause it to oscillate with a different amplitude. These oscillations are called conservative since there is no damping and energy is conserved. Volumes in phase space are constant. This is shown in Figure 8.2. A MATLAB program for simulating the pendulum is shown below.

```
% MATLAB program for simulating the pendulum using double Halijak's
% integrator.
axis([-4 4 -4 4]);
T = 0.08
% Initial Conditions
xss = 3;
xd = 0;
xs = xss + T * xd + 0.5 * T * T * (-sin(xss));
plot(xs, (xs-xss)/T,'.')
hold
% Main Loop
for n = 1:300
    f = - sin(xs)
```

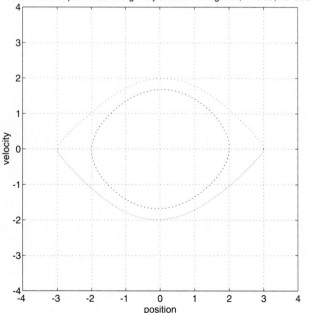

Simulation of pendulum using Halijak double integrator, T=0.08, x0=2&3

Figure 8.2 Simulation of the pendulum using Halijak double integrator, $T = 0.08$, $x0$ = 2 and 3.

```
      x = 2 * xs - xss + T * T * f;
      xss = xs;
      xs = x;
      xd = (xs - xss) / T;
      plot(xs,xd,'.')
end
xlabel('position'),ylabel('velocity')
title('Simulation of the pendulum using Halijak double integrator,
      T = 0.08, x0 = 2 & 3')
```

Unlike conservative oscillations, a limit cycle attracts trajectories. Any slight perturbation will still be attracted back to the limit cycle. Equivalently, volumes in phase space
contract. A good example of a limit cycling system is van der Pol's oscillator:

$$\ddot{x} + (x^2 - 1)\dot{x} + x = 0 \tag{8–11}$$

Here, the damping is negative when x is small, indicating instability. When x is large, however, the damping is positive, indicating stability. Thus the trajectory is basically stuck in
the middle and a continuous limit cycle occurs. This is shown in Figure 8.3, and a
MATLAB program for producing the simulation is shown below.

```
% MATLAB program for simulating van der Pol's equation
% using Euler's method.
```

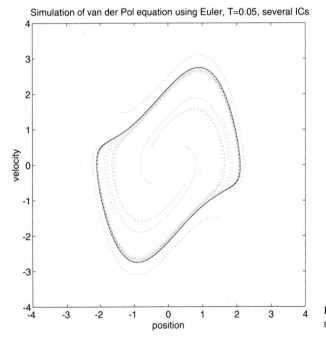

Figure 8.3 Simulation of van der Pol system using Euler, $T = 0.05$, several ICs.

```
axis('square')
axis([-4 4 -4 4]);
plot(0,0,'.')
hold
T = 0.05
% Initial Condition Loop
for x10 = -1.5:1:1.5
    for x20 = -1.5:1:1.5
        x1 = x10;
        x2 = x20;
% Main Loop
        for n = 1:200
            f1 = x2;
            f2 = -x1 - (x1 * x1 - 1) * x2;
            x1 = x1 + T * f1;
            x2 = x2 + T * f2;
            plot(x1,x1,'.')
        end
    end
end
xlabel('position'),ylabel('velocity')
title('Simulation of van der Pol equation using Euler, T = 0.05,
    several ICs')
```

When second-order systems are forced by a periodic forcing function, several frequency dependent phenomena are possible. These include subharmonics, superharmonics, jump resonances, and entrainment, among others. Since a time-dependent input can simply be considered to be the function of another state of a system, for example, $\dot{x} = 1$, these are also subsets of the next section.

8.5 THIRD-ORDER SYSTEMS

Third-order systems display all of the above behavior in addition to some more interesting features. It should be noted that as the dimension of the phase space increases, the possible dimension of the attractors and repellers also increases. The limiting factor is that the dimension of the attracting-repelling objects must be strictly less than the dimension of the phase space. Thus the first object that logically follows is a two-dimensional attractor-repeller. A two-dimensional object basically looks like a donut and can be thought of as a limit cycle that is being forced at a different frequency. Trajectories are then attracted to, or repelled from, this object, which is usually referred to as a torus.

An interesting, and unexpected, phenomenon that can also exist in third- and higher-order continuous-time systems is called chaos. Although difficult to explain briefly, chaos usually is caused by the tangled interaction of several unstable repellers. The resulting trajectory, which basically is being repelled from everywhere, is called a chaotic, or strange, attractor. The dimension of this object is usually noninteger, for example, dim = 2.6, and is called fractal. At any given cross section on the attractor, the flow is converging in one direction and diverging in the other, so that volumes in phase space are still getting smaller. However, since the flow is diverging in one direction, it is usually folded back over onto itself, and the attractor can resemble taffy being stretched and folded on a machine. Some authors have referred to this behavior as sensitive dependence on initial conditions. We prefer to describe chaos as the apparent random motion demonstrated by a completely de-terministic system. It should also be remembered that the chaotic behavior, just like a limit cycle, is steady-state behavior. That is, although the system is oscillating with no apparent period, the trajectories have converged to their attractor and steady-state chaotic behavior is observed.

A simple system that behaves in this way is the chaotic oscillator of Cook (1986):

$$x''' + x'' + x' - a(x - x^3) = 0 \qquad\qquad (8\text{--}12)$$

As the parameter a is increased, the system goes through a series of period doubling limit cycle bifurcations and eventually becomes chaotic (this is discussed further in the next sec-tion). Some of this behavior is shown in Figure 8.4. It is truly remarkable that the interesting behavior of such simple systems went unnoticed until relatively recently. The MATLAB program for this simulation is given below.

```
% MATLAB program for simulating Cook's scroll
% using Euler's method.
axis('square')
```

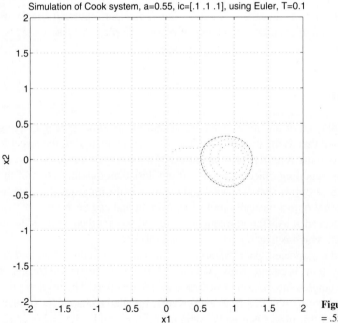

Simulation of Cook system, a=0.55, ic=[.1 .1 .1], using Euler, T=0.1

Figure 8.4a Simulation of Cook system, a = .55, ic = [.1 .1 .1], using Euler, T = 0.1.

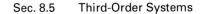

Simulation of Cook system, a=0.7, ic=[.1 .1 .1], using Euler, T=0.1

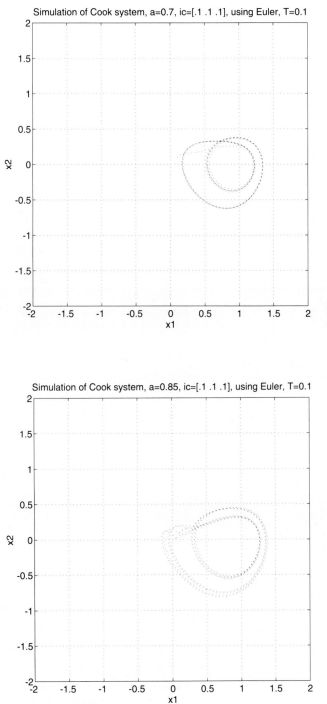

Figure 8.4b Simulation of Cook system, *a* = .7, *ic* = [.1 .1 .1], using Euler, *T* = 0.1.

Simulation of Cook system, a=0.85, ic=[.1 .1 .1], using Euler, T=0.1

Figure 8.4c Simulation of Cook system, *a* = .85, *ic* = [.1 .1 .1], using Euler, *T* = 0.1.

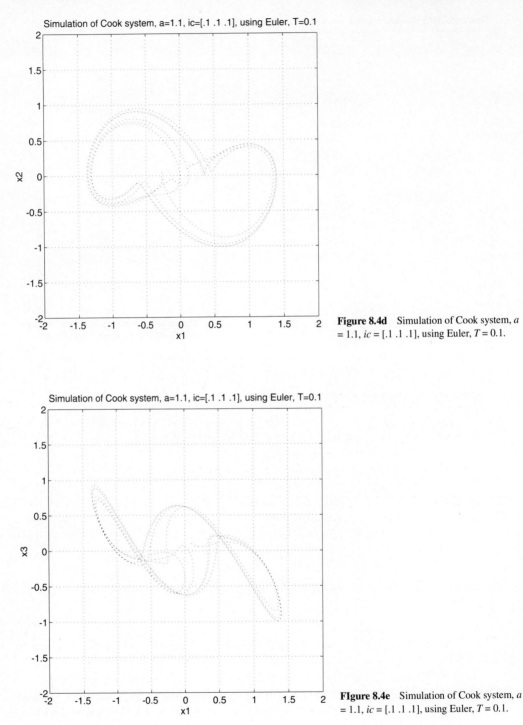

Simulation of Cook system, a=1.1, ic=[.1 .1 .1], using Euler, T=0.1

Figure 8.4d Simulation of Cook system, a = 1.1, ic = [.1 .1 .1], using Euler, $T = 0.1$.

Simulation of Cook system, a=1.1, ic=[.1 .1 .1], using Euler, T=0.1

FIgure 8.4e Simulation of Cook system, a = 1.1, ic = [.1 .1 .1], using Euler, $T = 0.1$.

```
axis([-2 2 -2 2]);
plot(0,0,'.')
hold
T = 0.1;
a = 0.55;
% Initial Conditions
x1 = 0.1;
x2 = 0.1;
x3 = 0.1;
% Main Loop
for n = 1:800
    f1 = x2;
    f2 = x3;
    f3 = -x3 - x2 + a * (x1 - x1^3);
    x1 = x1 + T * f1;
    x2 = x2 + T * f2;
    x3 = x3 + T * f3;
    plot(x1,x2,'.')
end
xlabel('x1'),ylabel('x2')
title('Simulation of Cook system, a = 0.55, ic = [.1 .1 .1], using
Euler, T = 0.1')
```

With respect to chaos, this complicated deterministic phenomenon is sometimes confused with the effect of stochastic, or noisy, signals. Although it can be analyzed as if it were a noisy signal, chaos is a completely deterministic feature of a nonlinear system and, at this time, is not known to be related to the unpredictable behavior resulting from physically occurring randomness. Alternatively, if we wished to simulate a system being forced by a stochastic signal, that signal is represented in the computer by a random number generator. A continuous-time stochastic signal is usually represented by its probabilistic properties such as mean and variance; although a chaotic signal can use these properties as one form of measure, they do not characterize the system. It should be noted in simulating a continuous-time noisy input signal that the sampling of this noise requires its variance to be divided by the simulation timestep to preserve the same level of excitement in the system as the continuous-time noise. This is not surprising when it is remembered that the frequency spectrum of white noise is identical to that of an impulse. In Chapter 4, it was demonstrated that the impulse must also be similarly scaled to maintain its excitement level.

8.6 HIGHER-ORDER SYSTEMS

For systems of order higher than 3, all the above behavior can be displayed. In addition, higher-dimensional tori can also exist, which have oscillations at several different frequencies simultaneously. Furthermore, hyperchaos can exist. Hyperchaos is very similar to chaos, except that now the divergence on the trajectory can occur in more than one direction. As long as the overall phase space volumes are contracting, all but one of the directions on the flow can be diverging.

An example of a hyperchaotic system is that of Rossler (1979):

$$
\begin{aligned}
\dot{x} &= -y - z \\
\dot{y} &= x + 0.25y + w \\
\dot{z} &= 3 + xz \\
\dot{w} &= -0.5z + 0.05w \\
x_0 &= -20, \; y_0 = 20, \; z_0 = 20, \; w_0 = 20
\end{aligned}
\tag{8-13}
$$

It should be remembered that not all nonlinear systems will display these phenomena. However, we should be aware of their existence so that we will recognize the behavior when it is encountered.

8.7 DISCRETE-TIME SYSTEMS

With discrete systems the situation is much more complicated. Only a first-order difference equation is necessary for chaotic behavior. Thus a first-order difference equation can display all of the phenomena that a third-order differential equation can. Although this can be a problem, it will be beneficial in that first-order difference equations are fairly easy to understand. A thorough discussion of the very popular logistic equation follows to give some intuition into the more complicated nonlinear behavior. It is important to get a good feel for how this system works since everything else will behave similarly, but will only be more complicated. Also since we will be simulating continuous-time nonlinear systems in discrete time, it turns out that each and every simulation of a nonlinear system contains the potential for complicated or chaotic behavior, whether the continuous-time system does or not. Thus, this discussion is also important so that the designer can recognize numerical chaos from that displayed by a real system.

Before introducing the logistic equation, it is important to present some information about first-order, discrete linear systems. Consider

$$
x_{n+1} = f(x_n) = ax_n
\tag{8-14}
$$

for positive a. This equation can be represented graphically, as in Figure 8.5. In steady state, for any discrete system, $x_{n+1} = x_n = x_\infty$. Thus, for this system, $x_\infty = 0$. From the figure, this can be found from the intersection of the $f(x)$ curve with the regression line at 45 degrees, $x_{n+1} = x_n$. This also occurs at $x = 0$. Notice that the dynamics of this system can be determined graphically, simply by iterating from $f(x)$ to the regression line as shown in the figure. Thus it can be seen that the origin is an attractor when $0 < a < 1$ and that the origin is a repeller when $a > 1$. Mathematically, this can be determined by computing the system Jacobian (df/dx), evaluating it at the fixed points, and then computing its eigenvalues. These eigenvalues are like the discrete system poles. As long as the poles are less than unity in magnitude, the small perturbation system about the given fixed point will be stable. This then determines the stability of the fixed point. When $a < 0$, Figure 8.6 arises. It should be observed that the transient response is an oscillation. When $a < -1$, the oscillation becomes unstable. Again, this can be determined either graphically or mathematically.

The logistic equation is now introduced to clarify some of the concepts involved with

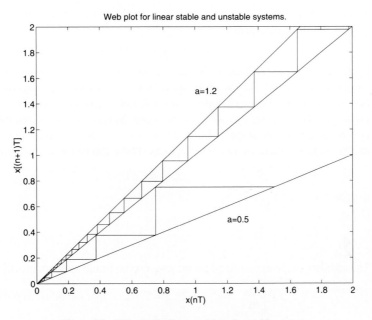

Figure 8.5   Web plot for linear stable and unstable system.

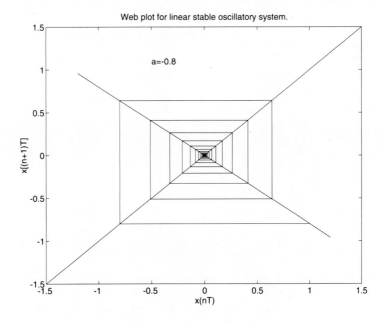

Figure 8.6   Web plot for linear stable oscillatory system.

nonlinear dynamics and also because it has become the somewhat prototypical nonlinear system. The logistic equation is

$$x_{n+1} = ax_n(1 - x_n), \qquad 0 \leq a \leq 4, 0 < x_0 < 1 \qquad (8\text{–}15)$$

Notice here that $f(x)$ on the right side of the equation is the equation of a parabola that passes through $x = 0$ and $x = +1$, concave down. As a increases, the concavity increases so that the hump in the parabola moves upward. Let us first consider $a < 1$, which is shown in Figure 8.7. Here the only intersection between the regression line and the parabola is at the origin, for $x > 0$. The slope of the line is a, and thus the origin is stable for $a < 1$. As a is increased past 1, there are two intersections. These can be found mathematically by solving

$$x_\infty = ax_\infty (1 - x_\infty)$$

to give

$$x_\infty = 0, \frac{a - 1}{a}$$

The stability of these two fixed points can also be found mathematically as

$$J = \left[\frac{\partial f}{\partial x} \right]_{x_\infty} = a (1 - 2x_\infty) = 2 - 2a + a \text{ at the nonzero point}$$

$$= a \text{ at the origin}$$

Thus it can be seen that the nonzero fixed point is stable for $1 < a < 3$. When $a = 2$, the stability of the point changes from indicating damped behavior to indicating oscillatory

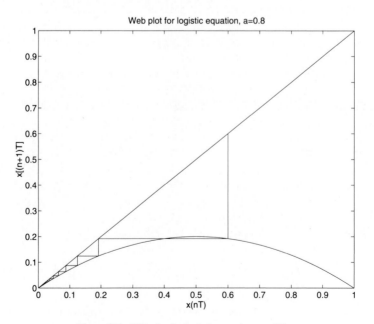

Figure 8.7 Web plot for logistic equation, $a = 0.8$.

behavior. At $a = 3$, the oscillatory attractor becomes marginally stable. What happens as a is increased further can best be understood by considering the second map, $f(f(x))$, plotted in Figure 8.8. Notice as a passes through 3 that one fixed point suddenly becomes three fixed points of the second map. The one in the center is unstable, whereas the other two are stable. This manifests itself in the time response as a limit cycle. The previously existing fixed point is usually referred to as a period-1 cycle. The new limit cycle is referred to as a period-2 limit cycle. Thus we have observed our first period doubling change in system structure, or bifurcation. Since the number of fixed points plotted versus a for the second map looks somewhat like a trident, this structure change is often called a pitchfork bifurcation. As a increases further, we must consider the stability of the fourth map, $f(f(f(f(x))))$.

What is observed is another period doubling, a process that continues to repeat itself with the higher maps until period infinity is reached around $a = 3.57$. After this, the trajectories have no stable place to go, and the result is chaos. The trajectory is essentially repelled from everywhere and ends up on what is now called a chaotic attractor. As a is increased further, other fundamental periods occasionally become stable, such as period 3, found from the third map, $f(f(f(x)))$, around $a = 3.83$. These orbits then go through their own period doubling routes to chaos as a increases. Amazingly enough, it turns out that every integer has its own little window of period doubling to chaos somewhere after $a = 3$ and before $a = 4$. The overall structure of the resulting steady-state behavior is shown in a bifurcation diagram, as in Figure 8.9. This is easily created for a discrete system by doing a simulation for several values of a, say, $a = 0$ to 4 in steps of 0.01, giving 400 simulations. The initial value of x is always chosen as some "unusual" number, such as 1/9, or an irrational number, such as $\pi/10$. For a given a, the simulation is run for about 100 steps without plotting anything (in the dark) to eliminate any transient. The simulation is then run for about 100 more steps, plotting all the values of x in the a-x bifurcation diagram (in the light). After this is completed, a is increased to its next value, and the process is repeated. For $a < 3.5$, most of these points will be plotted on top of one another. For $a > 3.5$, however, some interesting distribution of points on the resulting attractor occurs.

Notice that it is easy to see intriguing patterns running through the bifurcation diagram in the chaotic region. It should be noted that obtaining bifurcation diagrams for continuous-time systems is somewhat different than for discrete-time systems. For the latter, every point is plotted for a given parameter. For continuous-time systems, plotting every point would result in a solid dark line that would give information only about the boundaries. Instead, the value of the output should only be plotted whenever something convenient happens. For example, the output could be plotted only when the slope of the output variable changes sign. In this case, it may be necessary to plot more than one of the states in a bifurcation diagram to get the complete representation of the attractor.

Now we have gained some insight about chaotic systems, it is amazing to see that this period doubling to chaos occurs in many other systems, including continuous-time systems. For example, consider Rossler's (1976) equations:

$$\dot{x} = -y - z$$
$$\dot{y} = x + 0.2y \qquad\qquad (8\text{--}16)$$
$$\dot{z} = 0.2 + xz - cz$$

As c is increased, this system goes through the period doubling route to chaos.

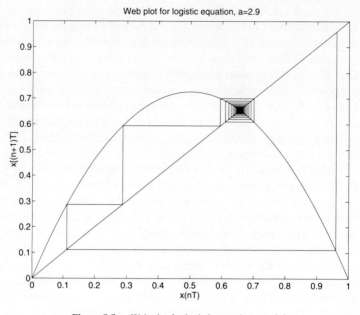

Figure 8.8a   Web plot for logistic equation, $a = 2.9$.

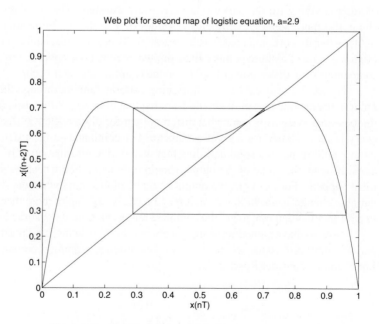

Figure 8.8b   Web plot for second map of logistic equation, $a = 2.9$.

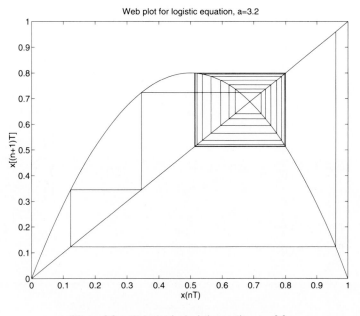

Figure 8.8c   Web plot for logistic equation, $a = 3.2$.

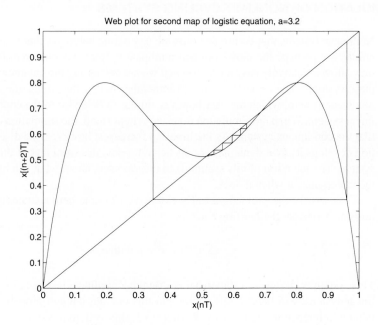

Figure 8.8d   Web plot for second map of logistic equation, $a = 3.2$.

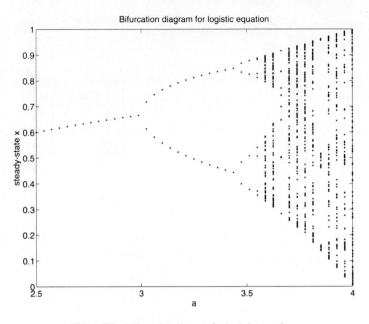

Figure 8.9   Bifurcation diagram for logistic equation.

8.8 SIMULATION OF NONLIMIT CYCLING SYSTEMS

Nonlinear systems that do not demonstrate any exotic behavior, that is, possess only fixed points, are perhaps the most common nonlinear systems encountered. It is fortunate that they allow a relatively easy analysis when we are discussing the behavior of simulations of these systems. The important thing to remember is that the simulation is a discrete-time nonlinear system, which the user hopes is similar in behavior to the continuous-time nonlinear system. When the simulation timestep is kept small, the simulation does often behave like the continuous system. As the timestep increases, however, the discrete-time simulation develops its own dynamics and its behavior deviates from that of the continuous-time system. The intention of this section is to inform users about what can happen and to help them recognize it when it does.

To clarify the problem, a fairly extensive example demonstrates the resulting phenomena. Consider the Duffing equation:

$$\ddot{x} + B\dot{x} + x^3 = u \text{ with } u = -1 \qquad (8\text{--}17)$$

This system is chosen as a somewhat prototypical nonlinear test equation. Clearly, all nonlinear systems will behave differently, but this one has enough properties to demonstrate what will happen in many cases. With $u = -1$, this system has one fixed point at $x = -1$, $\dot{x} = 0$ and it attracts all the phase space; that is, its basin of attraction is the entire phase plane. Also observe that as B increases from zero, the nonlinear system changes its damping. This is important when attempting to form a stability region, as is shown later.

The first thing to consider in simulating a nonlinear system is the basin, or basins, of attraction. This system has one fixed point whose basin of attraction is the entire plane. When simulated, however, the size of the basin for the simulation varies somewhat inversely proportional to T. In fact, for $\lambda T < -1$ (λ represents the poles of the linearized system), the basin is roughly 2 units by 2 units and has become a fractal basin, containing many self-similar islands. An example is shown in Figure 8.10, which was created by choosing a B and T and running the simulation by using each pixel point as an initial condition. Thus the basin represents over 10,000 separate simulations. The stable region in black was obtained by letting the simulation run for a large number of steps, say, 500. If blowing up is not detected in this period, the pixel is made black. The other shadings were obtained by setting the pixel to the shade associated with the iteration number when the simulation output blew up, that is, became larger than 100 in magnitude. Thus, the conclusion is, when initial conditions are far away from the expected steady states, users must be careful to choose T small enough so that the simulation stays in the correct basin.

The next thing to consider is the simulation time response as a function of T. To obtain this, the user must specify a B and some initial condition. The initial condition, however, must remain inside the basin of attraction as T is increased. Thus a good choice for this analysis is an initial condition close to the fixed point in question. Then a simulation can be

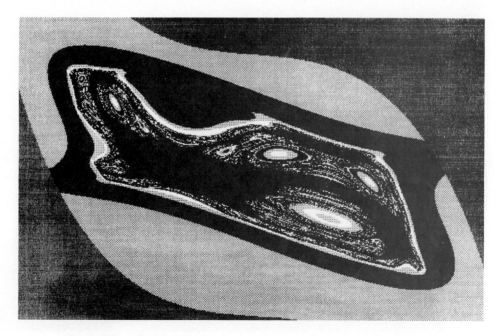

Figure 8.10a   Dark, fractial region near center represents the system equation 8–17 basin of attraction using Euler's method with $\lambda T = -1.5 \pm j0.8$.

Figure 8.10b Enlargement of left center of Figure 8.10a.

Figure 8.10c Enlargement of lower center of Figure 8.10b.

Figure 8.10d   Enlargement of upper center of Figure 8.10c.

run for several values of T to create a bifurcation diagram. For the given system, this is shown in Figure 8.11, using an initial condition of $x = -0.99$ and $\dot{x} = 0$. It can be observed that the single fixed point exists as expected until about $T = 0.6$. After that, a torus forms, which then develops several interesting bifurcations until $T = 0.95$, when instability occurs. It should be remembered that since this is a second-order system, the steady states all exist in a two-dimensional phase plane. Thus, the true bifurcation diagram is a three-dimensional object, with T being the third dimension. The object basically resembles a roselike flower, in that it bifurcates from a single steady-state point to form interesting petallike structures. Some cross sections of this object are given in Figure 8.12. It should be noted that the nonlinearity has effectively saved the simulation from blowing up as it passed the usual linear stability boundary. This is fortunate in that the simulation will continue to run. This is unfortunate, however, in that the simulation typically demonstrates high-frequency chaotic behavior in this region, which can be mistaken for the true system dynamics.

  For linear systems, the stability region has been shown to be our basic tool for understanding given integrators. The utility of the linear stability region carries over to nonlinear systems if the user is careful in interpreting them. For example, if a given nonlinear system is highly dissipative, Euler's method will work sufficiently well for it. This is true even if the system is limit cycling or chaotic. As long as there is sufficient damping in the system, Euler's method will work fine for a sufficiently small T. When the system is not highly dissipative, however, a better method should be used, particularly one that has good stability properties on the imaginary axis.

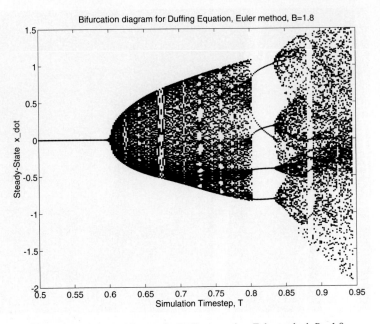

Figure 8.11   Bifurcation diagram for Duffing equation, Euler method, $B = 1.8$.

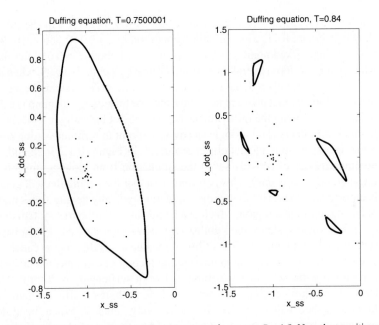

Figure 8.12   Steady-state phase plane attractors, $x'ss$ vs. xss, $B = 1.8$. Note the transitions from 1-torus (a) to 5-torus (b) to 5-chaos (c) to 1-chaos (d, e, f).

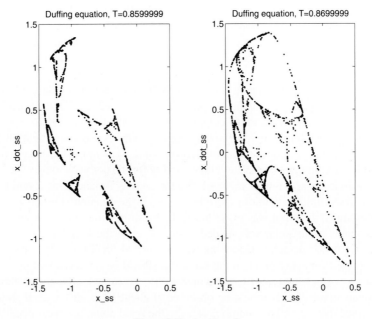

Figure 8.12   (continued)

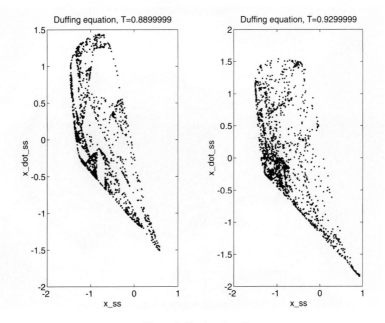

Figure 8.12   (continued)

Is there a nonlinear stability region? Well, not exactly, for two reasons. First, every nonlinear system behaves differently and thus requires its own analysis. Second, the behavior of a given simulation depends not only on the simulation timestep but also on initial conditions. Thus the true stability region for a given system is really a higher-dimensional object.

The particular system considered in this section is particularly illuminating, however. By changing B, the system damping can be changed, and then for each B, T can also be changed. By choosing initial conditions close to the fixed point at $x = -1$, a stability region can then be plotted in a linearized λT-plane. This is shown in Figure 8.13, with the usual Euler stability region superimposed on top of it. This figure was created by choosing a pixel point (y,z) in the λT-plane and then solving for the corresponding values of B and T in the linearized system model at $x = -1$. For the given λT-plane point, a simulation was run from the fixed initial conditions close to $x = -1$. If the simulation remained finite for, say, 500 iterations, the point was made black. If the simulation output became larger than 100 in magnitude, the point was shaded with respect to the iteration number when blowing up (> 100) was detected. Once this was completed for a given (y,z) point, the process was repeated for every pixel on the screen. It should be observed that the stability region is a fractal object, with fine filaments of stability stretching far into the unstable region. The two small stable islands should be observed at about $\lambda T = -0.5 \pm j2$. It should be further observed that for initial conditions close to the fixed point, the usual linear stability boundary is roughly still intact, with a buffer of chaotic stability of varying widths surrounding it.

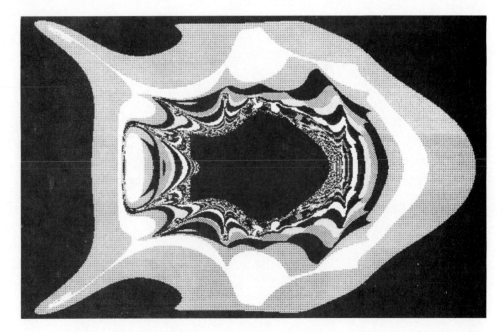

Figure 8.13a Dark central area is stable region in λT-plane of $\ddot{x} + b\dot{x} - x + x^3 = 0$, $x_0 = -.99$, $\dot{x}_0 = 0$ simulated with Euler's method.

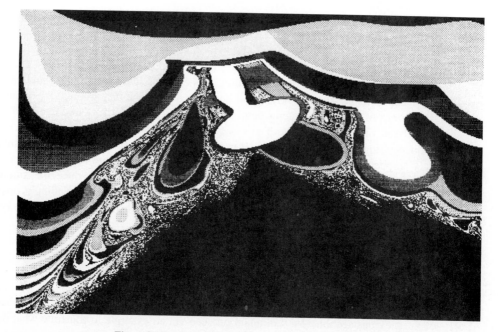

Figure 8.13b   Enlargement of upper center part of Figure 8.13a.

Figure 8.13c   Enlargement of lower right part of Figure 8.13b.

This is a phenomenon that appears to be universal for the simulation of nonlinear systems with any method.

The resulting general stability region for the simulation of nonlinear systems is given in Figure 8.14. Although it is somewhat qualitative in nature, as long as users are careful and recognize what the simulation is doing, it is not difficult to obtain good simulations of nonlinear systems. Alternatively, it has been our experience that around 98 percent of all simulations of nonlinear systems demonstrate chaotic behavior for large enough T, using most fixed timestep methods. The basic manifestation of this behavior is very high-frequency ringing in the simulation. If users see this in a simulation, they should immediately suspect the simulation, and not suggest that this is the true behavior of the given continuous-time system. Finally, when in doubt, users should reduce the simulation timestep by an order of magnitude and repeat the simulation. If the same thing is observed again, perhaps it is the true behavior. Users are reminded that they should not believe everything they see.

8.9 SIMULATION OF SYSTEMS WITH LIMIT CYCLES

The analysis of the simulation of limit cycling systems is even more complicated than that of the previous section. We will again proceed via an example to illustrate what occurs. Consider first the van der Pol system:

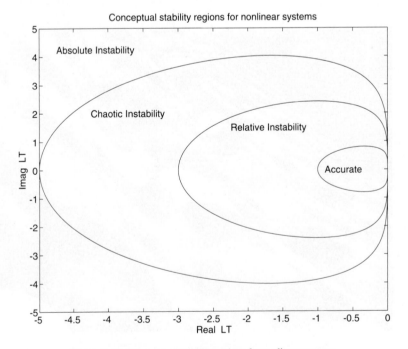

Figure 8.14   Conceptual stability regions for nonlinear systems.

$$\ddot{x} + (x^2 - 1)\,\dot{x} + x = 0 \tag{8-18}$$

A simulation of this equation is shown in Figure 8.3. Figure 8.15 shows the steady-state limit cycle behavior as T increases. Notice that as T is increased, little change in shape occurs before $T = 0.25$. For larger T, some bifurcation occurs, with instability setting in at approximately $T = 0.326$.

Similarly, consider the Rayleigh equation,

$$\ddot{x} + (\dot{x}^2 - 1)\dot{x} + x = 0 \tag{8-19}$$

which is qualitatively similar to the van der Pol system. The only difference is that the derivative is squared in the nonlinear damping rather than the position. Figure 8.16 shows the bifurcation structure of this system as T is increased. Notice that although they are apparently very similar, the bifurcation structure for the Rayleigh system is significantly different from that of the van der Pol system.

What can we learn from the two previous examples? One thing is that it is hard to predict what nonlinear systems are going to do. Another thing is the realization of when the simulation gets a life of its own rather than behaving like the original continuous-time system. Notice that as T is increased, the successive points on the limit cycle get farther and farther apart. When there is a significant visible difference between the values of the simulation output on successive timesteps, the simulation should be suspect. Furthermore, when interesting, visibly appealing patterns begin to emerge, the simulation should again be sus-

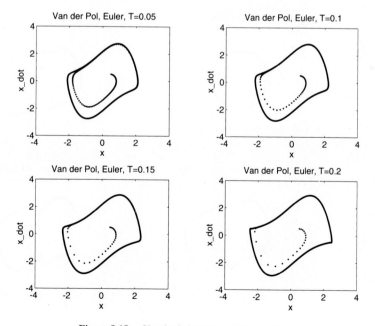

Figure 8.15a   Van der Pol equation, Euler method.

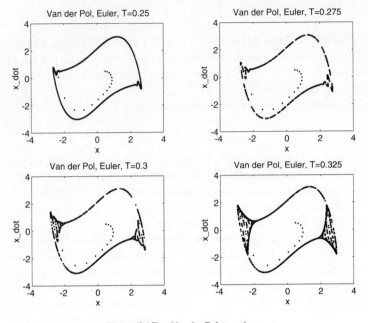

Figure 8.15b   Van der Pol equation.

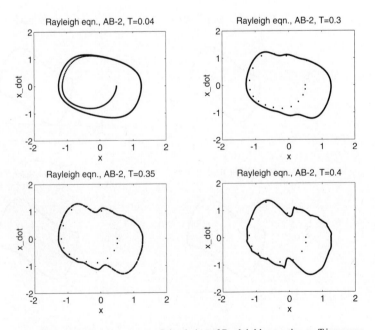

Figure 8.16   Bifurcation structure of simulation of Rayleigh's equation as T increases.

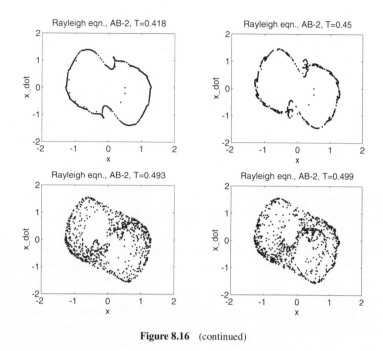

Figure 8.16   (continued)

pect. Also notice that for these systems, the onset of chaotic behavior is manifested in a somewhat lower frequency oscillation than for non–limit cycling systems.

8.10 SIMULATION OF CHAOTIC SYSTEMS

Simulation of chaotic systems is virtually a very difficult problem, fraught with dangerous pitfalls. One of the major problems is regional stiffness; that is, the continuous-time system would behave with very slow dynamics in most places, but when a trajectory approaches an unstable fixed point, one or more of the states can become very fast. Therefore, a variable-timestep Runge-Kutta algorithm is probably the best choice for a simulation. Otherwise, it may be possible to step across separating manifolds in the stiff regions, which can lead to apparent chaotic behavior in systems that may not be chaotic. It is nearly impossible to determine when this happens by observing the simulation output. An example of a system where this may occur is

$$\dddot{x} + \ddot{x} + \dot{x} - a \sin(x) = u \qquad (8\text{--}20)$$

For some a, near the fixed points at $x = n\pi$, trajectories can cross into the basin of $x = (n \pm 1)\pi$. Unfortunately, no one has yet determined the true behavior of this apparently simple system.

Another problem with chaotic systems is that the rule of reducing T and repeating the simulation does not exactly work because chaotic systems are sensitive to initial conditions. Thus if T is reduced, there is basically a different value of x at a given small time t. Not long into the future, therefore, the simulation output will be different from that of any other timestep or any other method. At the same time all of these simulations can be giving outputs that exist on approximately the same attractor as the true system.

Fortunately, many chaotic systems are significantly dissipative everywhere, which implies that simple methods, such as Euler's method, will work adequately. This is discussed further in Section 8.13.

8.11 SIMULATION OF SYSTEMS WITH DISCONTINUOUS NONLINEARITIES

Discontinuous nonlinearities can be a particularly difficult problem. The essence of the problem can be understood by considering Euler's method applied to any system with a discontinuous derivative, as shown in Figure 8.17. The simulation progresses as usual with a fixed timestep until the discontinuity is reached. At that point, the simulation steps over the discontinuity an unknown distance. If the discontinuity is close to the end of the timestep, little error is incurred. If the discontinuity is close to the beginning of the timestep, a large error is incurred. Unfortunately, this distance is unpredictable since the simulation cannot know exactly where the discontinuity is going to lie. Without any a priori consideration by the user, the time response of such a system will usually contain some high-

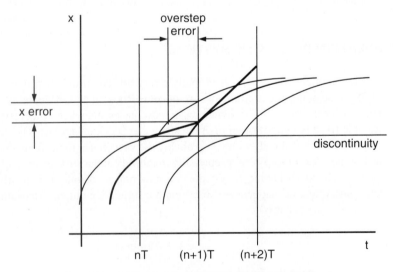

Stepping across a discontinuity with Euler's method.

Figure 8.17 Stepping across a discontinuity with Euler's method.

frequency ringing at the discontinuity, with magnitude proportional to the size of the timestep. As most engineering systems contain considerable dissipation, this ringing will usually decay rapidly, and the simulated trajectory will quickly approach the true trajectory. In systems containing little damping, however, this problem is magnified since the ringing in the simulation can be prolonged. Several specialized methods are available in the literature to deal with this problem; however, it is not too difficult to control the size of the ringing by using standard methods intelligently.

To reduce the size of the error incurred at a discontinuity, the simplest method is to use a smaller simulation timestep. Although this will increase the computation time, it is a very easy way to reduce this ringing. The next easiest way is to use a variable timestep Runge-Kutta method. In this case, the method monitors the size of the error. When the error becomes large, such as at a discontinuity, the simulation timestep will be reduced to an acceptable level. Following this, the timestep will slowly increase until it reaches the previous level. Clearly, variable timestep Runge-Kutta methods can be customized to efficiently reduce their timesteps at the discontinuity and then quickly enlarge them again. This exercise is left to the reader. The ringing at the discontinuity is particularly large for linear multistep methods because they use many previous values to compute the next output. Once the method steps across the discontinuity, it temporarily uses slopes from both sides. This effectively causes some confusion in the method, which is again manifested in the large ringing.

For example, consider

$$\dot{x} = x - sgn(x) \tag{8–21}$$

which is effectively given by the block diagram of Figure 8.18. The steady-state value of this system is easily found to be zero, when the magnitude of the initial condition is less than 1, by using the graphical methods of Section 8.2. This is simulated by using Euler's method with the timestep varying from $T = 0$ to $T = 1$. The resulting bifurcation diagram for the steady states is given in Figure 8.19. Notice that the output has only the steady-state value of zero when $T = 0$, which would take a very long time to simulate. For nonzero T, the bifurcation diagram indicates some sort of chaotic behavior, with instability occurring for $T > 1$. Interestingly enough, as T goes to zero, the bifurcation diagram takes on an infinite number of values, separated by small regions, which become infinitely close together. Hence, this system can never accurately be simulated by Euler's method or any other fixed timestep method. As a practical note, this is a good system for testing specific simulation packages for the handling of discontinuities.

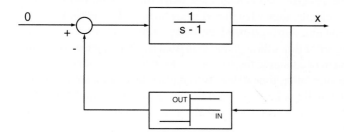

Figure 8.18   Example system containing a discontinuity.

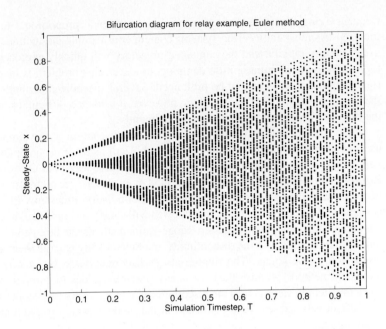

Figure 8.19 Bifurcation diagram for relay example with Euler method.

8.12 SIMULATION OF TIME-DELAY SYSTEMS

Linear or nonlinear systems containing a time delay are also very common. Their simulation, however, is not too difficult, although maybe more time consuming. Practically, time delays are usually given with only three or four-digit accuracy. To represent accurately a system containing time delays, we must first pick a simulation timestep that is a submultiple of all the time delays in the system. This sometimes requires an unnecessarily small timestep. In these situations, it is convenient, and not particularly inaccurate, to round the given time delays to a value that will allow a larger simulation timestep to be used. In any case, once T is chosen, the time delays can be simulated in two ways. One is actually to write out a discrete-time state space representation for all the delays and program it. Unfortunately, these representations can become very large, sometimes containing hundreds of states. Then, the alternative is to insert a for-next loop into the simulation for each time delay. Every time through the main simulation loop, these loops for the delays will transfer the data one step further back in time, somewhat like a bucket brigade. The reader is further cautioned that if the simulation timestep is changed, the delay loops are easy to lengthen or shorten by changing the number of times through the loop (that is, changing the size of the discrete-time equivalent system or changing the size of the discrete state vector).

Another alternative to simulating time delays is to replace them with some lumped approximation. This approach will not be considered further, as it is more in the realm of modeling. Furthermore, the discrete delays discussed above are almost always a more accurate approach.

As an example of the simulation of a time-delay system, consider a continuized version of the logistic equation (Hartley, Killory, De Abreu-Garcia, & AbuKhamseh 1990),

$$\varepsilon \dot{x}(t) = -x(t) + \alpha x(t-1)[1 - x(t-1)] \qquad (8\text{--}22)$$

where α is the usual logistic equation parameter, ε is some small number representing a singular perturbation, and the time delay is 1. This equation can also be represented in block diagram form, as in Figure 8.20. A BASIC program for simulating this system can be written, with a loop representing the time delay. If a simulation timestep of $T = 0.01$ is chosen, then there are 100 discrete-time states associated with this delay. Rather than build a large state matrix, these states are computed in the delay loop.

8.13 DIVERGENCE PRESERVING METHODS

One of the properties of limit cycling and chaotic systems is that they have negative divergence. This means that initial selected volumes in phase space, or state space, will contract as time evolves. Chaotic systems are special, in that there is a particular direction on the attractor where states are diverging. Blowing up is avoided, however, since the other directions in phase space are converging faster, and the global state space is distorted enough to ensure that the diverging directions are folded back onto the attractor. Thus, since all stable (not blowing up) dissipative systems have negative divergence, it is reasonable to require simulations of these systems to preserve divergence. This section presents some background for designing methods of this type.

Given a nonlinear, autonomous, continuous-time, n-vector system,

$$\dot{x} = f(x) \qquad (8\text{--}23)$$

its divergence is defined as

$$\text{Div}(f) = \nabla \cdot f = \frac{\partial f_1}{\partial x_1} + \frac{\partial f_2}{\partial x_2} + \dots \frac{\partial f_n}{\partial x_n} = \sum_{i=1}^{n} d_i \qquad (8\text{--}24)$$

Here each of these terms is similar to poles, characteristic exponents, or Lyapunov exponents (Thompson & Stewart, 1986). If this system were ideally discretized, its discrete divergence would be

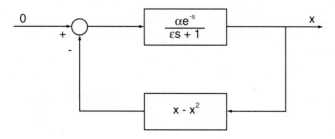

Figure 8.20 Example of a time-delay system, continuized logistic equation.

$$e^{(\nabla \cdot f)T} = \text{Exp}\left[T\frac{\partial f_1}{\partial x_1}\right]\text{Exp}\left[T\frac{\partial f_2}{\partial x_2}\right]\ldots\text{Exp}\left[T\frac{\partial f_n}{\partial x_n}\right] \qquad (8\text{--}25)$$

where now each of these terms is similar to discrete-time poles or characteristic multipliers (Thompson & Stewart). Thus any discrete-time approximation should preserve this localized discrete divergence through the computation of

$$\frac{\partial F_1}{\partial x_1} \cdot \frac{\partial F_2}{\partial x_2} \ldots \frac{\partial F_n}{\partial x_n} = m_1 m_2 \ldots m_n \qquad (8\text{--}26)$$

where the F's are the components of the resulting discretized state vector. It should be noted that this process would be useful in a variable timestep method for error control. This is clarified through an example.

Example 8.1

Here we consider divergence preservation for simulating the van der Pol oscillator:

$$\dot{x}_1 = x_2$$
$$\dot{x}_2 = -x_1 - (x_1^2 - 1)x_2$$

Its divergence is easily calculated to be

$$\text{div}(f) = d_1 + d_2 = 0 + (1 - x_1^2)$$

Discretizing this gives

$$e^{\text{div}(f)T} = e^{0T}\,e^{(1-x_1^2)T} = e^{(1-x_1^2)T}$$

Now, simulating the van der Pol system with Euler's method, we get the discrete-time system

$$x_{1,n+1} = x_{1,n} + Tx_{2,n}$$
$$x_{2,n+1} = x_{2,n} + T[-x_{1,n} - (x_{1,n}^2 - 1)x_2]$$

The local divergence of this is then

$$m_1 m_2 = (1)(1 + T(1 - x_1^2))$$

Comparison with the ideal gives

$$1 + T(1 - x_1^2) \approx e^{(1-x_1^2)T}$$

To determine the simulation timestep for a fixed timestep method, first decide on the range of possible values of x_1. Then determine an allowable error in the divergence. It is then possible to determine an acceptable T by plotting the error versus T. For this problem, x_1 is about unity on the limit cycle, but this is of no help since it gives zero for the error directly. Then $x_1 = 0$ should be considered since it apparently gives larger values of divergence. Then the absolute error in divergence is

$$|e^T - 1 - T|$$

T can be chosen here to make the error as small as possible. Choosing $T = 0.1$ gives an error of 0.00517. For variable timesteps, the original error expression can be monitored as x_1 changes. Then the local divergence can be controlled by considering this error and changing T in the same manner discussed in the variable timestep Runge-Kutta section in Chapter 6. The advantage here is that the variable timestep can be used directly with Euler's method, or any other desired method, without the requirement of a method-based error estimator. Here the error estimate comes from the local divergence error. Obviously, we cannot get something for nothing.

8.14 CONCLUSIONS

The purpose of this chapter is to provide some background information into the possible dynamics of nonlinear systems. The approach has been somewhat qualitative and intuitive, rather than highly mathematical, to give insight into the phenomena. Clearly, one chapter in any book cannot provide even a review of the vast area of nonlinear dynamics. It is hoped that this chapter is an introduction into this interesting and exciting area.

PROBLEMS

1. Simulate the system $x' = \tan(x)$. Discuss.
2. Simulate $x' = \sin(1/x)$, $x_0 = 0.00001$. Discuss.
3. Simulate $x' = x^3 + u$, with $u = 0.5$ for $t \geq 0$. Use Euler's method with a timestep of 1.55. Explain what you observe.

$$\boxed{9}$$

Multiple Integration

9.1 INTRODUCTION

In this chapter, we present methods for performing multiple integration. The need for double integration follows directly from some of the original motivating problems in classical physics. When we consider these mostly conservative systems, it becomes clear that methods designed for second-order systems containing no damping are necessary, and multiple integration is a natural extension. With this extension, it becomes possible to express, and obtain, all of the substitution methods from Chapter 3 in one compact formula. Using this formula, we can derive new and more accurate substitution methods.

Note that it has been necessary to wait until this point to discuss the multiple integration process applied to substitution methods so that the reader may easily grasp the results. This ability should now be possible, following the extensive discussion of linear multistep methods. Had these results been presented in Chapter 3, they would not have been so clear.

9.2 MOTIVATING PROBLEMS FOR DOUBLE INTEGRATION

Double integration is basically concerned with developing discrete approximations for the double integral, or $1/s^2$. As discussed in Chapter 3 dealing with operational substitution, one should be able to obtain better approximations to the double integral directly, rather than obtaining approximations to a single integral and squaring. It will be shown here that this is indeed the case.

The double integration problem was originally motivated by astronomers who were dealing with bodies moving in an inverse square field. Consider the system of Figure 9.1. This is basically the problem of a small mass, such as a planet, orbiting a much larger mass, such as the sun. In rectangular coordinates, the dynamics of this configuration can be found as

$$\ddot{x} = -\frac{MGx}{(x^2 + y^2)^{3/2}} \text{ and } \ddot{y} = -\frac{MGy}{(x^2 + y^2)^{3/2}} \tag{9-1}$$

Basically, this system has the form

$$\ddot{x} = f(x) \tag{9-2}$$

where f and x can be vectors. This form is typical of many astronomical and ballistic problems. It is reminiscent of the standard harmonic oscillator without damping, which is studied in introductory engineering courses:

$$\ddot{x} = -x + u \tag{9-3}$$

Even if $f(x)$ is nonlinear, no first derivatives appear explicitly, and the system is necessarily conservative, that is, marginally stable. This is easily seen in the undamped pendulum equation

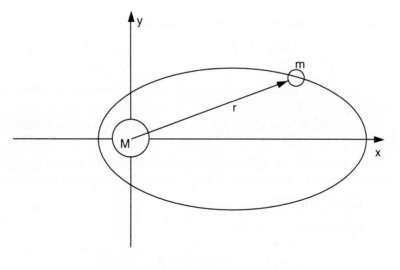

Astronomical central potential problem.

Figure 9.1   Astronomical central potential problem.

$$\ddot{x} = -\sin(x) + u \tag{9-4}$$

This form is essentially a state space representation; however, the term on the left is a second derivative. It could be put into the usual state space form containing first derivatives on the left:

$$\begin{aligned} \dot{x}_1 &= x_2 \\ \dot{x}_2 &= -\sin(x_1) + u \end{aligned} \tag{9-5}$$

This particular type of system, however, is very difficult to simulate with the single integration linear multistep methods or Runge-Kutta methods because it has zero damping, or zero divergence. Most of the methods presented in chapters 4, 5, and 6 will add a small amount of positive or negative damping through their local truncation error. Equivalently, their stability regions do not lie exactly on the imaginary λT-axis, and thus systems with purely imaginary poles will have their responses either converge to a point or diverge to infinity, both in an oscillatory fashion. Consequently, it is of considerable importance to develop methods that perform double integration directly and avoid such problems.

9.3 GENERAL LINEAR MULTISTEP DOUBLE INTEGRATOR

The linear k-step method for approximating the double integral can be found as (Lambert, 1973)

$$\sum_{j=0}^{k} \alpha_j x_{n+j} = T^2 \sum_{j=0}^{k} \beta_j \ddot{x}_{n+j}, \quad \text{again with } \alpha_k \equiv 1 \tag{9-6}$$

Again we can z-transform this equation and obtain

$$\frac{X(z)}{\ddot{X}(z)} = \frac{T^2(\beta_k z^k + \beta_{k-1} z^{k-1} + \ldots + \beta_1 z + \beta_0)}{z^k + \alpha_{k-1} z^{k-1} + \ldots + \alpha_1 z + \alpha_0} = \frac{T^2 \sigma(z)}{\rho(z)} \tag{9-7}$$

As with the single integral, the local truncation error can be defined as

$$\text{LTE} = \sum_{j=0}^{k} [\alpha_j x_{n+j} - T^2 \beta_j \ddot{x}_{n+j}] = C_0 x_n + C_1 T x_n^{(1)} + \ldots + C_q T^q x_n^{(q)} + \ldots \tag{9-8}$$

The C-coefficients are again related to coefficients of the method through term-by-term comparisons using Taylor series. These equations for the double integral are

$$C_0 = \alpha_0 + \alpha_1 + \ldots + \alpha_k$$

$$C_1 = \alpha_1 + 2\alpha_2 + \ldots + k\alpha_k$$

$$C_2 = \frac{1}{2!}[\alpha_1 + 4\alpha_2 + \ldots + k^2\alpha_k] - [\beta_0 + \beta_1 + \ldots + \beta_k] \tag{9-9}$$

$$C_q = \frac{1}{q!}[\alpha_1 + 2^q\alpha_2 + \ldots + k^q\alpha_k] - \frac{1}{(q-2)!}[\beta_1 + 2^{q-2}\beta_2 + \ldots + k^{q-2}\beta_k]$$

The method is said to have order p if

$$C_0 = C_1 = C_2 = \ldots = C_p = C_{p+1} = 0, \; C_{p+2} \neq 0 \qquad (9\text{-}10)$$

The method is said to be zero stable if no roots of $\rho(z)$ are larger than unity in magnitude and those with unity magnitude have multiplicity less than or equal to 2. Note here that we can now have two open-loop poles on the unit circle since this is a double integrator. The method is said to be consistent if it has order greater than or equal to one, $p \geq 1$. This implies that $C_0 = C_1 = C_2 = 0$ for the method to behave like a double integrator. The method is said to converge at a point if it is both consistent and zero stable. This can be shown in a manner similar to that for the single integrator. It is interesting to note that the first two of these equations guarantee two poles at $z = +1$. The third equation guarantees elimination of steady-state errors.

It is now instructive to derive the most accurate two-step explicit method. Two steps and explicit yield four unknowns, which then require four equations. From above, these are

$$C_0 = \alpha_0 + \alpha_1 + 1 = 0$$

$$C_1 = \alpha_1 + 2 = 0$$

$$C_2 = \frac{1}{2}[\alpha_1 + 4] - \beta_0 - \beta_1 = 0$$

$$C_3 = \frac{1}{6}[\alpha_1 + 8] - \beta_1 = 0$$

Solving these, we get $\alpha_1 = -2$, $\alpha_0 = 1$, $\beta_1 = 1$, and $\beta_0 = 0$. Thus the transfer function for the method is

$$H(z) = \frac{X(z)}{\ddot{X}(z)} = \frac{T^2 z}{z^2 - 2z + 1} = \frac{T^2 z}{(z-1)^2} \qquad (9\text{-}11)$$

This can also be recognized as the Halijak and z-transform approximation to $1/s^2$.

Example 9.1

To illustrate the use of double integrators, we now present a simulation of the pendulum equation. We use the Halijak double integrator because it is probably the simplest. A basic program for this simulation is given below. A simulation output is presented in Chapter 8 for this system.

```
T = 0.01
xss = 0.1                              Initial position (rads)
xd = 0                                 Initial velocity
xdd = -sin(x)                          Initial acceleration
xs = xss + T*xd + (1/2)*T*T*xdd        Position after one timestep
for n = 2 to 1000000                   Main loop
        fx = -sin(x)                   Acceleration
        x = 2*xs - xss + T*T*fx        Integration
        xss = xs                       Regress xs
        xs = x                         Regress x
        xd = (xs - xss)/T              Estimate velocity
```

```
                    pset(100 + 50*x,100 - 50*xd)        Plot phase plane
          next n
          end
```

9.4 WEAK STABILITY FOR DOUBLE INTEGRATION

To study the closed loop properties of these methods, it is first necessary to define a linear test equation. A convenient choice is

$$\ddot{x} = \omega^2 x + u \tag{9–12}$$

This continuous system will first be studied. Its characteristic equation is

$$\Delta(s) = s^2 - \omega^2 \tag{9–13}$$

If $\omega^2 > 0$ this system will have a pole $\pm\omega$ and will represent an unstable system. If $\omega^2 < 0$, the system will have a pole at $\pm j\omega$ and will represent a marginally stable conservative system. Clearly, most of the interest will be in the marginally stable systems.

Using this test equation to close the loop around the integrator, we get the following closed-loop transfer function:

$$G(z) = \frac{X(z)}{U(z)} = \frac{T^2 \sigma(z)}{\rho(z) - (\omega T)^2 \sigma(z)} \tag{9–14}$$

To understand the system behavior, consider the Halijak method derived earlier. Then

$$G(z) = \frac{T^2 z}{(z-1)^2 - (\omega T)^2 z} = \frac{T^2 z}{z^2 - [2 + (\omega T)^2]z + 1} \tag{9–15}$$

We will now consider the closed-loop characteristic equation. Clearly it is desirable to use a root locus analysis, where here $(\omega T)^2$ acts like gain and should be negative for a conservative system. The root locus is obtained in Figure 9.2. It should be observed that the root locus goes around the z-plane unit circle and corresponds to a marginally stable system, as it should. When $(\omega T)^2 = -4$, the root trajectories come together at $z = -1$ just before one goes to minus infinity. This is the maximum value of $(\omega T)^2$ for stability. It should be remembered that this property of the root locus going around the unit circle should be maintained for double integrators.

9.5 STABILITY REGIONS FOR DOUBLE INTEGRATORS

The major tool for studying single integrators is the stability region. This can also be done here, but it turns out not to be quite so interesting. Forming the stability region for the Halijak double integrator by using

$$(\omega T)^2 = \frac{\rho(z)}{\sigma(z)} \text{ with } z = e^{j\theta} \text{ and } \theta = 0 \to 2\pi \tag{9–16}$$

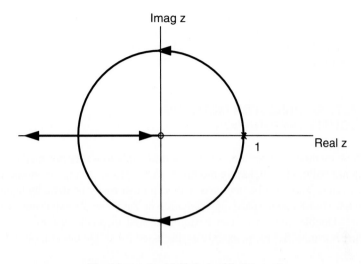

Root locus for Halijak double integrator

Figure 9.2   Root locus for Halijak double integrator.

gives the stability region of Figure 9.3. Notice that the boundary locus goes directly out of the negative real axis of the $(\omega T)^2$-plane. It then reaches -4 and turns around to return to zero. This implies that only negative real values of $(\omega T)^2$ are possible for stability of the simulation. This is reasonable since complex values of $(\omega T)^2$ would imply complex frequencies squared, which does not make any apparent physical sense. Alternatively, if the

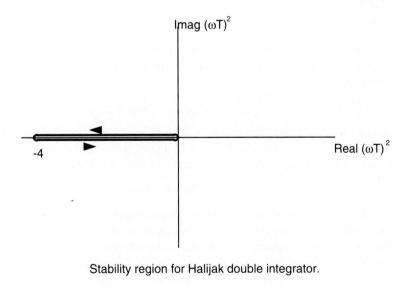

Stability region for Halijak double integrator.

Figure 9.3   Stability region for Halijak double integrator.

stability region leaves the negative real $(\omega T)^2$-axis and encircles it in a counterclockwise direction, the implication is that the root locus would move inside the unit circle and yield a damped response for those values of $(\omega T)^2$. Conversely, a clockwise encirclement would imply instability.

9.6 PREDICTOR-CORRECTOR METHODS FOR DOUBLE INTEGRATION

These methods can be directly extended to predictor-corrector methods and used in any of the normal modes, including modifiers. Also, practical implementation requires that starting values must still be obtained. This can most easily be done by using single integrators on a first-order state space representation or Taylor series expansions for the first timestep.

Double integrators can be implemented in predictor-corrector form, or predictor-corrector-modifier form, exactly as discussed for single integration in Chapter 4.

9.7 TRIPLE INTEGRATION

We do not derive triple integrators explicitly here; however, we consider the behavior of continuous-time third-order systems of the form

$$\dddot{x} = f(x) \tag{9–17}$$

Assume, for the purpose of this discussion, that $f(x)$ is in the form of a linear test function, giving

$$\dddot{x} = \lambda x + u \tag{9–18}$$

The question to be addressed here is this: How does this system behave as λ is changed. This is most easily addressed through a root locus analysis. For $\lambda > 0$, the characteristic equation is

$$\Delta(s) = s^3 - |\lambda| = 0 \tag{9–19}$$

and root locus of Figure 9.4 results. Remember that s is a complex variable and that the roots of this equation are complex, not all at $\lambda^{1/3}$. Notice that there is always an unstable branch on the positive real s-axis. Alternatively, for $\lambda < 0$, the characteristic equation is

$$\Delta(s) = s^3 + |\lambda| = 0 \tag{9–20}$$

and the root locus of Figure 9.5 results. Notice that there are always two branches of the root locus in the right-half plane. It is thus observed that no linear system of order higher than second, and containing no intermediate derivatives between the given order and zero, can be stable. This follows from the fact that all of the root loci for these higher-order (say, mth) systems will have two-root locus branches at an angle from the positive real s-axis of π/m, or 0. Thus $m = 2$ is the largest value to keep the angle greater than, or equal to, $\pi/2$.

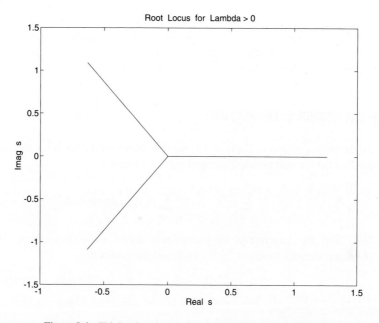

Figure 9.4   Third-order system with no damping terms, lambda > 0.

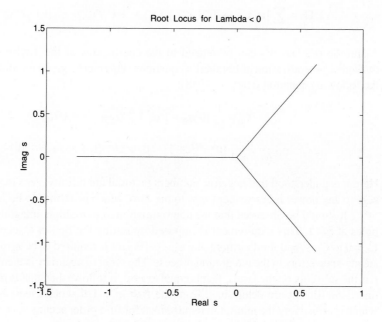

Figure 9.5   Third-order system with no damping terms, lambda < 0.

Since multiple integration is thus limited to second-order systems for stability, no stability analysis is presented for specific multiple integrators here. The next section contains a generalization that will allow us to return to substitution methods possessing greater sophistication than those we visited before.

9.8 HIGHER ORDER INTEGRATION

From the discussion in Section 9.3, we can infer that the form for the general linear k-step method for the mth multiple integral can be written as

$$\sum_{j=0}^{k} \alpha_j x_{n+j} = T^m \sum_{j=0}^{k} \beta_j x_{n+j}^{(m)}, \text{ again with } \alpha_k \equiv 1 \qquad (9\text{--}21)$$

Note that the superscript in parentheses refers to m-derivatives. The corresponding z-domain transfer function for the multiple integral is then

$$\frac{X(z)}{X^{(m)}(z)} = \frac{T^m(\beta_k z^k + \beta_{k-1} z^{k-1} + \ldots + \beta_1 z + \beta_0)}{z^k + \alpha_{k-1} z^{k-1} + \ldots + \alpha_1 z + \alpha_0} = \frac{T^m \sigma(z)}{\rho(z)} \qquad (9\text{--}22)$$

If $\beta_k = 0$ the method is explicit; otherwise the method is implicit. Also, if $k > m$, spurious discrete-time poles exist. Clearly this method will be associated with a local truncation error, defined as usual

$$\text{LTE} = \sum_{j=0}^{k} [\alpha_j x_{n+j} - T^m \beta_j x_{n+j}^{(m)}] = C_0 x_n + C_1 T x_n^{(1)} + \ldots + C_q T^q x_n^{(q)} + \ldots \qquad (9\text{--}23)$$

Again the α's and β's can be related to the coefficients of the Taylor series expansion through a generalization of Lambert's equations. This can be given for all k and for all m by the following equation (Hartley, 1988):

$$C_q = \frac{1}{q!} [0^q \alpha_0 + 1^q \alpha_1 + 2^q \alpha_2 + \ldots + k^q \alpha_k]$$

$$- \frac{1}{(q-m)!} [0^{q-m} \beta_0 + 1^{q-m} \beta_1 + 2^{q-m} \beta_2 + \ldots + k^{q-m} \beta_k] \qquad (9\text{--}24)$$

Here it is understood that negative numbers factorial are infinity; zero factorial is one; and zero to any power is zero except zero to the zero, which is defined to be 1.

It should be observed that the approximation to m-multiple integrals should have m-poles at $z = 1$. This requirement is imposed by setting the first m C_q-coefficients to zero: $C_0 = 0$ to $C_{m-1} = 0$. Furthermore, the C_m-coefficient is required to be zero to eliminate any steady-state errors in the integration process. The order of accuracy is then $p - m + 1$, where p is the subscript on the last C_q that is equal to zero. It follows then that m α-coefficients and one β-coefficient are determined from the first $m + 1$ C_q-coefficients being set to zero, which is effectively the minimum requirement for first-order accuracy, or consistency. Furthermore, given k, the maximum number of C_q's that can be set to zero is $2k + 1$, as this is the number of unknowns. This gives the maximum possible order of accuracy for a given k as $2k - m$. Clearly, a wide variety of methods are possible, containing any number of spurious roots.

9.9 SUBSTITUTION METHODS REVISITED

Using the general formula of the last section, we can now return to the substitution methods of Chapter 3 and get some sense of their accuracy. It is important to remember that the formula of the last section only determines the order of accuracy of a given multiple integrator and says nothing of how the particular substitution operator family will work in conjunction with one another for a particular system.

Returning to Table 3.1, we see that all of these methods have been substituted into the general formula of the last section. The resulting orders of accuracy for the particular integrators are given in Table 9.1. There it can be seen that the Madwed family is effectively no more accurate than the Tustin family. Additionally, we also know that the Tustin family simply iterates the trapezoidal rule, and thus is always stable for a stable system. This is not the case for the Madwed family. We would then conclude that the Tustin family should usually be used over the Madwed family. The order of accuracy for the Halijak family does not follow any particular pattern but should probably be considered superior to both the Tustin and Madwed families with respect to accuracy. The explicit z-transform family is the most accurate explicit method possible, as a simple analysis of the C_q-coefficient equations shows. The Boxer-Thaler z-forms are the most accurate family possible and can be derived from the C_q-coefficient equations as the most accurate explicit family. The reader should be

TABLE 9.1   ACCURACY, ORDER, AND PRINCIPAL LOCAL TRUNCATION ERROR (LTE) FOR COMMON FAMILIES OF OPERATIONAL SUBSTITUTION METHODS

Method	$\dfrac{1}{s}$	$\dfrac{1}{s^2}$	$\dfrac{1}{s^3}$	$\dfrac{1}{s^4}$
Euler	order = 1 LTE = $1/2T^2x^{(2)}$	order = 1 LTE = $T^3x^{(3)}$	order = 1 LTE = $3/2T^4x^{(4)}$	order = 1 LTE = $2T^5x^{(5)}$
First difference	order = 1 LTE = $-1/2T^2x^{(2)}$	order = 1 LTE = $-T^3x^{(3)}$	order = 1 LTE = $-3/2T^4x^{(4)}$	order = 1 LTE = $-2T^5x^{(5)}$
Tustin	order = 2 LTE = $-1/12T^3x^{(3)}$	order = 2 LTE = $-1/6T^4x^{(4)}$	order = 2 LTE = $-1/4T^5x^{(5)}$	order = 2 LTE = $-1/3T^6x^{(6)}$
Madwed	order = 2 LTE = $-1/12T^3x^{(3)}$	order = 2 LTE = $-1/12T^4x^{(4)}$	order = 2 LTE = $-1/12T^5x^{(5)}$	order = 2 LTE = $-1/12T^6x^{(6)}$
Halijak	order = 1 LTE = $-1/2T^2x^{(2)}$	order = 2 LTE = $1/12T^4x^{(4)}$	order = 4 LTE = $1/240T^7x^{(7)}$	order = 2 LTE = $1/12T^6x^{(6)}$
Explicit z-transform	order = 1 LTE = $-1/2T^2x^{(2)}$	order = 2 LTE = $1/12T^4x^{(4)}$	order = 4 LTE = $1/240T^7x^{(7)}$	order = 4 LTE = $-1/720T^8x^{(8)}$
Implicit z-transform	order = 1 LTE = $-1/2T^2x^{(2)}$	order = 2 LTE = $1/12T^4x^{(4)}$	order = 4 LTE = $1/240T^7x^{(7)}$	order = 4 LTE = $-1/720T^8x^{(8)}$
Boxer-Thaler	order = 2 LTE = $-1/2T^3x^{(3)}$	order = 4 LTE = $-1/240T^5x^{(5)}$	order = 4 LTE = $1/240T^7x^{(7)}$	order = 6 LTE = $1/3024T^{10}x^{(10)}$

cautioned, again, that this accuracy is based only on consideration of the open loop local truncation error, and little can be said concerning the accuracy of a given closed-loop situation.

Now that the familiar substitution methods of Chapter 3 have been considered, we wonder if there may be any other useful families of methods. Clearly many ideas are possible with various pole-zero or coefficient constraints, but only one idea is considered here to yield an explicit and implicit family. The new methods presented, the Filicky-Hartley families, are based on improving the accuracy of the lower-order substitution methods. It was shown in Chapter 3 that the major source of error in a closed-loop simulation using substitution methods lay in the approximation to the lower-order terms, in particular, $1/s$. Thus, by using the general formula from the previous section, the order of the lower integrators can be raised to that of the highest integrator, simply by adding some extra steps, that is, spurious poles. Unfortunately, in doing so, maximum accuracy can not be obtained because of the requirement for zero stability. Thus the Filicky-Hartley families provide, for a given mth-order continuous-time system, the most accurate explicit methods with all of the spurious open-loop poles constrained to be at the origin (Table 9.2) and the most accurate implicit methods with all of the spurious poles constrained to be at the origin (Table 9.3). Notice in these tables that there is necessarily a different family for each given m. These methods have been applied to several systems in Filicky (1990) and have been shown to be capable of providing superior results to the usual substitution methods when applied to systems. Again, this improvement is obtained by reducing the errors associated with the least accurate terms, thereby increasing the order of the resulting discrete-time system from m to something less than that obtained by a normal linear multistep method of Chapter 4.

9.10 CONCLUSIONS

This chapter has presented methods for performing multiple integration. It has been shown that there is one simple compact formula relating the method coefficients to the local truncation error for any number of multiple integrals. With this formula it is possible to reconsider the classical substitution methods and provide families of new substitution methods that improve on these earlier results.

PROBLEMS

1. For the system $x^{(3)} = \lambda x + u$, determine the simplest explicit linear multistep method (for $1/s^3$) that is most accurate with respect to local truncation error. Compare the weak stability of this method with the actual system stability. How does this compare with Halijak's method?

2. The following system represents a small object that is gravitationally attracted to a much larger body (i.e., this is the Kepler problem) in which the mass of the smaller object has been neglected: $x'' = -x/(x^2 + y^2)^{3/2}$, $y'' = -y/(x^2 + y^2)^{3/2}$, $x(0) = 2$, $x'(0) = 0$, $y(0) = 0$, $y'(0) = -0.4$. Simulate this system for several orbits, using the method of Halijak. Be careful how you start the simulation.

TABLE 9.2 IMPLICIT OPERATIONAL METHODS

	One-Step Method	Two-Step Method	Three-Step Method	Four-Step Method	Five-Step Method
$\dfrac{1}{s}$	$\dfrac{T(z+1)}{2(Z-1)}$ order = 2 LTE ≈ $\frac{1}{12}T^3 y^{(3)}$	$\dfrac{T(5z^2+8z-1)}{12(z-1)}$ order = 3 LTE ≈ $\frac{1}{24}T^4 y^{(4)}$	$\dfrac{T(9z^3+19z^2-5z+1)}{24z^2(z-1)}$ order = 4 LTE ≈ $\frac{19}{720}T^5 y^{(5)}$	$\dfrac{T(251z^4+646z^3-264z^2+106z-19)}{720z^3(z-1)}$ order = 5 LTE ≈ $\frac{27}{1440}T^6 y^{(6)}$	$\dfrac{T(475z^5+1427z^4-798z^3+482z^2-173z+27)}{1440z^4(z-1)}$ order = 6 LTE ≈ $\frac{863}{60480}T^7 y^{(7)}$
$\dfrac{1}{s^2}$		$\dfrac{T^2(z^2+10z+1)}{12(z-1)^2}$ order = 3 LTE ≈ $\frac{1}{12}T^5 y^{(5)}$	$\dfrac{T^2(z^2+10z+1)}{12(z^2-2z+1)}$ order = 4 LTE ≈ $\frac{1}{240}T^6 y^{(6)}$	$\dfrac{T^2(147z^4+192z^3+762z^2-528z+147)}{720z^2(z^2-2z+1)}$ order = 5 LTE ≈ $\frac{3}{20}T^7 y^{(7)}$	$\dfrac{T^2(18z^5+209z^4+4z^3+14z^2-6z+1)}{240z^3(z-1)^2}$ order = 6 LTE ≈ $\frac{19.87}{720}T^8 y^{(8)}$
$\dfrac{1}{s^3}$			$\dfrac{T^3 z(z+1)}{2(z-1)^3}$ order = 4 LTE ≈ $\frac{1}{240}T^7 y^{(7)}$	$\dfrac{T^3(z^4+116z^3+126z^2-4z+1)}{240(z^4-3z^3+3z^2-z)}$ order = 5 LTE ≈ $\frac{7}{360}T^8 y^{(8)}$	$\dfrac{T^3(.00625z^5+.4729z^4+.54583z^3-.03749z^2+.01458z-.002089)}{z^2(z^3-3z^2+3z-1)}$ order = 6 LTE ≈ $\frac{17}{12096}T^9 y^{(9)}$
$\dfrac{1}{s^4}$				$\dfrac{T^4(-z^4+124z^3+474z^2+124z-1)}{720(z^4-4z^3+6z^2-4z+1)}$ order = 5 LTE ≈ $\frac{185}{432}T^9 y^{(9)}$	$\dfrac{T^4(-.00225z^5+.17697z^4+.6479z^3+.18365z^2-.007648z+.001374)}{z^5-4z^4+6z^3-4z^2+z}$ order = 6 LTE ≈ $\frac{631}{403200}T^{10} y^{(10)}$
$\dfrac{1}{s^5}$					$\dfrac{T^5 z(z^3+11z^2+11z+1)}{(z-1)^5}$ order = 6, LTE ≈ $\frac{1}{6048}T^{11} y^{(11)}$

TABLE 9.3 EXPLICIT OPERATIONAL METHODS

	One-Step Method	Two-Step Method	Three-Step Method	Four-Step Method	Five-Step Method
$\dfrac{1}{s}$	$\dfrac{T}{(Z-1)}$ order = 1 LTE $\approx \frac{1}{2}T^2 y^{(2)}$	$\dfrac{T(3z-1)}{z(2z-2)}$ order = 2 LTE $= \frac{5}{12}T^3 y^{(3)}$	$\dfrac{T(23z^2-16z+5)}{12z^2(z-1)}$ order = 3 LTE $= \frac{9}{24}T^4 y^{(4)}$	$\dfrac{T(55z^3-59z^2+37z-9)}{24z^3(z-1)}$ order = 4 LTE $\approx \frac{251}{720}T^5 y^{(5)}$	$\dfrac{T(1901z^4-2774z^3+2616z^2-1274z+251)}{720z^4(z-1)}$ order = 5 LTE $\approx \frac{475}{1440}T^6 y^{(6)}$
$\dfrac{1}{s^2}$		$\dfrac{T^2 z}{(z-1)^2}$ order = 2 LTE $= \frac{1}{12}T^4 y^{(4)}$	$\dfrac{T^2(13z^2-2z+1)}{12z^2(z^2-2z+1)}$ order = 3 LTE $= \frac{1}{12}T^5 y^{(5)}$	$\dfrac{T^2(14z^3-5z^2+4z-1)}{12z^2(z^2-2z+1)}$ order = 4 LTE $= \frac{53}{720}T^6 y^{(6)}$	$\dfrac{T^2(1.246z^4-.7331z^3+.80797z^2-.3997z+.0791)}{z^3(z-1)^2}$ order = 5 LTE $= \frac{9}{504}T^7 y^{(7)}$
$\dfrac{1}{s^3}$			$\dfrac{T^3 z(z+1)}{2(z-1)^3}$ order = 4 LTE $\approx \frac{1}{240}T^7 y^{(7)}$	$\dfrac{T^3(z^2+z)}{2(z-1)^3}$ order = 4 LTE $\approx \frac{1}{240}T^8 y^{(8)}$	$\dfrac{T^3(.5042z^4+.48331z^3+.02504z^2-.01667z+.004176)}{z^2(z^3-3z^2+3z-1)}$ order = 5 LTE $\approx \frac{242}{403202}T^8 y^{(8)}$
$\dfrac{1}{s^4}$				$\dfrac{T^4 z(z^2+4z+1)}{6(z-1)^4}$ order = 4 LTE $\approx \frac{1}{720}T^8 y^{(8)}$	$\dfrac{T^4(.16527z^4+.67224z^3+.1583z^2+.005575z-.001392)}{z^5-4z^4+6z^3-4z^2+z}$ order = 5 LTE $\approx \frac{49}{362880}T^9 y^{(9)}$
$\dfrac{1}{s^5}$					$\dfrac{T^5 z(z^3+11z^2+11z+1)}{24(z-1)^5}$ order = 6, LTE $\approx \frac{1}{6048}T^{11} y^{(11)}$

288

3. The Filicky substitution method is a hybrid between substitutional methods and linear multistep methods. Its explicit form for second-order systems is $1/s \approx (T/2)(3z - 1)/(z^2 - z)$, $1/s^2 \approx T^2 z/(z - 1)^2$. Apply this substitution method to the longitudinal dynamics of the X-15 aircraft and obtain an impulse response. The transfer function is given as $H(s) = (9.772s + 2.756)/(s^2 + 0.890s + 8.833)$. How many discrete poles are there?

4. The differential equation for the deflection, y, of a beam due to loads, u, is given by $d^4 y/dx^4 = u(x)$. For a simply loaded beam, $u = -\delta(x)/2 + \delta(x - 0.5) - \delta(x - 1)/2$, $y_0=0$, $y_0' = -1/16$, $y_0'' = 0$, $y_0^{(3)} = 1/2$. Discuss the single integration and multiple integration approaches to computing the deflection of the beam as a function of x.

10

Concluding Discussion

10.1 COMPETING CRITERIA

The purpose of simulating a system is to determine its response to a set of initial conditions or to a prescribed input sequence. The simulation results may be used to study the behavior of the system or to evaluate some component, for example, a new controller. To provide meaningful results, the simulation obviously must be accurate, at least to some sufficient degree. However, accuracy is not the only consideration. If we assume that the system being simulated is itself inherently stable, the simulation must also be stable. In fact, if it is not, accuracy has no meaning. Thus, simulation stability, as well as accuracy, is a concern.

There is still one other factor to consider, and that is computation time. If the simulation needs to be done in real time, all function evaluations, integration computations, and overhead operations must actually be performed within the prescribed timestep. Although "real time" is not a precisely defined range of numbers, the dynamics of the system equations impose an upper limit on the integration timestep in order to obtain an accurate simulation. Once an integration method is chosen, these dynamics also impose an upper limit on T to achieve a stable simulation. In addition, in real-time simulation there is some hardware-imposed lower limit on the simulation timestep. Even with batch simulation, computation time is still a concern, even if it is not so important. Excessively slow simulations are frustrating for the user, and round-off error increases with the number of computations.

The trade-offs among accuracy, stability, and computation time are subtle and not always appreciated by someone unfamiliar with the details of digital simulation. For example, to improve the accuracy of a simulation, the chosen integration algorithm could be replaced with one having a higher order of accuracy. However, we have seen that the regions of absolute stability for linear multistep integration methods shrink as the orders of accuracy increase. Therefore, if the simulation timestep is not reduced accordingly, the expected accuracy increase may not be realized; the simulation may even become unstable. In the case of real-time simulation, the reduction in timestep may not be possible, and so intelligent decisions need to be made concerning the problem specifications and the simulation parameter values. The simulation designer must learn to organize effectively and to manipulate knowledge about accuracy, stability, computation time, and their interrelationships.

As a simple example, consider a scalar system described by the homogeneous differential equation

$$\frac{dy(t)}{dt} = -100y(t) = f[y(t), t] \tag{10--1}$$

that is being solved for some initial condition by numerical integration, using the second-order Adams-Bashforth (AB-2) method:

$$y_{n+2} = y_{n+1} + \frac{T}{2}(3f_{n+1} - f_n) \tag{10--2}$$

We know that the AB-2 method has an interval of absolute stability for real eigenvalues of $(-1, 0)$ and that the local truncation error is proportional to T^3. Thus, T would have to be smaller than 0.01 seconds for the simulation to be absolutely stable. To achieve accuracy, relative stability is also desired, so a reasonable choice for T might be 0.005 seconds. Assume that when $T = 0.005$ seconds, the simulation is stable but still not sufficiently accurate. One approach is to use the fourth-order Adams-Bashforth (AB-4) method instead, which has a local truncation error proportional to T^5:

$$y_{n+4} = y_{n+3} + \frac{T}{24}(55f_{n+3} - 59f_{n+2} + 37f_{n+1} - 9f_n) \tag{10--3}$$

However, this algorithm has an interval of absolute stability on the real axis of $(-0.3, 0)$. Therefore, T would have to be less than 0.003 seconds for absolute stability, and even smaller for relative stability. If the reduction in step size were not done, the simulation would be not only less accurate but also unstable. In real-time applications, this reduction in step size, from 0.005 to perhaps 0.0015, might not be possible because of hardware constraints, and therefore other methods would have to be investigated. In any event, it should be realized that the total computation time would increase as a result of the need for more accuracy, this increase being due both to the smaller T required for stability and to the slightly more complex algorithm that would be solved at each timestep. Although this is a trivial example, which admits an analytic solution, the interaction among simulation accuracy, stability, and computation time is clearly demonstrated. Each of these three factors must be considered when planning a simulation.

10.2 REVIEW OF BASIC APPROACHES

Two major categories for digital simulation of continuous-time systems are presented in this text. The first consists of engineering approaches to discretizing a continuous-time transfer function. The z-transform and data reconstruction methods of Chapter 2 and the operational substitution technique of Chapter 3 fit into this category. The most commonly used data reconstructor is the zero-order hold. The z-transform of the series combination of zero-order hold and original transfer function yield a discrete-time transfer function that has the same step response as the continuous-time system. With the operational substitution methods, a continuous-time transfer function is converted to a discrete-time transfer function by substituting an expression in the variable z for each power of the variable s. The resulting transfer function could then be converted into a difference equation to be iteratively solved for the output variable. With the matrix design procedure presented in Chapter 3, the substitution process is fairly simple. However, not all substitution models guarantee that a stable continuous-time transfer function will map into a stable discrete-time transfer function, so care must be exercised in choosing the value of T. Unfortunately, some of the more accurate substitution methods fail to guarantee the mapping of stable continuous-time systems into stable discrete-time systems. If the system being simulated has a closed-loop block diagram, the choice of substitution method is important since some do not yield a discrete-time model that is closed-loop realizable. The primary application area with either of the engineering approaches is for linear time-invariant systems, which can be represented by transfer functions in the Laplace variable s. Time-varying systems can sometimes be handled by assuming that the coefficients are "frozen" for the purpose of creating the transfer function.

The second category of simulation techniques is a numerical analysis approach, which involves the numerical integration of the system's differential equations. This requires approximating the mathematical integration process, represented by $1/s$ in continuous time, by an expression in discrete time. The differential equation in its normal form is evaluated, and its solution is numerically obtained from the computations required by the discrete-time expression approximating the integration process. Since the numerical integration technique does not modify the system's differential equation in any way, it can be applied to nonlinear and time-varying systems with no further approximations required.

Two subcategories of models for numerical integration are presented and discussed in detail. The first, linear multistep methods, are introduced in Chapter 4. In their simplest form (explicit methods), only one function evaluation per timestep is needed. Simulation results can be generated quickly and easily. Unfortunately, the explicit methods suffer from poor accuracy and small stability regions compared to implicit or predictor-corrector methods. Implicit methods generally have to be iterated at each timestep to obtain an accurate solution, and predictor-corrector methods require at least two function evaluations per timestep. Modifiers can be added to the predict-correct cycle, improving accuracy still further, but at the cost of increased computation time. One problem with linear multistep methods in the accuracy-stability trade-off is the decreasing size of the stability region with an increasing order of accuracy. As seen in the example in Section 10.1, the value of T may have to be reduced when one integration operator is replaced by another one of higher accuracy. A catch-22 situation can arise in trying to satisfy constraints on accuracy, stability, and computation time.

The other class of numerical integration methods that are discussed are the Runge-Kutta methods. These are all one-step methods; accuracy is achieved by increasing the number of function evaluations that must be computed for each timestep. Variable timestep methods are available that alter the value of T during the simulation to maintain some measure of the simulation error within a specified bound. Because of the multiple function evaluations per timestep, Runge-Kutta methods are not generally considered suitable for real-time simulation; and although they can generate very accurate results, the user must be willing to wait to get them. One nice thing about Runge-Kutta methods is the increase in the size of the stability region with an increasing order of accuracy. This at least reduces part of the problem in the trade-off among the three competing criteria on the choice of T. The stability region for a Runge-Kutta method requiring a particular number of function evaluations can be further enlarged by choosing the order of accuracy less than the maximum possible for the method.

10.3 SYSTEM CHARACTERISTICS

If the system being simulated is linear and time-invariant, the most important characteristics affecting the simulation are its eigenvalues. The system is inherently stable if all the eigenvalues have negative real parts. If this is so, the initial condition response decays to zero with time. The simulation should provide an accurate reproduction of that response, and simulation stability is the obvious first requirement in obtaining it. For the linear time-invariant system, we know that a necessary and sufficient condition for simulation stability is that all the λT products lie within the region of absolute stability for the particular integration method being used.

If all of the eigenvalues fall within a narrow range of values, achieving stability is not a particular problem. The value of T needed for accuracy will be close to the value needed for stability. Excessively long computation times will not be needed in this case. However, as we see in Chapter 7, problems arise when the system is stiff. It that case, very long simulations may have to be run because of the small value of T and the slowness of the important dynamics. Design techniques like the MSRP or inverse Nyquist array may alleviate part of the problem. If the system has slowly decaying high-frequency oscillations, a long simulation with a small T is the only way to get an accurate simulation.

If the system has one or more eigenvalues with positive real parts, it is inherently unstable. The initial condition response will become unbounded with time. To be accurate, the simulation must mimic this response; thus, the simulation will also be unstable in an absolute sense. Accuracy would be achieved by ensuring that the value of T is chosen to provide relative stability with the integration method used.

From this discussion, we have seen that for the linear time-invariant system, the eigenvalue locations, both in an absolute and a relative sense, are the primary system characteristics affecting the simulation. For this type of system, the eigenvalues are constants; therefore, decisions about the integration method and timestep can be based on information that is available before the simulation is run. For nonlinear or time-varying systems, the situation is not as nice. The eigenvalues of the system Jacobian only give stability information in a local sense, that is, close to the point about which the system was linearized. How useful this information is depends on how smoothly and slowly the eigenvalues change over time. As we see in Chapter 8, some nonlinear systems can behave very erratically, even chaotically.

For example, the response of the discrete-time version of the simple logistic equation

$$x_{k+1} = 4x_k (1 - x_k), \qquad x_0 = \frac{\pi}{10} \tag{10–4}$$

to the initial condition $\pi/10$ is shown in Figure 10.1a. Each circle represents the value of x_k at the corresponding point in time. The equilibrium value is 0.75, shown by the horizontal line in the figure. It is clear that the response is not settling to the equilibrium value; rather, it has an apparent random motion. Figure 10.1b shows the sequence of eigenvalues for the system. Each circle represents the coefficient in the linearization of the system equation, evaluated at the appropriate value of x_k. The two horizontal lines at ±1 in the figure repre-

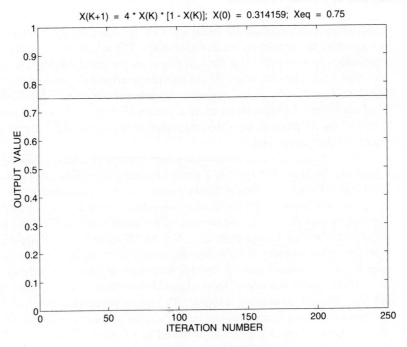

Figure 10.1a   Iteration Number.

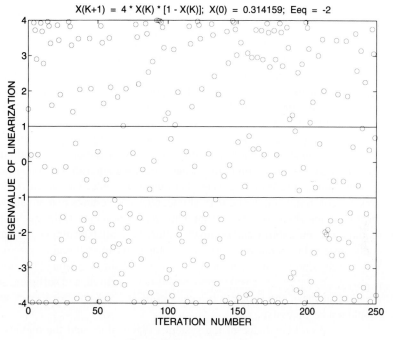

X(K+1) = 4 * X(K) * [1 - X(K)]; X(0) = 0.314159; Eeq = -2

Figure 10.1b   Iteration Number.

sent the real-axis limits of the unit circle. When the eigenvalues are between the lines, the system is locally stable; when they are outside the lines, the system is unstable at that point. It should be clear that knowledge of the eigenvalue at any single point in time does not characterize the system behavior well at all. Thus, for nonlinear or time-varying systems, much more information than just the eigenvalues is needed. Knowledge of how the particular system should respond is essential for obtaining an accurate simulation of the system.

10.4 CHOOSING A METHOD AND *T*

The first question generally asked by a person faced with a simulation problem is, "What value do I use for *T*?" This is indeed a valid question; unfortunately there is not a simple answer. There is no "magic" value for the simulation timestep any more than there is one integration algorithm or operational substitution method that is always the right one to use. Even when the integration method is restricted to available software, perhaps a Runge-Kutta method on the mainframe computer, the value of *T* must be chosen wisely. As we have discussed in the previous sections, the choice of simulation method and timestep are bound up with the characteristics of the system being simulated and in the performance requirements imposed on the simulation.

The decision about which integration algorithm to use and the proper value of *T* are obviously important considerations. Our general premise is that for many applications a user cannot blindly pick a simulation method and/or timestep and get acceptable results. The user must choose an integration method and integration timestep that will produce results that meet the user's requirements for accuracy, stability, and computation time. What needs to be developed by the simulation designer is an understanding of how an experienced worker in digital simulation approaches and solves a specific problem, that is, on the thought processes, analytic tools, and heuristic information that are used to produce an acceptable simulation.

For complex processes, particularly those having to be simulated in real time, the choices of simulation method and simulation timestep are often difficult to make. For real-time simulation, computation time of the simulation algorithm becomes perhaps as critical as its accuracy. Trade-offs between them must be made. The specified computation time imposes constraints on the numerical integration algorithms that are allowable. Decisions must be made about whether or not a standard simulation technique can meet all of the requirements on stability and accuracy within the available computation time. If not, a specially designed integration algorithm developed for the particular system may be needed. Whether the simulation is to be done in real time or not, selection of the simulation timestep is always an important consideration. Stability is required, and sufficient accuracy must be obtained to make the simulation meaningful; and yet computer costs and elapsed time should not be excessive.

Based on knowledge of the system's eigenvalues and the regions of absolute and relative stability for a particular integration method, the maximum usable integration timestep for that method can be chosen. The product of the simulation timestep and the magnitude of each of the eigenvalues (which has a negative real part) must lie inside the closed region of the stability boundary. As we see in the example in Section 10.1, for the system described by equation 10–1 the maximum allowed timesteps for the AB-2 and AB-4 methods are 0.01 and 0.003, respectively. To obtain relative stability, these numbers have to be reduced by roughly a factor of 2. Thus, the increased accuracy of the AB-4 method comes at a cost of more than a threefold increase in the total number of computations.

Although this may appear to contradict the discussion in the opening paragraphs of this section, the following general guideline is presented: *Use as simple an integration method and as large a value for the timestep as possible, consistent with satisfying the performance constraints on the simulation.* Often, this reduces to this advice: Use Euler's method whenever possible, with an initial choice for the timestep given by

$$T = \frac{1}{5|\lambda_{max}|} \tag{10-5}$$

Euler's method is easy to program; has only one function evaluation per timestep; needs no additional starting values; requires no iteration; and has a reasonably large stability region, at least for an explicit linear multistep method. Its major drawback, of course, is that it is only first-order accurate. However, in some cases that may be enough. What this means is that the engineer does not need to use a four-stage fourth-order Runge-Kutta method with a very small timestep if the only purpose of the simulation is to get a rough picture of the response of a system whose eigenvalues have a small range. The value of *T* in equation 10–5 provides reasonable accuracy and ensures that an exponentially decaying continuous-time signal will not be simulated by an oscillating response. A quick check on the adequacy of the simulation is to run the simulation a second time with a smaller *T*, for example, *T*/2. If there is no appreciable difference between the two simulations, the original one should be satisfactory. If there is a major difference, the smaller *T* or a more accurate method may be required for the application. The process of halving *T* and rerunning the simulation may be repeated until a sufficiently accurate simulation is achieved or until the computation time becomes excessive.

A method with a higher order of accuracy can be used if the need for the extra accuracy is indicated and the additional computational burden can be tolerated. Various simulation methods can be investigated in light of the performance objectives and stability constraints. If one or more of the standard methods are acceptable, one of them can be based on convenience or familiarity. Some Runge-Kutta and predictor-corrector-modifier methods have the capability of adaptively changing the timestep to maintain a prescribed accuracy. However, these methods typically have long computation times and thus generally are not practical for real-time simulation. They can be used when extreme accuracy is required and computation time is not a serious consideration.

Probably the first step toward choosing a method and timestep is to determine the system eigenvalues. If the system is stiff, special precautions may have to be taken. The inverse Nyquist array can be used to check for diagonal dominance, which might allow *T* to be based on the slow dynamics. A rule of thumb for the value of *T* to provide an accurate simulation would be five to ten times smaller than the reciprocal of the absolute value of the important eigenvalue. Equation 10–5 illustrates this, using the factor of 5. Stability of the fast dynamics could be achieved by using integration operators designed by the SRP procedure, by using the first backward difference method, or by using reduced-order Runge-Kutta methods. Alternatively, the MSRP procedure could be used to design the complete integration algorithm. The computation time would be larger with this approach since the MSRP integrator's matrices are not diagonal. If the system is not stiff, the timestep can still be based on the rule of thumb mentioned above.

Unfortunately, there is not an easy mapping of simulation accuracy specifications into a selection of order of accuracy. However, once an initial choice of *T* has been made, the problem is partly solved. For a linear multistep method, the region of relative stability is roughly one-half radially the size of the absolute stability region. Thus, the designer would want to place the λT products in the right half of the stability region. Once *T* is chosen, the λT products are known, and they can be superimposed on the stability regions of several integration methods. This will allow an easy determination of whether or not absolute and relative stability can be achieved with each method. For example, if equation 10–5 is used to choose *T*, the largest λT product will be –0.2. This is well within the region of relative stability for the AB-2 method but not for the AB-4 method. The reduction of *T* by 25 percent so that $\lambda T = -0.15$ would allow a relatively stable simulation from the AB-4 method also. The AB-4 method would certainly provide a more accurate simulation than the AB-2, using the same value for T. It should be noted that the differences in accuracy will appear in the simulation of the system's transient response. If the system input is constant, all the integrators will provide the correct steady-state response.

We conclude this section by summarizing the steps in choosing an integration method and the value of the timestep. If T can be selected by the simulation designer, an initial choice for the value should be in accordance with equation 10–5. That should provide an accurate representation of the continuous-time system's response. With that value of T, Euler's method or another explicit linear multistep method can be studied from a stability point of view to determine if the integration will be relatively stable or at least not severely detuned. A trial simulation can be run to see whether or not accuracy requirements are satisfied. If the form of the actual response is unknown, the simulation can be rerun with a smaller T and/or a method with a higher order of accuracy. Runge-Kutta methods can be used in place of linear multistep methods. They are self-starting and have larger stability regions for a given order of accuracy. Another alternative for achieving accuracy is to use a predictor-corrector method, with or without modifiers. If the system being simulated is stiff, special methods will be needed unless the small value of T and the resulting large number of computations are acceptable. If the value of the timestep is fixed by external constraints, the choice of integration method is more restricted. A method will have to be chosen that has a stability region enclosing all the λT products, which are also fixed. In this situation, a specially designed integration method, using perhaps the MSRP approach, may be needed.

Whether the system being simulated is stiff or nonstiff, linear or nonlinear, and whether the simulation is to be done in real time or not, the engineer should realize that there is a certain amount of trial and error associated with obtaining an acceptable simulation. Knowing how the system is supposed to respond is essential; otherwise, there is no way to tell if the simulation results are correct. This is especially true for nonlinear systems, as was pointed out in Chapter 8.

10.5 AVAILABLE SOFTWARE

Throughout the text, simulation programs have been provided with the examples. Some of these have been in the BASIC programming language; others have been in MATLAB. The actual code for a simulation can be written in any programming language. Typical languages are FORTRAN, C, PASCAL, or BASIC. The arithmetic operations needed for most simulations are just additions and multiplications, plus trigonometric or transcendental functions needed for function evaluations. Instructions for these operations would be provided by any programming languages. The choice of which language to choose would be based on availability, personal preference, speed of execution, and company policy.

Special simulation programming languages are also available. These languages make the programming of the actual computations transparent to the user. Many use an interactive block diagram format for inputting the system equations and have excellent graphics capabilities for plotting the output responses.

In addition to the software that actually executes the simulation, good control design software is very convenient during the design process. Plots of stability regions, root loci, and Gershgorin circles are powerful tools in evaluating integration methods and in developing new integration techniques. Software packages developed for the design of control systems are probably the most convenient for this type of work. These include MATLAB, Program CC, Control C, and others. Engineers engaged in serious simulation design should avail themselves of this type of software.

PROBLEMS

1. Indicate the method that you generally prefer to use and discuss your choice.
2. Compare and contrast the general accuracy, stability, computation time, and utility properties of linear multistep methods, Runge-Kutta methods, and operational calculus (substitution). Be complete and concise.
3. If you observed very high frequencies in your simulation, what might you suspect the cause to be? How might you fix this problem?

A

References

BEALE, G. O., & HARTLEY, T. T. (1987, May). Stability considerations: Numerical methods and control theory equivalences. *IEEE Trans. Ind. Elect.,* 34 (2), 180–187.

BEALE, G. O., & KAWAMURA, K. (1989). Coupling symbolic and numerical computation for intelligent simulation. In *Knowledge-based approach to system diagnosis, supervision, and control,* pp. 137–152, ed. S. Tzafestas. New York: Plenum Press.

BOXER, R. & THALER, S. (Jan. 1956). A simplified method of solving linear and nonlinear systems. *Proc. IRE,* 44, 89–101.

BRENNAN, T. C. & LEAKE, R. J. (Nov. 1975). Simplified simulation models for control studies of turbojet engines. *NASA Technical Report EE-757 (N77–10061).*

BUTCHER, J. C. (1987). *The numerical analysis of ordinary differential equations.* Chichester, England: Wiley.

CHICATELLI, S. P. (1989). *Graphical stability methods for numerical integration of ordinary differential equations.* M.S. thesis, University of Akron, Akron, OH.

COOK, P. A. (1986). *Nonlinear dynamical systems,* Englewood Cliffs, NJ: Prentice-Hall.

DAHLQUIST, G. (1956). Convergence and stability in the numerical integration of ordinary differential equations. *Math. Scand.,* (4), 33–53.

DAHLQUIST, G. (1963). A special stability problem for linear multistep methods. *BIT,* 3, 27–43.

DANIELI, C. J., & KROSEL, S. M. (1979, February). Generation of linear dynamic models from a nonlinear simulation. *NASA Tech. Paper 1388.*

D'AZZO, J. J., & HOUPIS, C. H. (1981). *Linear control system analysis and design.* New York: McGraw-Hill.

DE ABREU-GARCIA, J. A., & HARTLEY, T. T. (1990). Multistep matrix integrators for real-time simulation. In *Control and dynamic systems,* Vol. 38. San Diego: Academic Press.

FEHLBERG, E., (1968). Classical Fifth, Sixth, Seventh, and Eighth Order Runge-Kutta Formulas with Stepsize Control. *NASA TR R-287.*

FILICKY, D. J. (1990). *Higher order accurate operational substitution methods for real-time simulation.* M.S. thesis, University of Akron, Akron, OH.

GEAR, C. W. (1971). *Numerical initial value problems in ordinary differential equations.* Englewood Cliffs, NJ: Prentice-Hall.

GOODWIN, G.C., & SIN, K.S. (1984). *Adaptive filtering, prediction, and control.* Englewood Cliffs, NJ: Prentice-Hall.

HARTLEY, T. T. (1984, December). *Parallel methods for the real-time simulation of stiff nonlinear systems.* Ph.D. dissertation, Vanderbilt University, Nashville, TN.

HARTLEY, T. T. (1988). Operational transforms via time domain optimization. *Transactions of the Society for Computer Simulation,* 5 (3), 221–228.

HARTLEY, T. T., & BEALE, G. O. (1987, November). Matrix integrators for real-time simulation. *Proc. of IEEE IECON 1987,* pp. 63–69, Cambridge MA.

HARTLEY, T. T. KILLORY, H. DE ABREU-GARCIA, J. A., & ABU-KHAMSEH, N. (August 1990). Analysis and reduction of an infinite dimensional chaotic system. *Proceedings of the 33rd Midwest Symposium on Circuits and Systems,* Calgary, Can.

HENRICI, P. (1962). *Discrete variable methods in ordinary differential equations.* New York: Wiley.

HOROWITZ, I. M. (1963). *Synthesis of feedback systems.* New York: Academic Press.

JORDAN, D. W. & SMITH, P. (1987). *Nonlinear ordinary differential equations,* 2nd ed. Oxford University Press.

JURY, E. I. (1958). *Sampled data control systems.* New York: Wiley.

JURY, E. I., (1964). *Theory and application of the z-transform method.* Malabar, FL: Krieger.

JURY, E. I. & CHAN, O. W. C. (Sept. 1973). Combinatorial rules for some useful transformations. *IEEE Trans. Cir. Thy.,* CT-20(5), 476–480.

JURY, & LEE, (1967).

KAILATH, T. (1980). *Linear systems.* Englewood Cliffs, NJ: Prentice-Hall.

KREYZIG, E. (1988). *Advanced engineering mathematics,* 6th ed. New York: Wiley.

KUO, B. C. (1980). *Digital control systems.* New York: Holt, Rinehart & Winston.

KUO, B. C. (1992). *Automatic control systems.* Englewood Cliffs, NJ: Prentice-Hall.

LAMBERT, J. D. (1973). *Computational methods in ordinary differential equations.* New York: Wiley.

LAPIDUS, L., & SEINFELD, J. H. (1971). *Numerical solution of ordinary differential equations.* New York: Academic Press.

MACIEJOWSKI, J. M. (1989). *Multivariable feedback design.* Wokingham, Eng.: Addison-Wesley.

MARINCHEK, B. (1992). *Graphical stability for predictor-corrector-modifier methods.* M.S. thesis, University of Akron, Akron, OH.

MERSON, R. H. (1957). An operational method for the study of integration processes. *Proc. Symp. Data Processing, Weapons Research Establishment,* Salisbury, S. Australia.

MIDDLETON, R. H., & GOODWIN, G. C. (1990). *Digital control and estimation, A unified approach.* Englewood Cliffs, NJ: Prentice-Hall.

ORTEGA, J. M., & POOLE, W. G. (1981). *An introduction to numerical methods for differential equations.* Mansfield, MA: Pitman.

OPPENHEIM, A. V. & SCHAFER, R. W. (1989). *Discrete-time signal processing.* Englewood Cliffs, NJ: Prentice-Hall.

PAYNTER, H. (1989). The differential analyzer as an active mathematical element. *Proc. American Control Conference,* pp. 1863–1872, Pittsburgh.

PHILLIPS, C. L., & HARBOR, R. D. (1991). *Basic feedback control systems.* Englewood Cliffs, NJ: Prentice-Hall.

PHILLIPS, C. L., & NAGLE, H. T. (1990). *Digital control systems analysis and design,* 2nd ed. Englewood Cliffs, NJ: Prentice-Hall.

RABINER, L. R. & GOLD, B. (1975). *Theory and application of digital signal processing.* Englewood Cliffs, NJ: Prentice-Hall.

RAGAZZINI, J. R., & FRANKLIN, G. F. (1958). *Sampled data control systems.* New York: McGraw-Hill.

RALSTON, A., & RABINOWITZ, P. (1978). *A first course in numerical analysis,* 2nd ed. New York: McGraw-Hill.

ROSKO, J. S.(1972). *Digital simulation of physical systems.* Reading, MA: Addison-Wesley.

ROSSLER, O. E. (1976). An equation for chaos. *Phys. Lett.,* 57A, 397.

ROSSLER, O. E. (1979). An equation for hyperchaos. *Phys. Lett.,* 71A, 155–157.

SCHWARTZ, M. (1970). *Information transmission, modulation, and noise,* 2nd ed. New York: McGraw-Hill.

SHAMPINE, L. F. & GORDON, M. K. (1975). *Computer solution of ordinary differential equations: The initial value problem.* San Francisco: Freeman.

SMITH, J. M. (1987). *Mathematical modeling and digital simulation for engineers and scientists,* 2nd ed. New York: Wiley-Interscience.

THALER, G. J. (1989). *Automatic control systems.* St. Paul, MN: West Publishing.

THOMAS, B. (1968, April). The Runge-Kutta methods. *Byte,* 11 (4).

THOMPSON, J. M. T., & STEWART, H. B. (1986). *Nonlinear dynamics and chaos.* New York: Wiley.

TUSTIN, A. (May 1947). A method of analyzing the behavior of linear systems in terms of time series. *Journal IEE,* 94, Pt. II-A.

VAN DE VEGTE, J. (1990). *Feedback control systems.* Englewood Cliffs, NJ: Prentice-Hall.

VAN VALKENBERG, M. E. (1974). *Network analysis,* 3rd ed. Englewood Cliffs, NJ: Prentice-Hall.

VERNER, J. H. (1978). Explicit Runge-Kutta methods with estimates of the local truncation error. *SIAM J. Numer. Anal.,* 15, 772–790.

Stability Region Plots

TABLE B.1 COEFFICIENTS OF AM METHODS

k	α_k	α_{k-1}	β_7	β_6	β_5	β_4	β_3	β_2	β_1	β_0	p	C_{p+1}
1	1	−1							1	0	1	−1/2
1	1	−1							1/2	1/2	2	−1/12
2	1	−1						5/12	8/12	−1/12	3	−1/24
3	1	−1					9/24	19/24	−5/24	1/24	4	−19/720
4	1	−1				251/720	646/720	−264/720	106/720	−19/720	5	−27/1440
5	1	−1			$\dfrac{475}{1440}$	$\dfrac{14427}{1440}$	$\dfrac{-798}{1440}$	$\dfrac{482}{1440}$	$\dfrac{-173}{1440}$	$\dfrac{27}{1440}$	6	$\dfrac{-863}{60480}$
6	1	−1		$\dfrac{19087}{60480}$	$\dfrac{65112}{60480}$	$\dfrac{-46461}{60480}$	$\dfrac{37504}{60480}$	$\dfrac{-20211}{60480}$	$\dfrac{6312}{60480}$	$\dfrac{-863}{60480}$	7	$\dfrac{-1375}{120960}$
7	1	−1	$\dfrac{36799}{120960}$	$\dfrac{139849}{120960}$	$\dfrac{-121797}{120960}$	$\dfrac{123133}{120960}$	$\dfrac{-88547}{120960}$	$\dfrac{41499}{120960}$	$\dfrac{-11351}{120960}$	$\dfrac{1375}{120960}$	8	$\dfrac{-339533}{3628800}$

TABLE B.2 COEFFICIENTS OF AB METHODS

k	α_k	α_{k-1}	β_6	β_5	β_4	β_3	β_2	β_1	β_0	p	C_{p+1}
1	1	−1							1	1	1/2
2	1	−1						3/2	−1/2	2	5/12
3	1	−1					23/12	−16/12	5/12	3	9/24
4	1	−1				55/24	−59/24	37/24	−9/24	4	251/720
5	1	−1			1901/720	−2774/720	2616/720	−1274/720	251/720	5	475/1440
6	1	−1		4277/1440	−7923/1440	9982/1440	−7298/1440	2887/1440	−475/1440	6	19087/60480
7	1	−1	198721/60480	−447288/60480	705549/60480	−688256/60480	407139/60480	−134472/60480	19087/60480	7	36799/120960

TABLE B.3 OTHER COMMONLY USED EXPLICIT LMM

	k	α_{k-1}*	α_{k-2}	β_6	β_5	β_4	β_3	β_2	β_1	β_0	p	C_{p+1}
Midpoint	2	0	−1						2	0	2	1/3
Krogh A	4	−1/2	−1/2				119/48	−99/48	69/48	−17/48	4	161/480
Krogh B	4	−4/7	−3/7				103/42	−88/42	61/42	−15/42	4	85/252
Krogh	5	1/31	−32/31			22321/7440	−21774/7440	24216/74440	−12034/7440	2391/7440	5	13861/44640
Krogh	6	11/12	−23/12		62249/17280	−62255/17280	101430/17280	−76490/17280	30545/17280	−5079/17280	6	5977/20736
Krogh	7	−1	0	2578907/604800	−2454408/604800	5615199/604800	−5719936/604800	3444849/604800	−1149048/604800	164117/604800	7	21691/80640

*For all of the methods, α_k is 1.

TABLE B.4 COEFFICIENTS OF BD METHODS

k	α_6	α_5	α_4	α_3	α_2	α_1	α_0	β_k	p	C_{p+1}
1						1	−1	1	1	−1/2
2					1	−4/3	1/3	2/3	2	−2/94
3				1	−18/11	9/11	−2/11	6/11	3	−3/22
4			1	−48/25	36/25	−16/25	3/25	12/25	4	−12/125
5		1	−300/137	300/137	−200/137	75/137	−12/137	60/137	5	−10/137
6	1	−360/147	450/147	−400/147	225/147	−72/147	10/147	60/147	6	−20/343

The Runge Kutta methods are presented in the standard Butcher matrix form. To conserve space, the C vectors are not given. Recall that the C vector is obtained as the horizontal or row sum of the rows of the A matrix. The B vectors are stated with the order of the approximation given in parenthesis, that is, $B(3) = 1/8 \; 1/5 \; 0 \; 1/2$ would indicate that use of this B vector with the corresponding A matrix would yield a third-order approximation. For embedded methods, the row number of the embedded method is given, followed by its order. For example, A row 3 (2) indicates that if the third row is used as a B vector, a second-order method will result.

Consider the Runge Kutta A matrix below in the format of the appendix (Merson, 1957).

A =					
0	0	0	0	0	These numbers make up the A matrix of this Runge
1/3	0	0	0	0	Kutta method.
1/6	1/6	0	0	0	
1/8	0	3/8	0	0	
1/2	0	−3/2	2	0	

A row 5 (4)					The fifth row is an embedded method.[1]
B (4) =					Using this B vector and the above A matrix will
1/6	0	0	2/3	1/6	produce a fourth-order approximation.[2]
B (5) =					Using this B vector and the above A matrix will
1/10	0	3/10	2/5	1/5	produce a fifth-order approximation.[3]

[1] A row 5 (4) indicates that using $B = [1/2,0,−3/2,2,0]$ with the first four rows or stages of A will produce a fourth-order Runge Kutta approximation.
[2] $B (4) = [1/6,0,0,2/3,1/6]$ indicates that this B with the rows/stages of A produce a fourth-order Runge Kutta approximation.
[3] Similarly $B (5) = [1/10,0,3/10,2/5,1/5]$ indicates that this B will produce a fifth-order Runge Kutta approximation.

RUNGE KUTTA METHODS

MERSON (1957)

$A =$

0	0	0	0	0
$\dfrac{1}{3}$	0	0	0	0
$\dfrac{1}{6}$	$\dfrac{1}{6}$	0	0	0
$\dfrac{1}{8}$	0	$\dfrac{3}{8}$	0	0
$\dfrac{1}{2}$	0	$\dfrac{-3}{2}$	2	0

A row 5 (4)

B (4) =

$$
\begin{array}{ccccc}
\dfrac{1}{6} & 0 & 0 & \dfrac{2}{3} & \dfrac{1}{6}
\end{array}
$$

B (5) =

$$
\begin{array}{ccccc}
\dfrac{1}{10} & 0 & \dfrac{3}{10} & \dfrac{2}{5} & \dfrac{1}{5}
\end{array}
$$

BUTCHER (1987, p. 300)

A =

$$
\begin{array}{cccc}
0 & 0 & 0 & 0 \\[6pt]
\dfrac{1}{2} & 0 & 0 & 0 \\[6pt]
0 & \dfrac{1}{2} & 0 & 0 \\[6pt]
0 & 0 & 1 & 0
\end{array}
$$

A row 4 (2)

B (4) =

$$
\begin{array}{cccc}
\dfrac{1}{6} & \dfrac{1}{3} & \dfrac{1}{3} & \dfrac{1}{6}
\end{array}
$$

BUTCHER (1987, p. 301)

A =

$$
\begin{array}{ccccc}
0 & 0 & 0 & 0 & 0 \\[6pt]
\dfrac{1}{4} & 0 & 0 & 0 & 0 \\[6pt]
\dfrac{-9}{4} & 3 & 0 & 0 & 0 \\[6pt]
\dfrac{1}{18} & \dfrac{5}{12} & \dfrac{1}{36} & 0 & 0 \\[6pt]
\dfrac{7}{9} & \dfrac{-5}{3} & \dfrac{-1}{9} & 2 & 0
\end{array}
$$

A row 5 (4)
B (4) =

$\dfrac{1}{6}$	0	0	$\dfrac{2}{3}$	$\dfrac{1}{6}$

BUTCHER (1987, p. 302)

A =

0	0	0	0	0
$\dfrac{2}{7}$	0	0	0	0
$\dfrac{-8}{35}$	$\dfrac{4}{5}$	0	0	0
$\dfrac{29}{42}$	$\dfrac{-2}{3}$	$\dfrac{5}{6}$	0	0
$\dfrac{1}{6}$	$\dfrac{1}{6}$	$\dfrac{5}{12}$	$\dfrac{1}{4}$	0

A row 5 (3)
B (4) =

$\dfrac{11}{96}$	$\dfrac{7}{24}$	$\dfrac{35}{96}$	$\dfrac{7}{48}$	$\dfrac{1}{12}$

BUTCHER (1987, p. 302)

A =

0	0	0	0	0	0	0
$\dfrac{1}{4}$	0	0	0	0	0	0
$\dfrac{1}{8}$	$\dfrac{1}{8}$	0	0	0	0	0
0	$\dfrac{-1}{2}$	1	0	0	0	0
$\dfrac{13}{200}$	$\dfrac{-299}{1000}$	$\dfrac{78}{125}$	$\dfrac{13}{50}$	0	0	0

$\dfrac{548}{7475}$	$\dfrac{688}{2875}$	$\dfrac{572}{2875}$	$\dfrac{-88}{575}$	$\dfrac{132}{299}$	0	0
$\dfrac{37}{312}$	0	$\dfrac{4}{33}$	$\dfrac{8}{9}$	$\dfrac{-100}{117}$	$\dfrac{575}{792}$	0

A row 7 (4)
B (5) =

$\dfrac{41}{520}$	0	$\dfrac{58}{165}$	$\dfrac{16}{135}$	$\dfrac{50}{351}$	$\dfrac{575}{2376}$	$\dfrac{1}{15}$

FEHLBERG (1970)

A=

0	0	0
$\dfrac{1}{2}$	0	0
$\dfrac{1}{256}$	$\dfrac{255}{256}$	0

A row 3 (1)
B (2) =

$\dfrac{1}{512}$	$\dfrac{255}{256}$	$\dfrac{1}{512}$

FEHLBERG (1970)

A =

0	0	0	0
$\dfrac{1}{4}$	0	0	0
$\dfrac{-189}{800}$	$\dfrac{729}{800}$	0	0
$\dfrac{214}{891}$	$\dfrac{1}{33}$	$\dfrac{650}{891}$	0

A row 4 (2)
B (3) =

$\frac{41}{162}$	0	$\frac{800}{1053}$	$\frac{-1}{78}$

FEHLBERG (1970)

A =

0	0	0	0
1	0	0	0
$\frac{1}{4}$	$\frac{1}{4}$	0	0
$\frac{1}{2}$	$\frac{1}{2}$	0	0

A row 4 (2)
B (3) =

$\frac{1}{6}$	$\frac{1}{6}$	$\frac{2}{3}$	0

FEHLBERG (1970)

A =

0	0	0	0	0
$\frac{1}{4}$	0	0	0	0
$\frac{4}{81}$	$\frac{32}{81}$	0	0	0
$\frac{57}{98}$	$\frac{-432}{343}$	$\frac{1053}{686}$	0	0
$\frac{1}{6}$	0	$\frac{27}{52}$	$\frac{49}{156}$	0

A row 5 (3)
B (4) =

$\frac{43}{288}$	0	$\frac{243}{416}$	$\frac{343}{1872}$	$\frac{1}{12}$

FEHLBERG (1970)

$A =$

0	0	0	0	0
$\dfrac{2}{7}$	0	0	0	0
$\dfrac{77}{900}$	$\dfrac{343}{900}$	0	0	0
$\dfrac{805}{1444}$	$\dfrac{-77175}{54872}$	$\dfrac{97125}{54872}$	0	0
$\dfrac{79}{490}$	0	$\dfrac{2175}{3626}$	$\dfrac{2166}{9065}$	0

A row 5 (3)
B (4) =

$\dfrac{229}{1470}$	0	$\dfrac{1125}{1813}$	$\dfrac{13718}{81585}$	$\dfrac{1}{18}$

FEHLBERG (1970)

$A =$

0	0	0	0	0	0
$\dfrac{2}{9}$	0	0	0	0	0
$\dfrac{1}{12}$	$\dfrac{1}{4}$	0	0	0	0
$\dfrac{69}{128}$	$\dfrac{-243}{128}$	$\dfrac{135}{64}$	0	0	0
$\dfrac{-17}{12}$	$\dfrac{27}{4}$	$\dfrac{-27}{5}$	$\dfrac{16}{15}$	0	0
$\dfrac{65}{432}$	$\dfrac{-5}{16}$	$\dfrac{13}{16}$	$\dfrac{4}{27}$	$\dfrac{5}{144}$	0

A row 5 (2)

B (4) =

$$\frac{1}{9} \qquad 0 \qquad \frac{9}{20} \qquad \frac{16}{45} \qquad \frac{1}{12} \qquad 0$$

B (5) =

$$\frac{47}{450} \qquad 0 \qquad \frac{12}{25} \qquad \frac{32}{225} \qquad \frac{1}{30} \qquad \frac{6}{25}$$

FEHLBERG (1969)

A =

$$
\begin{array}{cccccccc}
0 & 0 & 0 & 0 & 0 & 0 & 0 & 0 \\[6pt]
\frac{1}{6} & 0 & 0 & 0 & 0 & 0 & 0 & 0 \\[6pt]
\frac{4}{75} & \frac{16}{75} & 0 & 0 & 0 & 0 & 0 & 0 \\[6pt]
\frac{5}{6} & \frac{-8}{3} & \frac{5}{2} & 0 & 0 & 0 & 0 & 0 \\[6pt]
\frac{-8}{5} & \frac{144}{25} & -4 & \frac{16}{25} & 0 & 0 & 0 & 0 \\[6pt]
\frac{361}{320} & \frac{-18}{5} & \frac{407}{128} & \frac{-11}{80} & \frac{55}{128} & 0 & 0 & 0 \\[6pt]
\frac{-11}{640} & 0 & \frac{11}{256} & \frac{-11}{160} & \frac{11}{256} & 0 & 0 & 0 \\[6pt]
\frac{93}{640} & \frac{-18}{5} & \frac{803}{256} & \frac{-11}{160} & \frac{99}{256} & 0 & 1 & 0
\end{array}
$$

A row 6 (2)
A row 8 (2)
B (5) =

$$\frac{31}{384} \qquad 0 \qquad \frac{1125}{2816} \qquad \frac{9}{32} \qquad \frac{125}{768} \qquad \frac{5}{66} \qquad 0 \qquad 0$$

$B(6) =$

$$\frac{7}{1408} \quad 0 \quad \frac{1125}{2816} \quad \frac{9}{32} \quad \frac{125}{768} \quad 0 \quad \frac{5}{66} \quad \frac{5}{66}$$

VERNER (1978)

$A =$

$$
\begin{array}{llllllll}
0 & 0 & 0 & 0 & 0 & 0 & 0 & 0 \\[6pt]
\dfrac{1}{18} & 0 & 0 & 0 & 0 & 0 & 0 & 0 \\[10pt]
\dfrac{-1}{12} & \dfrac{1}{4} & 0 & 0 & 0 & 0 & 0 & 0 \\[10pt]
\dfrac{-2}{81} & \dfrac{4}{27} & \dfrac{8}{81} & 0 & 0 & 0 & 0 & 0 \\[10pt]
\dfrac{40}{33} & \dfrac{-4}{11} & \dfrac{-56}{11} & \dfrac{54}{11} & 0 & 0 & 0 & 0 \\[10pt]
\dfrac{-369}{73} & \dfrac{72}{73} & \dfrac{5380}{219} & \dfrac{-12285}{584} & \dfrac{2695}{1752} & 0 & 0 & 0 \\[10pt]
\dfrac{-8716}{891} & \dfrac{656}{297} & \dfrac{39520}{891} & \dfrac{-416}{11} & \dfrac{52}{27} & 0 & 0 & 0 \\[10pt]
\dfrac{3015}{256} & \dfrac{-9}{4} & \dfrac{-4219}{78} & \dfrac{5985}{128} & \dfrac{-539}{384} & 0 & \dfrac{693}{3328} & 0
\end{array}
$$

A row 6 (2)
A row 8 (2)
$B(5) =$

$$\frac{3}{80} \quad 0 \quad \frac{4}{25} \quad \frac{243}{1120} \quad \frac{77}{160} \quad \frac{73}{700} \quad 0 \quad 0$$

$B(6) =$

$$\frac{57}{640} \quad 0 \quad \frac{-16}{65} \quad \frac{1377}{2240} \quad \frac{121}{320} \quad 0 \quad \frac{891}{8320} \quad \frac{2}{35}$$

LAWSON (1966)

A =

0	0	0	0	0	0
$\dfrac{1}{2}$	0	0	0	0	0
$\dfrac{3}{16}$	$\dfrac{1}{16}$	0	0	0	0
0	0	$\dfrac{1}{2}$	0	0	0
0	$\dfrac{-3}{16}$	$\dfrac{3}{8}$	$\dfrac{9}{16}$	0	0
$\dfrac{1}{7}$	$\dfrac{4}{7}$	$\dfrac{6}{7}$	$\dfrac{-12}{7}$	$\dfrac{8}{7}$	0

A row 6 (2) =
B (5) =

$\dfrac{7}{90}$	0	$\dfrac{16}{45}$	$\dfrac{2}{15}$	$\dfrac{16}{45}$	$\dfrac{7}{90}$

NYSTROM (LAMBERT, 1973)

A =

0	0	0	0	0	0
$\dfrac{1}{3}$	0	0	0	0	0
$\dfrac{4}{25}$	$\dfrac{6}{25}$	0	0	0	0
$\dfrac{1}{4}$	-3	$\dfrac{15}{4}$	0	0	0
$\dfrac{2}{27}$	$\dfrac{10}{9}$	$\dfrac{-50}{81}$	$\dfrac{8}{81}$	0	0
$\dfrac{2}{25}$	$\dfrac{12}{25}$	$\dfrac{2}{15}$	$\dfrac{8}{75}$	0	0

A row 4 (2)
B (5) =

| $\dfrac{23}{192}$ | 0 | $\dfrac{125}{192}$ | 0 | $\dfrac{-27}{64}$ | $\dfrac{125}{192}$ |

HUTA (LAMBERT, 1973)

A =

0	0	0	0	0	0	0
$\dfrac{1}{9}$	0	0	0	0	0	0
$\dfrac{1}{24}$	$\dfrac{1}{8}$	0	0	0	0	0
$\dfrac{1}{6}$	$\dfrac{-1}{2}$	$\dfrac{2}{3}$	0	0	0	0
$\dfrac{-5}{8}$	$\dfrac{27}{8}$	-3	$\dfrac{3}{4}$	0	0	0
$\dfrac{221}{9}$	-109	$\dfrac{289}{3}$	$\dfrac{-34}{3}$	$\dfrac{1}{9}$	0	0
$\dfrac{-61}{16}$	$\dfrac{113}{8}$	$\dfrac{-59}{6}$	$\dfrac{-11}{8}$	$\dfrac{5}{3}$	$\dfrac{1}{16}$	0
$\dfrac{358}{41}$	$\dfrac{-2079}{82}$	$\dfrac{501}{41}$	$\dfrac{417}{41}$	$\dfrac{-227}{41}$	$\dfrac{-9}{82}$	$\dfrac{36}{41}$

A row 8 (2)
B (6) =

| $\dfrac{41}{840}$ | 0 | $\dfrac{9}{35}$ | $\dfrac{9}{280}$ | $\dfrac{34}{105}$ | $\dfrac{9}{280}$ | $\dfrac{9}{35}$ | $\dfrac{41}{840}$ |

LAWSON (1967)

A =

A11 =	0.0000000000000000000000000000
A12 =	0.0000000000000000000000000000
A13 =	0.0000000000000000000000000000
A14 =	0.0000000000000000000000000000

A15 =   0.0000000000000000000000000

A16 =   0.0000000000000000000000000

A17 =   0.0000000000000000000000000

A21 =   0.2022766448981406349333370

A22 =   0.0000000000000000000000000

A23 =   0.0000000000000000000000000

A24 =   0.0000000000000000000000000

A25 =   0.0000000000000000000000000

A26 =   0.0000000000000000000000000

A27 =   0.0000000000000000000000000

A31 =   0.0758537418368027381000010

A32 =   0.2275612255104082143000040

A33 =   0.0000000000000000000000000

A34 =   0.0000000000000000000000000

A35 =   0.0000000000000000000000000

A36 =   0.0000000000000000000000000

A37 =   0.0000000000000000000000000

A41 =   1.3592822172833003172528910

A42 =   −5.2378857026288066156570600

A43 =   4.7536034853455062984041700

A44 =   0.0000000000000000000000000

A45 =   0.0000000000000000000000000

A46 =   0.0000000000000000000000000

A47 =   0.0000000000000000000000000

A51 =   −0.3210920022580216847152800

A52 =   1.6513531279223823812908960

A53 =   −0.9052866767637204932799910

A54 =   0.0750255510993597967043750

A55 =   0.0000000000000000000000000

A56 =   0.0000000000000000000000000

A57 =   0.0000000000000000000000000

A61 =   0.2923218393493635657197980

A62 =   −0.7482693860898295165224370

A63 =   0.5924708449664850399864190

A64 =   −0.0395545388491436203024900

A65 =   0.0280312406231245311187110

A66 =   0.0000000000000000000000000

A67 =   0.0000000000000000000000000

A71 =   −20.6627618949040851886373680

A72 =   63.8523209463321187432479580

A73 =   −74.1517509476888342486158630

A74 =   0.8641176443733843952193490

A75 =   14.5054816592948237061933360

A76 =   16.5925925925925925925925880

A77 =   0.0000000000000000000000000000

A row 7 (2)

B (6) =
B1 =   0.0142857142857142857142860
B2 =   0.0000000000000000000000000000
B3 =   0.0000000000000000000000000000
B4 =   0.2708994708994708994708990
B5 =   0.4296296296296296296296300
B6 =   0.2708994708994708994708990
B7 =   0.0142857142857142857142860

Note: In the following stability region plots, δr is 0.1. In plots with solid lines, δθ is 10°; in plots with dotted lines, δθ is 2.5° for linear multistep methods (including predictor-corrector and predictor-corrector-modifier methods), and δθ is 5° for Runge-Kutta methods.

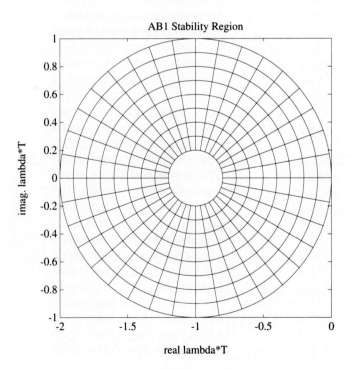

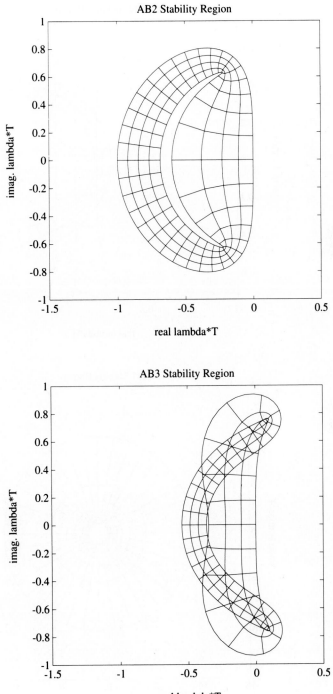

AB4 Stability Region

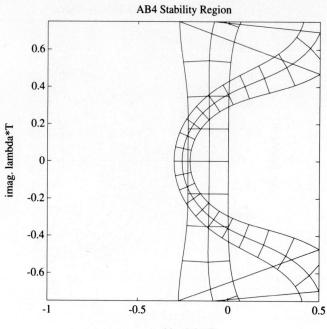

AM1 Stability Region

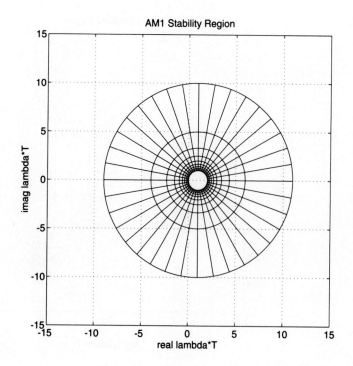

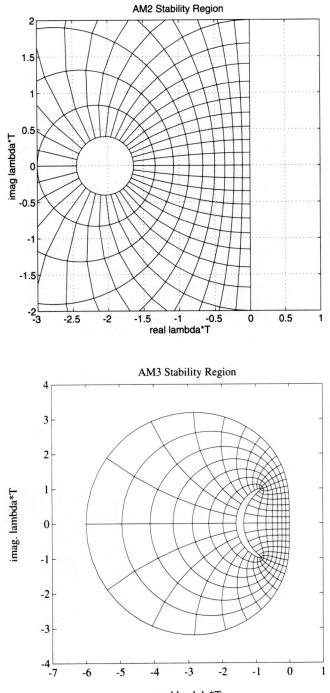

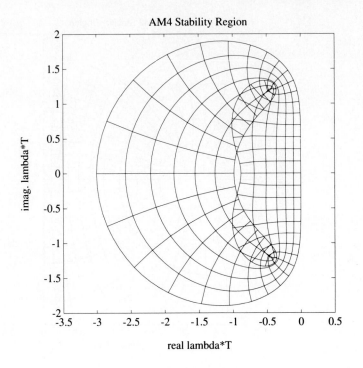

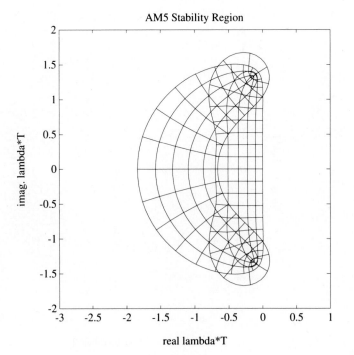

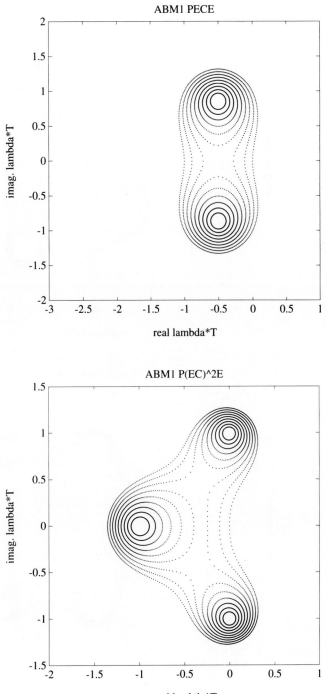

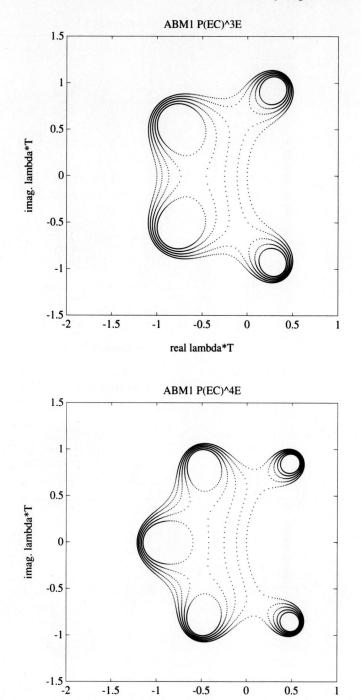

ABM1 P(EC)^3E

ABM1 P(EC)^4E

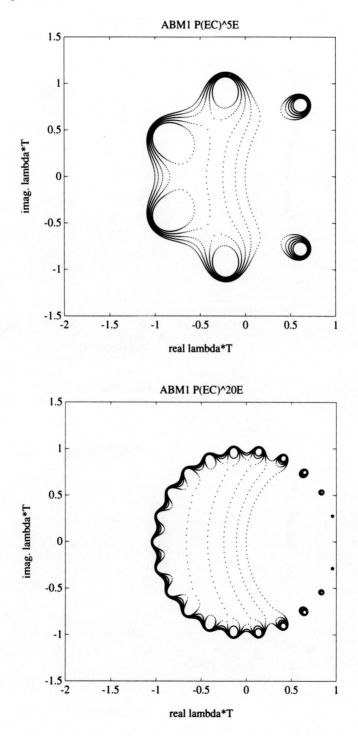

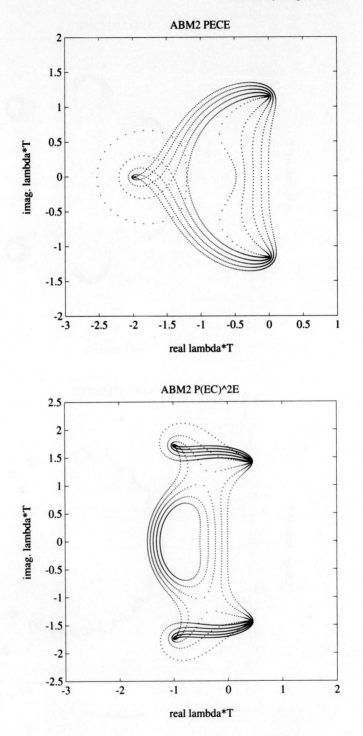

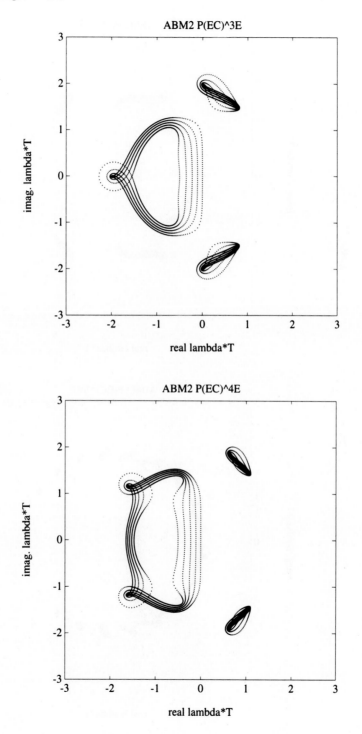

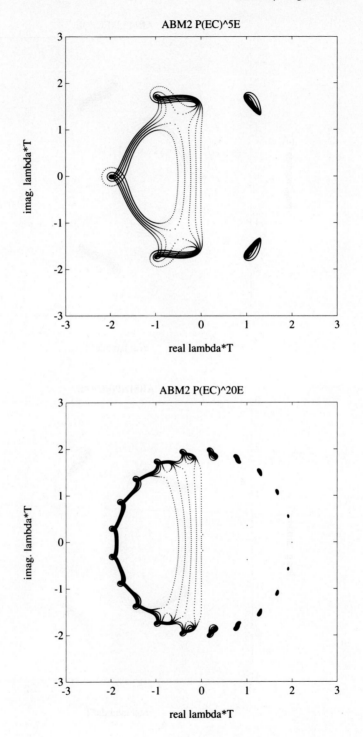

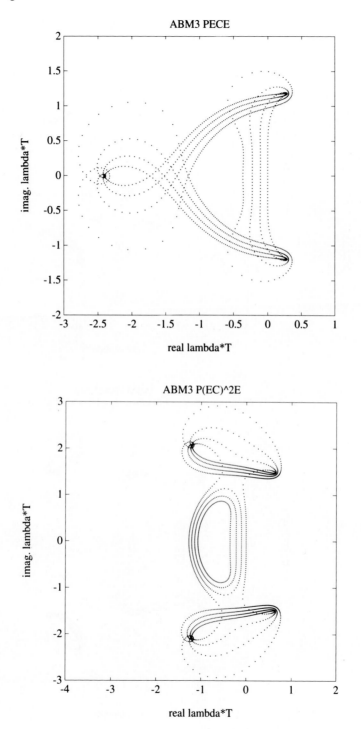

ABM3 PECE

ABM3 P(EC)^2E

ABM3 P(EC)^3E

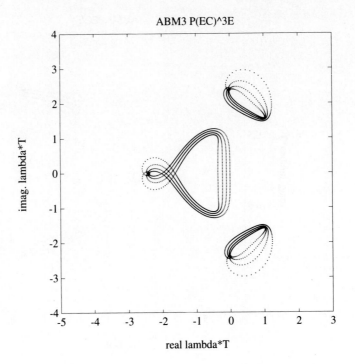

real lambda*T

ABM3 P(EC)^4E

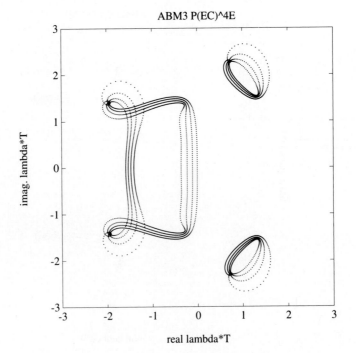

real lambda*T

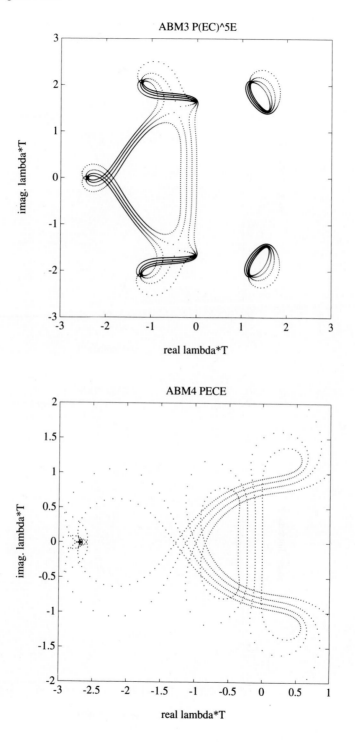

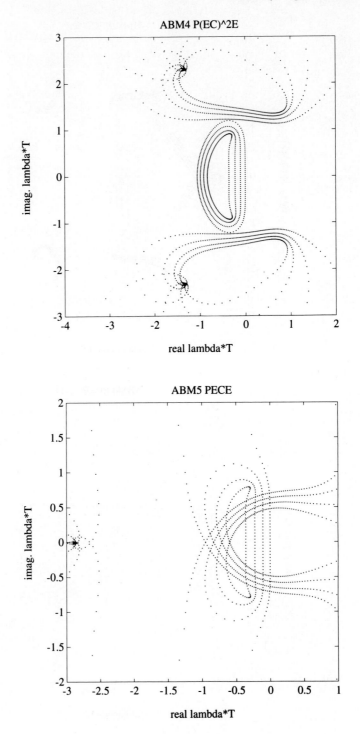

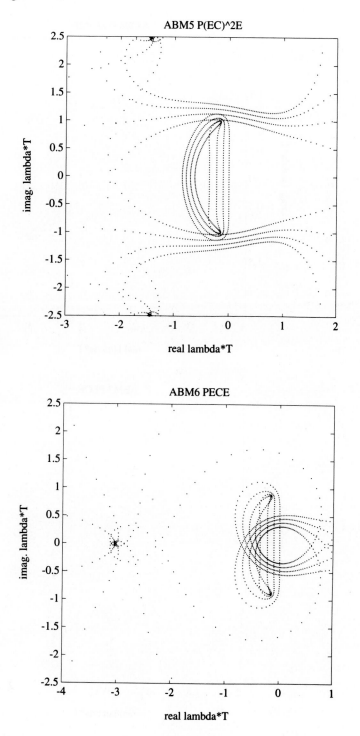

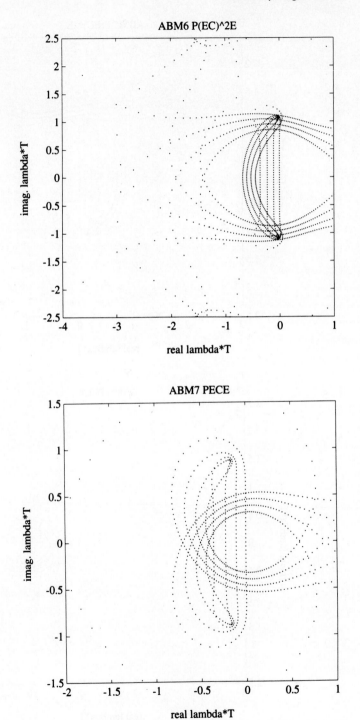

ABM6 P(EC)^2E

ABM7 PECE

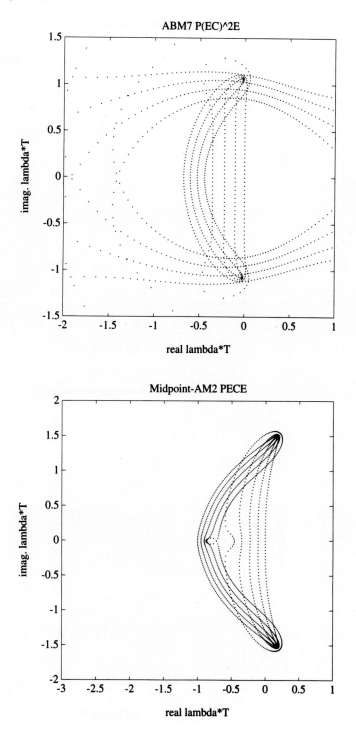

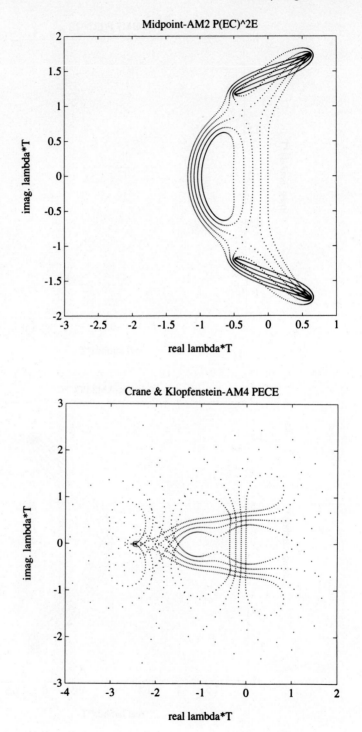

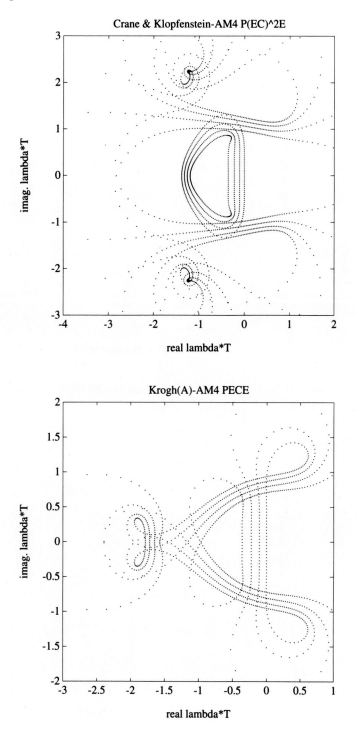

Crane & Klopfenstein-AM4 P(EC)^2E

Krogh(A)-AM4 PECE

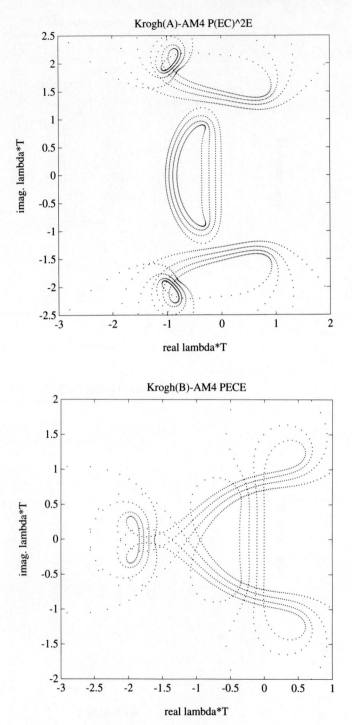

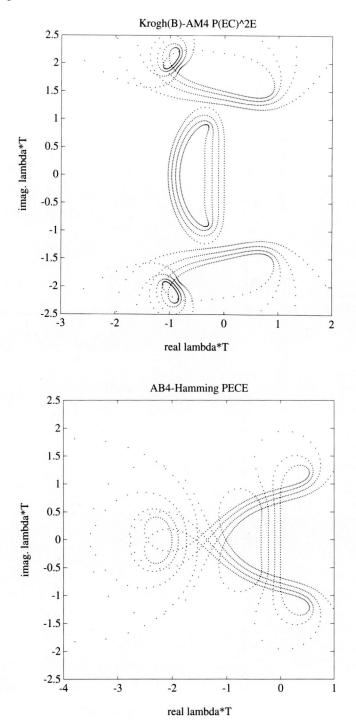

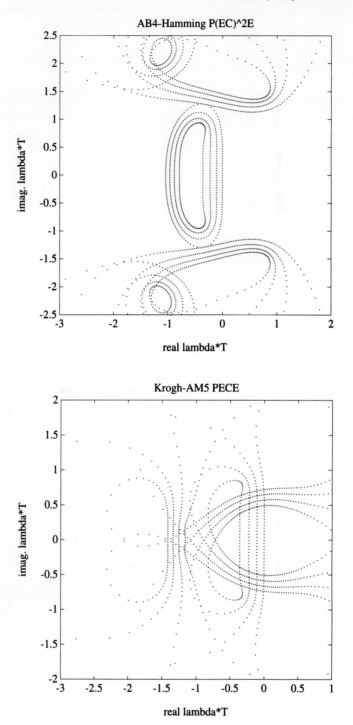

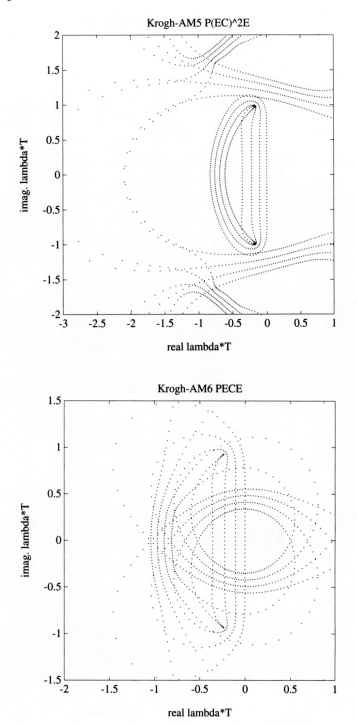

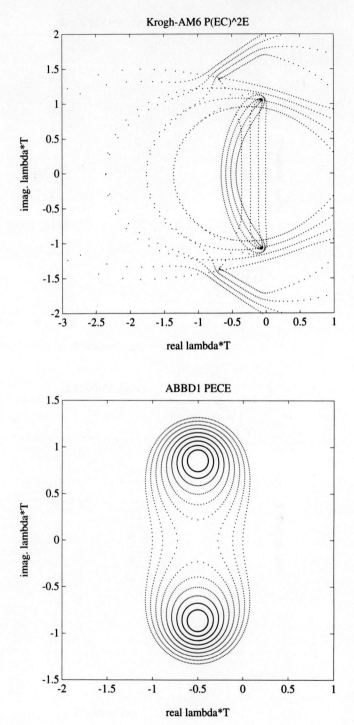

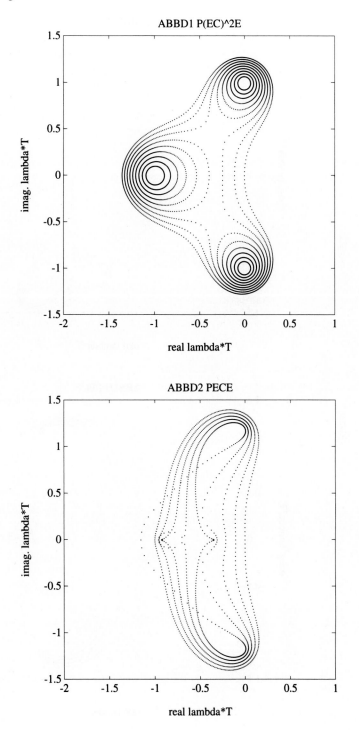

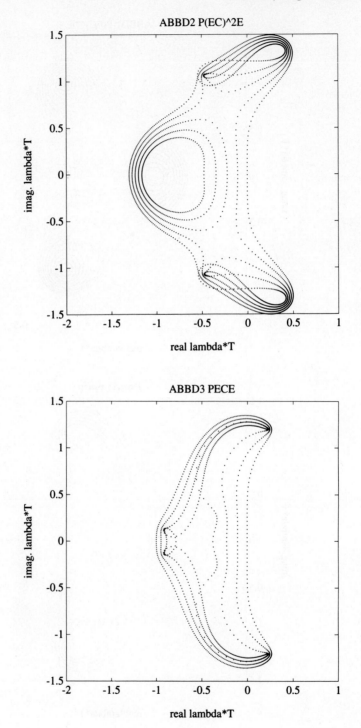

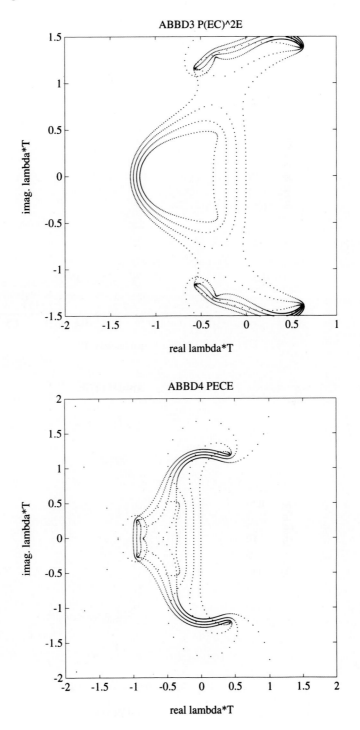

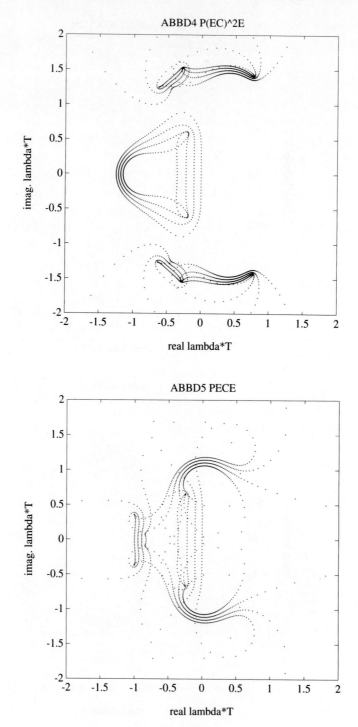

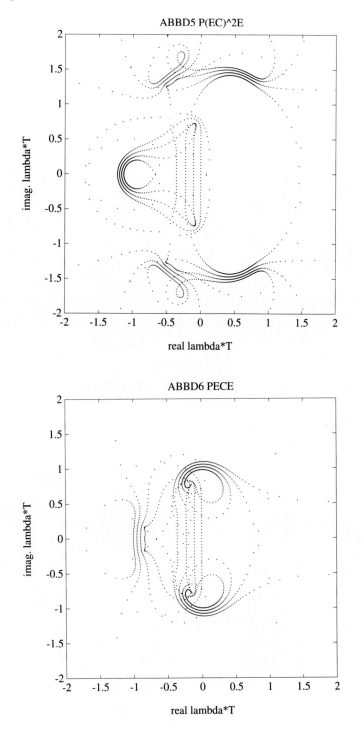

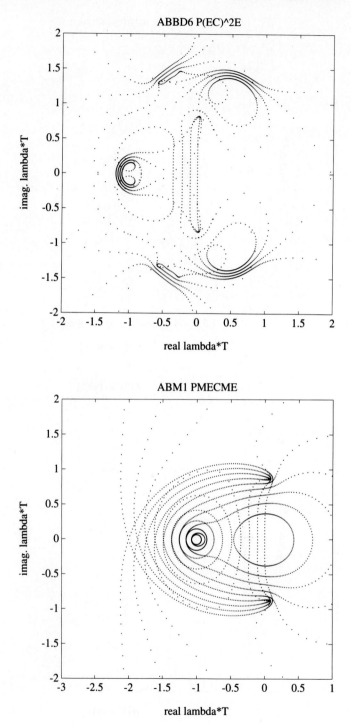

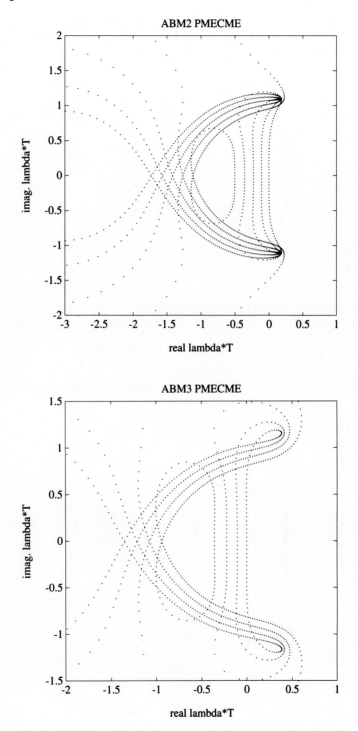

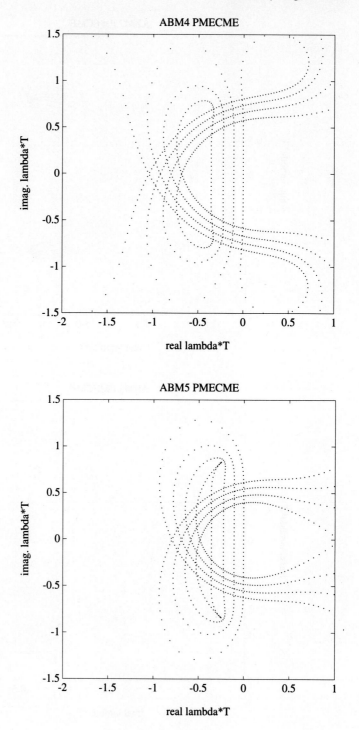

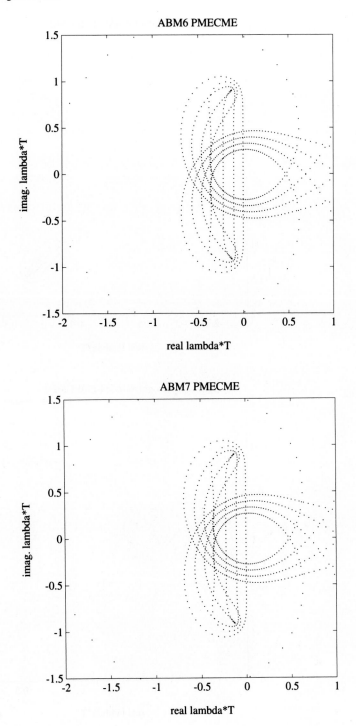

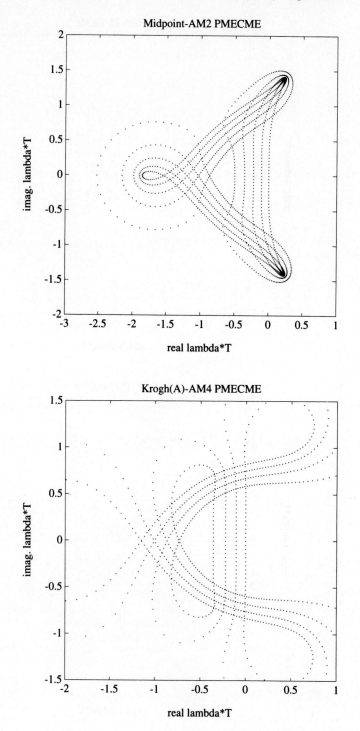

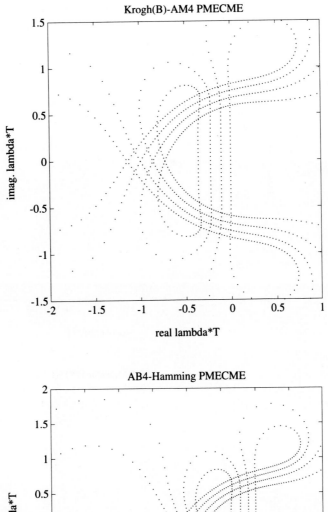

Krogh(B)-AM4 PMECME

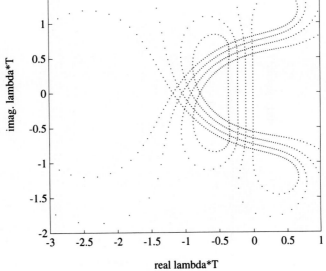

AB4-Hamming PMECME

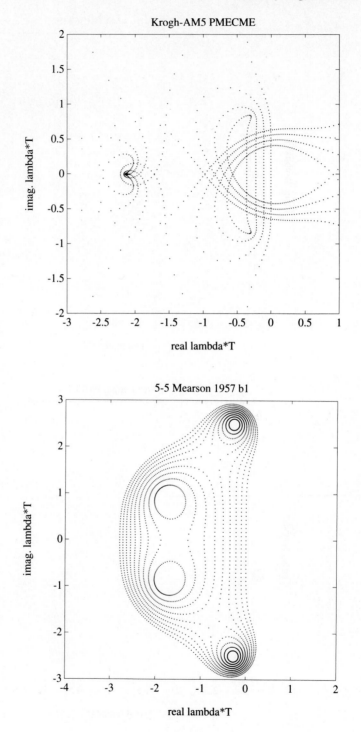

Krogh-AM5 PMECME

5-5 Mearson 1957 b1

5-5 Mearson 1957 b2

5-5 Mearson 1957 b3

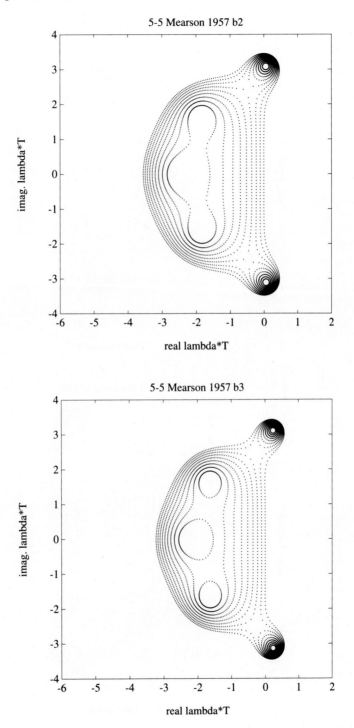

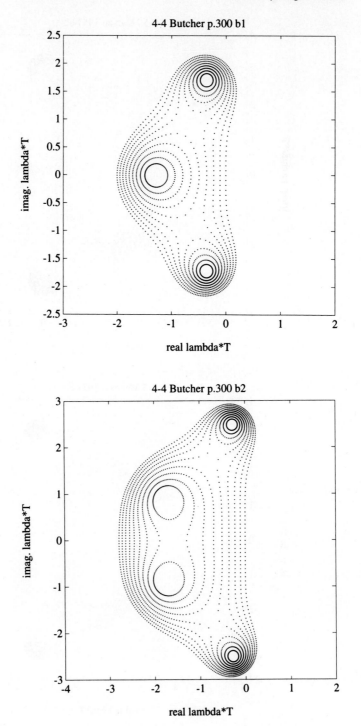

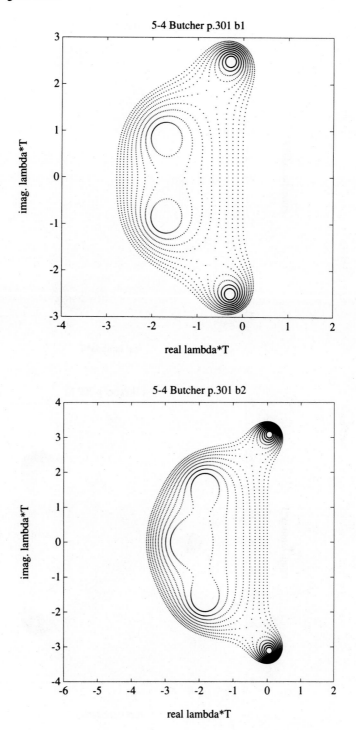

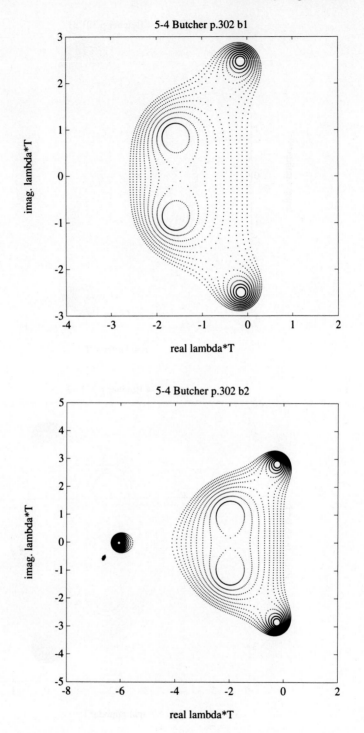

5-4 Butcher p.302 b1

5-4 Butcher p.302 b2

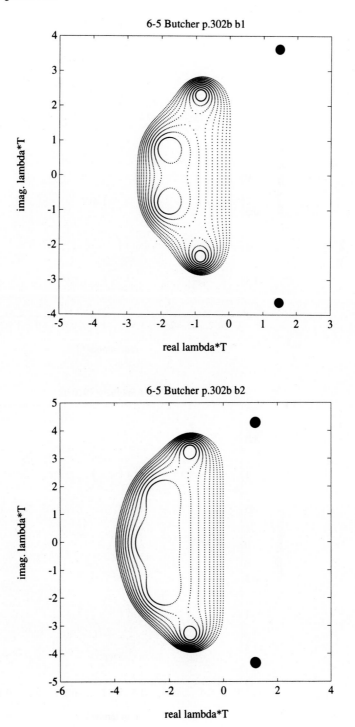

6-5 Butcher p.302b b1

6-5 Butcher p.302b b2

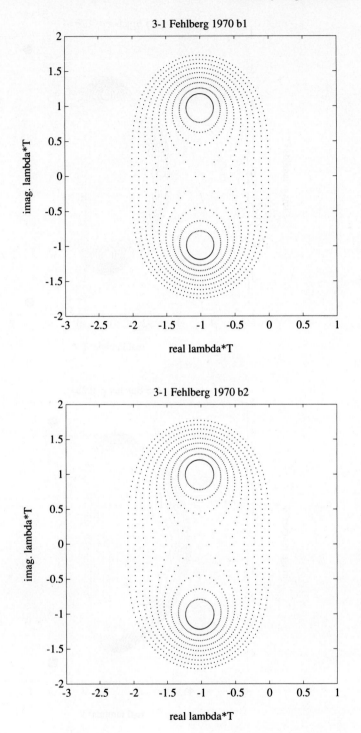

3-1 Fehlberg 1970 b1

3-1 Fehlberg 1970 b2

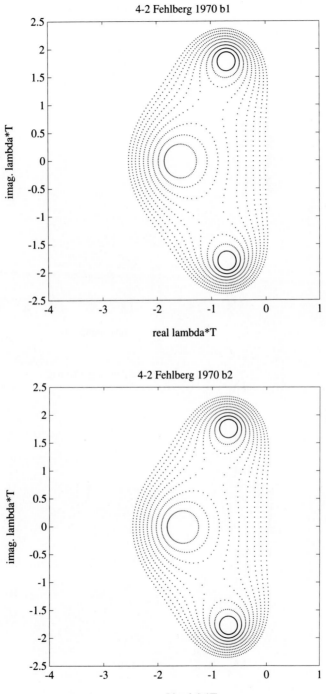

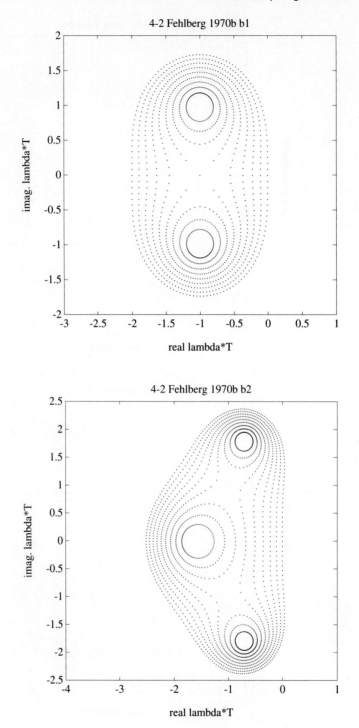

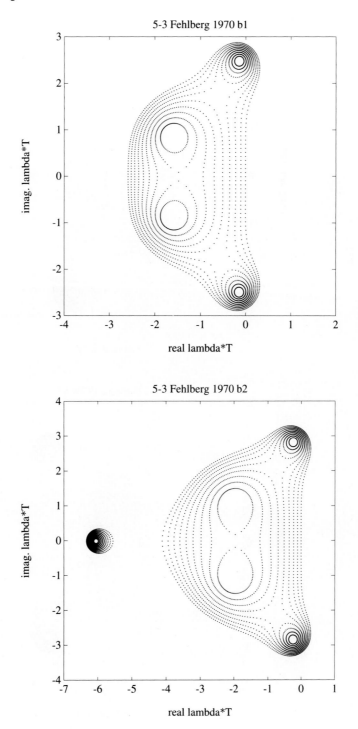

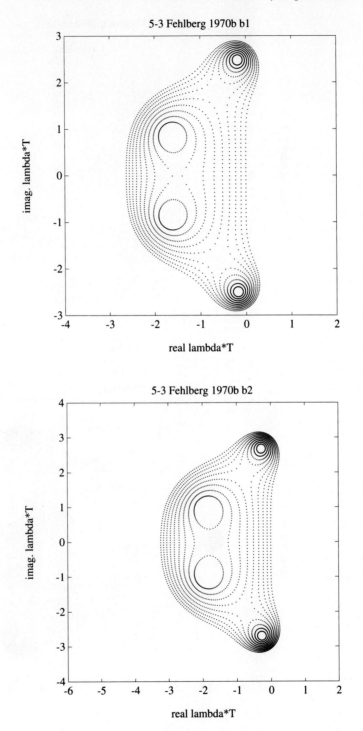

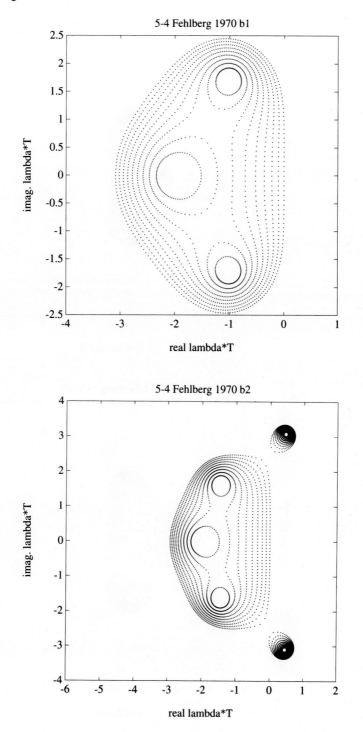

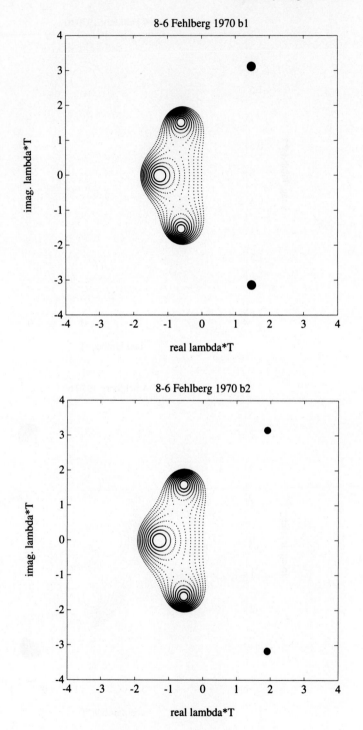

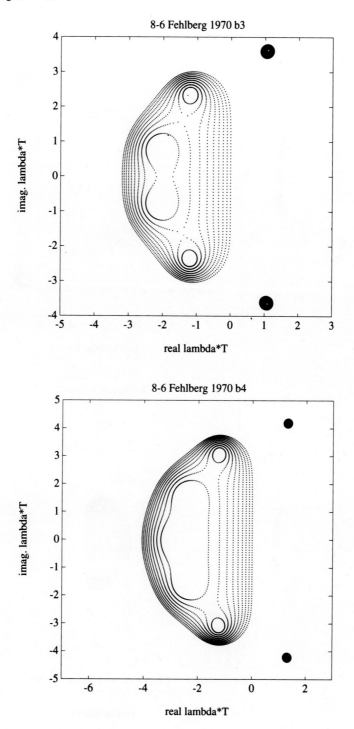

8-6 Fehlberg 1970 b3

8-6 Fehlberg 1970 b4

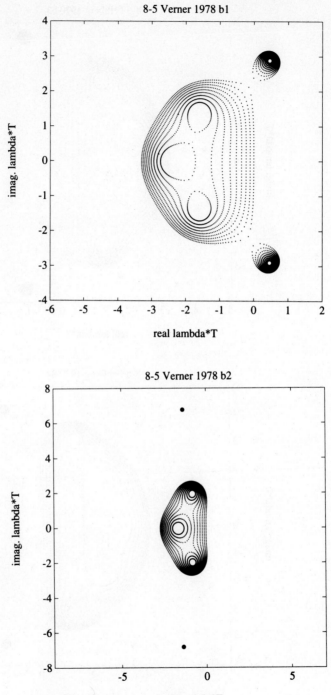

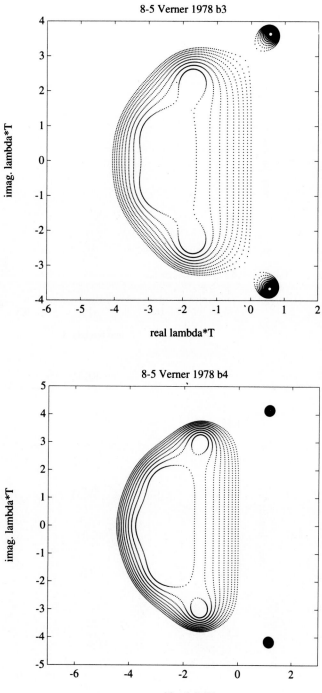

RK1

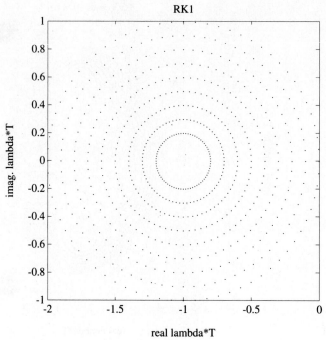

RK2

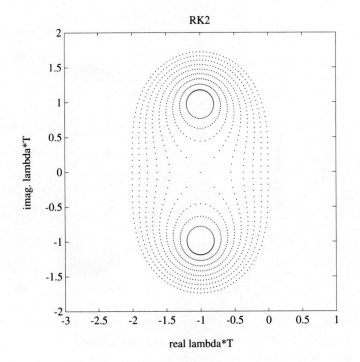

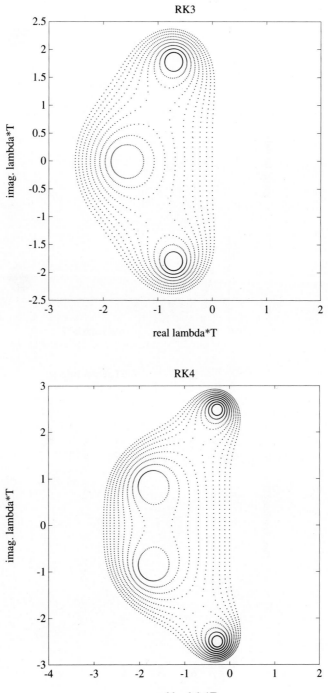

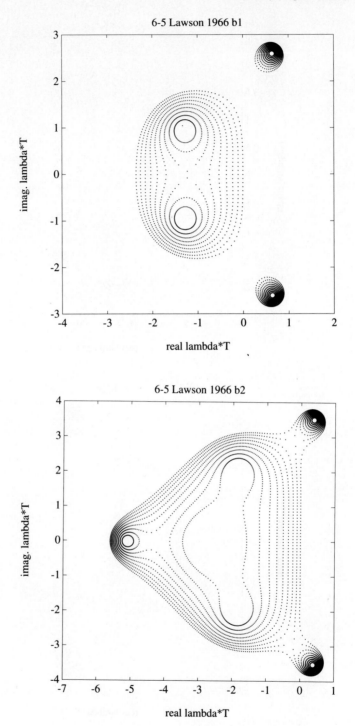

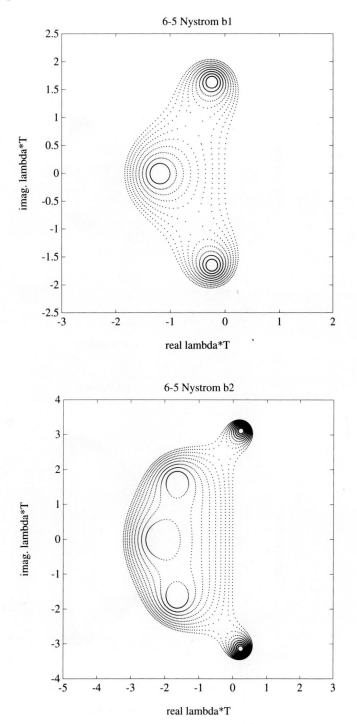

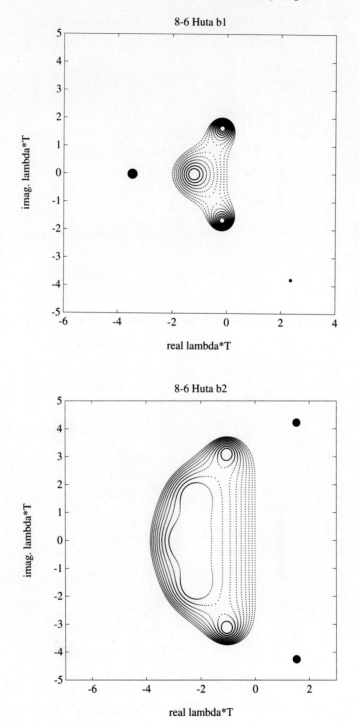

Listing of Provided Software

PROGRAMS IN THIS APPENDIX

ltregion—MATLAB program to generate stability regions for linear multistep methods.

rkstab—MATLAB program to generate stability regions for Runge-Kutta methods.

pcstab—MATLAB program to generate stability regions of linear multistep methods in predictor-corrector pairs in P(EC)^{m}E mode.

pcmstab—MATLAB program to generate stability regions of linear multistep methods in predictor-corrector pairs in PMECME mode.

```
function [x,y] = ltregion(num,den,rads)
%                    [x,y] = ltregion(num,den,rads)
%
% This function maps the unit circle in the Z-plane into the
% Lambda*T plane through the integrator 1/s = num/den
% at all radii specified in rads. If rads is omitted, [1] is used.
% The 'x' and 'y' returned are the real and imaginary parts
% of the mapping, respectively.
```

```
% Written by Stephen Chicatelli
%              University of Akron
%              Akron, Ohio
%              9/25/87
%
% Modified 10/15/87
%
if nargin <3, rads = [1]; end
j = (-1)^(1/2);
a = 0:pi/72:2*pi;
for count = 1:1:length(rads);
  z = rads(count) .*exp(j*a);
  n = polyval(num,z);
  d = polyval(den,z);
  resp(:,count) = (d./n)';
end
x = real(resp);
y = imag(resp);


function [x,y] = rkstab(stages,rs);
%
% Usage: function [x,y] = rkstab(stages, rs);
%
%       Where stages is the number of stages in the stability region
%       of the Runge-Kutta desired.
%       If stages > 4, it will prompt you for the stability polynomial.
%       'x' and 'y' are the real and imaginary parts of the mapping,
%       respectively
%
% Written by Stephen Chicatelli
%              University of Akron
%              Akron, Ohio
%              1/27/88
%
if nargin == 1, rs = [1]; end
j = (-1)^(1/2);
a = 0:pi/36:2*pi;
la = length(a);
m = ones(la,1);
    if stages <= 4,
      for b = stages:-1:1,
        vec(:,stages-b+1) = m./fctrl(b);
      end
    else
if length(stages) == 1,
    vect = input('Enter the stability polynomial ');
    for b = 1:stages+1,
      vec(:,b) = vect(b) .* m;
    end
```

```
else
  for b = 1:length(stages),
      vec(:,b) = stages(b).*m;
  end
  stages = length(stages)-1;
end
for c = 1:length(rs),
    r = [];
    z = [rs(c).*exp(j.*a)]';
    vec(:,stages+1) = m - z;
    for b = 1:la
        r = [r;roots(vec(b,:))];
    end
    x = [x real(r)];
    y = [y imag(r)];
end

function [x,y] = pcstab(predict_num,predict_den,correct_num,...
correct_den,rads,m);
% Usage: [x,y] = pcstab(predict_num,predict_den,correct_num,cor-
% rect_den,rads,m);
%
%         predict_num is the numerator of the predictor
%         predict_den is the denominator of the predictor
%         correct_num is the numerator of the corrector
%         correct_den is the denominator of the corrector
%
%         rads is the vector of radii to map through P(EC)^mE (optional)
%         the default value is 1
%
%         m is the number of (EC) steps to do (optional)
%         the default value is 1
%
% Written by Stephen Chicatelli
%            University of Akron
%            Akron, Ohio
%            10/28/87
%
clc
bk = correct_num(1);
if nargin <6,
    m = 1;
end
if nargin <5,
    rads = [1];
    lr = 1;
else
    lr = length(rads);
end
```

```
% loop over all radii

for counter = 1:lr,

  disp('Calculating results for radius');
  disp(rads(counter));
  z = rads(counter).* exp(sqrt(-1).*[eps:pi/72:pi]');
  lz = length(z);
  pn = polyval(predict_num,z);
  pd = polyval(predict_den,z);
  cn = polyval(correct_num,z);
  cd = polyval(correct_den,z);
  r = [];

% set up empty companion matrix

  if m == 1,
     r = sqrt(-1)*ones(2,lz);
     a = diag(ones(1,1),-1);
  else
     r = sqrt(-1)*ones(m+2,lz);
     a = diag(ones(1,m+1),-1);
  end

% loop over all values on unit circle

  for count = 1:lz,

% set up eigenvalue matrix characteristic equation

        vec = [];
        if m == 1,
           vec(1,1)     =  -bk*pn(count);
           vec(1,2)     =   bk*pd(count)-cn(count);
           vec(1,3)     =   cd(count);
           a(1,:)       =  -vec(2:3)./ vec(1);
           r(:,count)   =   eig(a);
        else
           vec(1,m+3)   =   bk^(m+1).*pn(count);
           vec(1,m+2)   =   bk^m.*(cn(count)-pn(count)-bk.*pd(count));
           vec(1,m+1)   =   bk^m.*(pd(count)-cd(count));
           vec(1,2)     =   vec(1,2) - cn(count);
           vec(1,1)     =   cd(count);
           vec(1,1:m+3) =   vec(1,m+3:-1:1);
           a(1,:)       =  -vec(2:m+3)./vec(1);
           r(:,count)   =   eig(a);
        end
```

```
    end
    results = [results,r.'];
  end
x = [real(results)];
y = [imag(results)];


function [x,y] = pcmstab(predict_num,predict_den,correct_num,...
correct_den,mp,mc,rads
%
% Usage: [x,y] = pcmstab(predict_num,predict_den,correct_num,cor-
% rect_den, ...... mp,mc,rads);
%
%          predict_num is the numerator of the predictor
%          predict_den is the denominator of the predictor
%          correct_num is the numerator of the corrector
%          correct_den is the denominator of the corrector
%
%          mp is the modifier for the predictor
%          mc is the modifier for the corrector
%
%          rads is the vector of radi to map through P(EC)^mE (op-
%          tional)
%          the default value is 1
%
% Written by Stephen Chicatelli
%             University of Akron
%             Akron, Ohio
%             9/25/92

clc
bk = correct_num(1);
if nargin <7,
   rads = [1];
   lr = 1;
else
   lr = length(rads);
end
%
% loop over all radi

for counter = 1:lr,

  disp('Calculating results for radius');
  disp(rads(counter));
  z = rads(counter).* exp(sqrt(-1).*[eps:pi/72:pi]');
  lz = length(z);
  pn = polyval(predict_num,z);
  pd = polyval(predict_den,z);
```

```
      cn = polyval(correct_num,z);
      cd = polyval(correct_den,z);
      r = [];

% set up empty companion matrix

      r = sqrt(-1)*ones(2,lz);
      a = diag(ones(1,1),-1);

% loop over all values on unit circle

      for count = 1:lz

% set up eigenvalue matrix characteristic equation

            vec = [];
            vec(1,1) = bk*pn(count) * ((mc+1) * z(count) - mp);
            vec(1,2) = z(count) * (mc * (cn(count) - pn(count)) + ...
            cn(count));
            vec(1,2) = vec(1,2) - bk*pd(count)*((mc+1)*z(count) ...
            - mp);
            vec(1,3) = -z(count) * (mc * (cd(count) - pd(count)) + ...
            cd(count));
            a(1,:) = -vec(2:3) ./ vec(1);
            r(:,count) = eig(a);
      end
      results = [results,r.'];
end
x = [real(results)];
y = [imag(results)];
```

D

Collection of Dynamic Systems

THE PREDATOR-PREY MODEL OF VOLTERRA

The predator-prey model of Volterra is concerned with only two species. Let t represent time, $x(t)$ represent the prey population and $y(t)$ the predator population. The dynamics of x and y are described in the following equations: $dx/dt = ax - bxy$, $dy/dt = -cy + dxy$. In these equations, $a,b,c,$ and d are constants with default values of 1.[1]

THE PREDATOR-PREY MODEL WITH FISHING

If we add terms that account for losses to a third species, the two equations above become $dx/dt = ax - bxy - fx$, $dy/dt = -cy + dxy - fy$. The coefficient f depends on the fishing effort.

THE F100 GAS TURBINE ENGINE

The sixth-order reduced state space model of the F100 gas turbine engine retains the following variables:

[1]C. M. Bender and S. A. Orszag, *Advanced mathematical methods for scientists and engineers* (New York: McGraw-Hill, 1978), pp. 179–183.

$$x_1 = \text{fan speed}$$
$$x_2 = \text{compressor speed}$$
$$x_3 = \text{augmentor pressure}$$
$$x_4 = \text{fan turbine inlet fast response}$$
$$x_5 = \text{fan turbine inlet slow response}$$
$$x_6 = \text{burner exit slow response}$$

The state matrices of the system, $\dot{x} = A_p x + B_p u$; $y = C_p x + D_p u$; $Ys = C_s x$ are shown below.[2]

$$A_p = \begin{bmatrix} -5.62 & 3.61 & -192 & 9.60 & 9.58 & -.832 \\ .291 & -3.09 & -122 & .066 & .686 & 13.0 \\ .087 & -.004 & -7.41 & -.014 & -.090 & -.070 \\ 3.83 & -2.27 & -591 & -50.0 & .771 & -10.1 \\ .170 & -.101 & -26.3 & -2.00 & -1.97 & -.449 \\ .006 & -.003 & -1.07 & .0002 & .003 & -.665 \end{bmatrix}$$

$$B_p = \begin{bmatrix} .467 & 237 & -126 & 13.1 & -35745 \\ .877 & -174 & 6.97 & -84.1 & -20138 \\ .012 & -95.0 & .535 & -.176 & -247 \\ 4.36 & -702 & 45.3 & 59.0 & 41611 \\ .194 & -31.2 & 2.01 & 2.62 & 1849 \\ .008 & -1.33 & .080 & .104 & 72.6 \end{bmatrix}$$

$$C_p = \begin{bmatrix} .923 & .030 & 180 & -.508 & -.271 & -.588 \\ .014 & .000003 & -.011 & 0 & 0 & 0 \\ .103 & -.064 & -17.9 & .004 & .044 & .032 \\ .00007 & .000007 & -.015 & .000003 & .00004 & .00002 \\ -.000007 & .00002 & .002 & -.000004 & -.000005 & -.00003 \end{bmatrix}$$

$$C_s = \begin{bmatrix} 1 & 0 & 0 & 0 & 0 & 0 \\ 0 & 1 & 0 & 0 & 0 & 0 \\ 0 & 0 & 1 & 0 & 0 & 0 \\ 0 & 0 & 0 & 1 & 1 & 0 \end{bmatrix}$$

$$D_p = \begin{bmatrix} .336 & -239 & 37.3 & 5.50 & -5615 \\ .0001 & .335 & .680 & -.00005 & -.013 \\ .130 & -22.7 & 1.31 & 1.43 & 1451 \\ .000003 & -.012 & -.006 & .0001 & .043 \\ -.000008 & .004 & -.0002 & -.006 & .839 \end{bmatrix}$$

[2]C. J. Daniele and S. M. Krosel, Generation of linear dynamic models from a digital nonlinear simulation, *NASA Tech. Paper 1388,* 1979.

DUAL-MASS/SINGLE-SPRING

The following system state space system describes two masses coupled by a spring:

$$x_1 = \text{position of mass 1}$$
$$x_2 = \text{position of mass 2}$$
$$x_3 = \text{velocity of mass 1}$$
$$x_4 = \text{velocity of mass 2}$$

$$
\begin{bmatrix} \dot{x}_1 \\ \dot{x}_2 \\ \dot{x}_3 \\ \dot{x}_4 \end{bmatrix} =
\begin{bmatrix}
0 & 0 & 1 & 0 \\
0 & 0 & 0 & 1 \\
-k/m_1 & k/m_1 & 0 & 0 \\
k/m_2 & -k/m_2 & 0 & 0
\end{bmatrix}
\begin{bmatrix} x_1 \\ x_2 \\ x_3 \\ x_4 \end{bmatrix} +
\begin{bmatrix} 0 \\ 0 \\ 1/m_1 \\ 0 \end{bmatrix} u
$$

The nominal mass-spring values are $m_1 = m_2 = k = 1$.[3]

VIBRATIONS OF A TAUT STRING

The following equations describe the fundamental transverse (x_1, x_2) and longitudinal (x_3) modes:

$$\ddot{x}_1 + w_0^2 x_1 + \beta(x_1^2 + x_2^2 - 2x_3^2)\, x_1 = f_1$$

$$\ddot{x}_2 + w_0^2 x_2 + \beta(x_1^2 + x_2^2 - 2x_3^2)\, x_2 = f_2$$

$$\ddot{x}_3 + w_1^2 x_3 - 2\beta\,(x_1^2 + x_2^2)x_3 = f_3$$

In the above equations $w_0^2 = \pi^2 T_0 / mL^2$, $w_1^2 = \pi^2 EA/mL^2$ are the transverse and longitudinal natural frequencies, and $\beta = 3\pi^4 T_0/8mL^4$. The parameter values are $L = 0.64m$, $f_1 = 188.5\text{Hz}$, $f_2 = 271.24\text{Hz}$, $f_3 = 364.37\text{Hz}$, $T_0 = 29.87\text{N}$, $E = 190GPa$, $A = 2.92E - 7m^2$, and $m = 2.315E - 3k$.[4]

HYPERCHAOTIC SYSTEM

The following state space system models a higher dimensional hysteresis generator of order four. The system contains one nonlinear element.

$$
\begin{aligned}
\dot{x} &= -z - w \\
\dot{y} &= r\,(2dy + z) \\
\dot{z} &= p(x - y) \\
e\dot{w} &= x - h(w)
\end{aligned}
$$

[3]B. Wie and D. S. Beernstein, A benchmark problem for robust control design, *Proc. 1990 American Control Conf.,* 1990, pp. 961–962.

[4]O. O'Reilly and P. J. Holmes, Non-linear, non-planer, and non-periodic vibrations of a string, *Jour. Sound and Vibrations,* 153(3), 1992, 413–435.

The nonlinear element $h(w)$ is

$$h(w) = \begin{cases} w - (1+n) & w \geq n \\ -w/n & -n < w < n \\ w + (1+n) & w \leq -n \end{cases}$$

The parameters of the system are $n = 1$, $r = 1$, $p = 14$, $0.52 < d < 1.04$, and $e \ll 1$ (0.02 nominal).[5]

THREE-FINNED TORPEDO

The following state space system ($\dot{x} = Ax + Bu, y = Cx$) models the behavior of a three-finned torpedo. The inputs are roll demand and yaw demand (u_1, u_2). The outputs are roll and yaw (y_1, y_2).[6]

$$A = \begin{bmatrix} 0 & 0 & 0 & 0 & 0 \\ 0 & -40 & 0 & 0 & 0 \\ 0 & 0 & -40 & 0 & 0 \\ 0 & 0 & 0 & -1.91 & 0 \\ 0 & 0 & 0 & 0 & -12.5 \end{bmatrix}, B = \begin{bmatrix} 191.5 & 46.7235 \\ 1 & 0 \\ 0 & 1 \\ 0 & 1 \\ 0 & 1 \end{bmatrix}$$

$$C = \begin{bmatrix} 1 & -191.5 & 5.1538 & -41.76 & -10.1102 \\ 0 & 0 & -37.4045 & 3.6239 & 33.7055 \end{bmatrix}$$

MAGNETOELASTIC BUCKLED BEAM

The second-order equation $\ddot{x} + \gamma\dot{x} - \alpha x + \beta x^3 = f\cos(\omega t)$ describes a periodically vibrating beam undergoing buckling by magnetic forces. The parameters of the system are $\alpha = \beta = 0.5$, $\omega = 0.8$, $\gamma = 0.15$, and $0.1 < f < 0.4$.[7]

AUTOMOBILE CLIMATE

The state space below ($\dot{x} = Ax + Bu, y = Cx$) models the compartment temperature of an automobile. The inputs are percent air mix and percent capacity of variable compressor (u_1, u_2). The output is compartment temperature y.[8]

[5]S. Toshimishi, An approach toward higher dimensional hysterisis chaos generators, *IEEE Trans. Cir. Sys.,* 37(3), 1990, 850–854.

[6]B. A. Stacey and M. A. Smith, "Multivariable control of a 3 fin torpedo, *Proc. 1989 American Control Conf.,* 1989, pp. 326–331.

[7]F. C. Moon, *Chaotic vibrations—An introduction for applied scientists and engineers,* (New York: Wiley, 1987), p. 77.

[8]T. Tabe, K. Matsui, T. Kakehi, and M. Ohba, Automotive climate control, *IEEE Control Sys. Mag,* October 1986, pp. 20–24.

$$A = \begin{bmatrix} -0.002 & 0.008 & 0.001 \\ 0.04 & -0.12 & -0.05 \\ -0.006 & 0.04 & -0.02 \end{bmatrix}, B = \begin{bmatrix} -0.0003 & 0.013 \\ -0.00003 & -0.013 \\ 0.0015 & 0.0018 \end{bmatrix}, C = [0.008 \quad 0.007 \quad 0.002]$$

SUPERSONIC AIRCRAFT

The following system describes a fictitious high-performance supersonic aircraft. The variables are

$$y_1 = \text{velocity}$$
$$y_2 = \text{flight path angle}$$
$$y_3 = \text{angle of attack}$$
$$y_4 = \text{angle of attack rate}$$
$$y_5 = \text{range}$$
$$y_6 = \text{altitude}$$
$$u = \text{elevator angle}$$

The system is described as follows:

$$\dot{y}_1 = -g \sin(y_2) - K_1 y_1^2 + K_2 \cos(y_3) \qquad y_1(0) = 599.594 \text{ ft/s}$$

$$\dot{y}_2 = \frac{1}{y_1}[-g \cos(y_2) + K_3 y_1^2 + K_2 \sin(y_3)] \qquad y_2(0) = 0$$

$$\dot{y}_3 = y_4 \qquad y_3(0) = 0$$

$$\dot{y}_4 = 1.311u - 0.806y_4 - 1.311y_3 \qquad y_4(0) = 0$$

$$\dot{y}_5 = y_1 \cos(y_2) \qquad y_5(0) = 0$$

$$\dot{y}_6 = y_1 \sin(y_2) \qquad y_6(0) = 100000 \text{ ft}$$

$$g = 32.2 \text{ ft/s}^2, K_1 = 5.326 \times 10^{-6} \text{ ft}^{-1},$$
$$K_2 = 1.915 \text{ ft/s}^2, K_3 = 8.957 \times 10^{-5} \text{ ft}^{-1}.$$

Index

A

A-stability, 103, 195–196
A(*a*)-stability, 197
A(θ)-stability, 197
AbuKhamseh, N., 273
Adams-Bashforth methods, 78–82,
    101–103, 126, 153, 303, 340–349
    AB-2, 67, 78, 81, 88, 101–102,
    143–147, 149–152, 193, 291, 317
    AB-3, 81, 102–3, 105–111, 317
    AB-4, 81, 103, 291, 318, 351
Adams-Bashforth-Moulton (PC) meth-
    ods, 153, 193
    stability regions, 321–333
    with modifiers stability regions,
    346–349
Adams-Moulton methods, 78–83, 127,
    153, 303
    AM-1 (see backward Euler), 81, 318

AM-2 (see trapezoidal rule), 81, 319,
    350
AM-3, 81, 320
AM-4, 81, 320, 350–351
AM-5, 320, 352
Adaptive methods, 124–125, 184–188,
    274
Associative methods, 30, 32–37
Attainable order
    linear multistep methods, 81, 86
    Runge-Kutta methods, 173
Attractor, 242, 244, 247
Automobile climate control model, 382

B

Backward Euler method, 81, 84, 89, 90,
    318
Backward difference operator, 125–126

Backward difference methods, 80, 84–85, 196, 304
  as corrector stability regions, 340–346
BASIC, 56, 70–73, 115–116, 164–165, 279, 298
Basin of attraction, 243, 259–261
Beale, G. O., 72, 76, 200, 201, 202, 210, 212, 300, 301
Bessel's equation, 53–54
Bifurcation diagram, 255–258, 261–264
  continuous-time systems, 258
Bilinear method (trapezoidal rule), 31, 35–37, 60
Block cascading property, 22
Boxer-Thaler method, 31, 37–38, 63, 300
Branch cuts, 145–147
Branch points, 145–147
Brennan and Leake jet engine model, 72, 212–216, 232–237, 300
Buckled beam model, 382
Bush, V., 5
Butcher, J. C., 165, 173, 188, 238, 300
  methods, 306–308, 354–357
  matrix notation, 304
Butterworth filter, 104–111, 133–134, 139–141, 142–144

C

Chaos, 57–58, 247–251
  discrete-time systems, 252–258
Chicatelli, S. P., 142, 300
Closed-loop integration, 86–97
Comparison of methods, 290–299
Complementary sensitivity, 130–132
Conformal map, 142
Conservative oscillation, 244–247
Crane-Klopfenstein method as predictor, 334–335
Consistency, 77–78, 81
Convergence, 81, 85
Cook scroll system, 57–58, 248–251, 300
Correcting to convergence, 118

D

D'Azzo, J. J., 98, 104, 131, 135, 225, 301
Dahlquist, G., 68, 81, 195, 300
Danieli, C. J., 243–244, 300, 380
Data reconstructors, 16–23
DeAbreu-Garcia, J. A., 210, 212, 273, 301
Delta function, 9, 76–77
Delta operator, 30
Detuned method, 29, 95
  severely, 95
Difference equations, 12–14, 18, 252–258
Differential analyzer, 5
Differential equations
  first order, 242
  second order, 244–247
  third order, 247–251, 282
  higher order, 251–252
Discontinuous systems, 270–272
Discrete-time systems, 8–13
  stability, 11
Disturbances, 130
Divergence, 273
Divergence preserving methods, 273–275
Double integration, 276–282
  corresponding Lambert's equation, 278–279
  linear multistep method for, 278–279
  predictor-corrector (-modifiers), 282
  stability regions, 280–282
  weak stability, 280
Dual-mass-single-spring system, 381
Duffing double scroll, 115–116, 123–124
Duffing's equation, 258–266

E

Eigenvalues (see poles), 96, 191, 224
Embedding method, 187
ENIAC, 5
Equation(s)
  Bessel's, 53–54
  Cook's scroll, 57–58, 248–251, 300

first characteristic, 69
first order differential, 242
first order difference, 252–258
Lambert's, 75, 77, 78, 152–153, 166
linear test, 86–87
Mathieu's, 54
nonlinear differential, 241–252, 258–275
Rayleigh's, 267–269
second characteristic, 69
second order differential, 244–247
stiff, 190–240
third order differential, 247–251, 282
van der Pol's, 246–247, 266–268, 274
Error
    constant, 73
    control, 124–125, 186–188
    estimation, 186
    round-off, 46
    truncation, 46, 73
Euler's method, 25–30, 61, 89–90,
        99–100, 142–143, 161–162,
        271–272, 274, 285
    backward, 89–90, 101
    choosing T for stability, 28–29
    forward looking rectangles, 26
    frequency response, 27
    improved, 162
    modified, 163
    multiple integration, 26
    root locus for linear test eqn., 28–29,
        89–90
    stability region, 99–100, 142–143, 316
    substitution method, 31
Existence, 67
Explicit method, 69, 114
    property, 89, 104
Explicit z-transform substitution method,
        31

F

F-100 jet engine, 379–380
Feedback configuration, 68–69

Fehlberg methods, 186, 188, 219, 301,
        308–312, 358–365
Filicky, J. D., 288, 301
Filicky-Hartley substitution methods,
        286–289
Final value theorem, 7–10, 76–77
First characteristic polynomial, 69
First difference substitution method
        (backward difference method), 31,
        32–35, 59–60, 285
First-order hold, 20
First-order low-pass filter hold, 19–20
Fixed point, 242, 254
Fixed point iteration, 113–114
Forward difference, 125–126
Franklin, G. F., 8, 19, 20, 302
Frequency response, 94, 129–160
Frozen transfer function, 53–56, 292

G

Gear, C. W., 193, 301
Gear's method, 238
Gershgorin bands, 224–238
Gold, B., 57, 302
Goodwin, G. C., 30, 187, 301, 302
Gordon, M. K., 125, 302

H

Halijak's method, 31, 62, 285
Hamming's method as corrector stability
        region, 337–338, 351
Harbor, R. D., 225, 302
Hardware-in-the-loop simulation,
        154–160
Harmonic Oscillator, 245, 277
Hartley, T. T., 72, 76, 93, 115, 123, 200,
        201, 202, 210, 212, 273, 284, 300,
        301
Hazen, 5
Helicopter model, 48–53

Henrici, P., 68, 78, 81, 301
Hermite's method, 78
Heun's method, 169
Holds, 16–23
   first-order, 20
   triangular, 18–20
   zero-order, 17–21, 25, 31, 48, 61, 156,
      161, 292
Horowitz, I. M., 129, 301
Houpis, C. H., 98, 104, 131, 135, 225, 301
Huta's method, 314, 372
Hyperchaos, 251–252, 381–382

I

Implicit methods, 69, 111–114
Implicit z-transform substitution method,
      31, 61
Improved Euler method, 162
Impulse invariance method, 58
Initial value theorem, 10
Initial value problem, 67
Input sampling, 15–16
Integrator
   ideal, 24, 99
   requirements for discrete-time equiva-
      lent, 77–78
Inverse complementary sensitivity, 135
Inverse Nyquist array, 224–238
Inverse Nyquist criterion, 137–139
Inverse Nyquist plot, 98, 135
Inverse sensitivity, 135–136
Inverse z-transform, 9

J

Jacobian matrix, 185, 242, 244
   discrete-time, 244, 252, 254
   input, 244
   system, 244
Jordan, D. W., 54, 301

Jury, E. I., 8–9, 11, 19, 20, 39, 40, 301
Jury's test, 11, 94

K

Kailath, T., 43, 104, 301
Kepler problem, 288
Kreyzig, E., 224, 301
Krogh's predictors, 304
   PC stability regions, 335–337, 338–340
   PCM stability regions, 350–351, 352
Krosel, S. M., 243–244, 300, 380
Kuo, B. C., 19, 225, 301

L

L-stability, 200
$L(k)$-stability, 201
Lambert, J. D., 67, 68, 73, 74, 75, 81,
      94, 114, 115, 118, 165, 166, 195,
      197, 200, 219, 278, 301, 313, 314
Lambert's equations, 75, 77, 78,
      152–153, 166
Lapidus, L., 25, 73, 238, 301
Laplace transform, 8, 11, 24, 292
   correspondence between s and z, 11,
      22, 37
   final value theorem, 10
   initial value theorem, 11
   time varying systems, 53–56
Lawson's method, 313, 314–315, 370
Limit cycle, 244–247
Linear differential equations, 67
Linear multistep methods, 66–160, 161,
      202–216, 276–289, 292–293
   absolute stability, 94
   Adams methods, 78–83
   backward difference methods, 80–81,
      84–85
   derivation, 73–76
   design example, 104–111

double integration, 278
difference equation, 68
error constant, 73
local truncation error, 73–75
maximal, 86
modifiers, 120–124
optimal method, 86
order of accuracy, 75
predictor-corrector methods, 114–120
predictor-corrector-modifier methods,
    120–124
properties, 81–86
stability regions, 98–104
transfer function, 68
Linear test equation, 86–87
Linearization, 243–244
Lipschitz constant, 67
Local truncation error, 46, 73
Logistic equation, 254–258, 273
Loop gain, 131
Low pass filters, 16–17, 19, 20
LTREGION, 373–374
Lyapunov exponents, 273

M

Maciejowski, J. M., 129, 131, 224, 301
Madwed's method, 31, 62–63, 285
Mapping of differentials, 57
    (see operational substitution)
Marinchek, B., 124, 301
Matched z-transform method, 58–59
Mathieu's equation, 54
MATLAB, 22, 59, 97, 107, 142, 212,
    245, 246, 248, 298
Matrix exponential, 21, 209–215
Maximal method, 86
Merson's method, 186, 188, 302,
    305–306, 352–353
Methods
  table of, 303–316
Middleton, R. H., 31, 302

Midpoint method, 304
    as predictor, 333, 334, 350
Milne's method, 122
Milne's device, 122
Modeling, 2
Modified Euler method, 163, 177
Modifiers, 120–124
Multiple integration, 276–289
    generalized Lambert's equations, 284
    transfer function for, 284
Multiplying polynomials using matrices,
    41
Multirate input sampling (MIS), 174–177

N

Nagle, H. T., 8, 11, 19, 94, 225, 302
Newton-Gregory expansion, 126
Noise, 95, 130, 251
    spurious, 95
    tuning, 95
Non-associative methods, 30, 37–39
Nonlinearities,
    mathematical, 241
    engineered, 241–242
Nonlinear differential equations, 241–275
Nonlinear difference equations, 252–258
Norms, 157–158, 186–187
Numerical integration, 292
Nyquist stability criterion, 131, 155, 225
Nyquist plot, 98, 131
Nystrom's method, 313–314, 371

O

One step methods, 161–162
Open-loop integration, 25, 76–77
Operational substitution methods, 24–65,
    69, 161, 285–288, 292
    associative, 30, 32–37
    choosing T, 31, 48

Operational substitution methods (*cont.*)
  closed-loop systems, 47–53
  error in using, 43–47
  error table, 285
  general procedure, 30–32
  matrix methods, 39–43, 59–64
  multi-element systems, 45–47
  non-associative, 31, 37–39
  nonlinear systems, 56–57
  table of methods, 31
  time-varying systems, 53–56
Oppenhiem, A. V., 57, 302
Optimal method, 86
Order of accuracy, 75
Ordinary differential equations (ODE's), 2
Ortega, J. M., 241, 302
Output sampling, 14–15

P

Parabolic hold, 20
Partial fraction expansions, 23, 24
Paynter, H., 5, 302
PCSTAB, 373, 375–377
PCMSTAB, 373, 377–378
Pendulum equation, 245–246, 277–278, 279
Period doubling route to chaos, 255–258
Phase variable form, 105
Philbrick, 5
Phillips, C. L, 8, 11, 19, 94, 225, 302
Planes (complex)
  λT, 98
  s-, 11–12
  $\hat{s}$-, 104, 107–111
  z-, 11–12
Poles, 11–13, 86–96, 131, 178–182
Poole, W. G., 241, 302
Predator-prey model, 379
  with fishing, 379
Predictor-Corrector methods, 114–120, 153

Predictor-Corrector-Modifier methods, 120–124, 153
Principal local truncation error, 73, 120–124
Principal root, 88, 94, 145, 178–182

Q

Qammar, H. Killory, 273, 301

R

Rabiner, L. R., 57, 302
Rabinowitz, P., 25, 69, 113, 165, 166, 302
Ragazzini, J. R., 8, 19, 20, 302
Ralston, A., 25, 69, 113, 165, 166, 302
Rayleigh's equation, 267–269
Real-time simulation, 291
Reconstruction, 16–17, 292
Relative stability, 94–95, 107
Repeller, 242, 244, 247
Riemann sheets, 143, 145
RKSTAB, 373, 374–375
Robust control, 129–134
  and the simulation problem, 148, 148–154
Rocket model, 54–56
Root locus, 89–92, 94–97, 178–182
Rosko, J. S., 8, 14, 19, 20, 39, 46–47, 69, 172, 302
Rossler, O. E., 252, 258, 302
Rossler's equations, 258
Round-off error, 46
Routh-Hurwitz test, 11, 94
Runge-Kutta methods, 125, 162–189, 216–224, 293
  coefficient constraints, 168, 172
  double integration, 278
  embedded methods, 187
  error estimation, 184–188, 274

Fehlberg, 186, 188, 219, 308–312, 358–365
general expression, 165–166
graphical interpretation, 162–164
Heun, 169
Huta, 314, 372
improved Euler, 162
input sampling, 174–177
Lawson, 313, 370
Merson, 186, 188, 305, 352–353
modified Euler, 163, 177
Nystrom, 313–314, 371
order of accuracy vs. stages, 173
transfer functions, 177–179
stability regions, 182–184, 368–369
stages, 162, 166
standard fourth order, 164–165
stiff systems, 216–224
variable timestep methods, 184–188
Verner, 186, 312, 366–367
weak stability, 178–182

S

Sampled data systems, 14
Sampling, 14
Schafer, R. W., 57, 302
Schur-Cohn test, 11, 94
Schwartz, M., 95, 302
Second characteristic polynomial, 69
Second-order hold, 20
Second-order low pass filter hold, 20
Seinfeld, J. H., 25, 73, 238, 301
Sensitivity, 130–134
Shampine, L. F., 125, 302
Signal-to-Noise ratio (SNR), 95, 107
Simpson's rule, 38, 91–92, 103
Simulation of,
chaotic systems, 269–270
discontinuous systems, 270–272
limit cycling systems, 266–269
non-limit cycling systems, 258–266
time delay systems, 272–273

Simulation method selection, 295–298
Simulation timestep selection, 2, 18, 295–298
Sin, K. S., 187
Smith, P., 54, 301
Smith, J. M., 20, 302
Spurious roots, 88, 94, 145, 161
Stability
A-, 103
$A(a)$-, 197–199
$A(\theta)$-197
absolute, 94
closed-loop, 94–97
discrete-time system, 11
L-, 200–202
open-loop, 76–77
relative, 94–95, 107
stiff, 197–199
weak, 94–97
zero-, 77
Stability regions, 98–104, 129–160, 190
double integration, 280–282
inside, 141–148
linear multistep methods, 98–104
nonlinear, 264–266
property, 98
Runge-Kutta methods, 182–184
table of, 316–372
Stability region placement methods, 202–216
implicit (SRPI), 207–209
matrix methods (MSRP), 209–216
scalar methods (SRP), 202–209
Starting a method, 107
State space representation, 66–69, 70, 96, 105, 111, 211, 243–244
linear, 21, 67, 70, 96, 105, 111, 211, 244
nonlinear, 67–69, 170, 243–244
Steady state error, 76–77
Step invariance method, 57
Stewart, H. B., 273, 302
Stiffness
definition, 191
ratio, 191

Stiff stability, 197–199
Stiff systems, 190–240
Stochastic systems, 251
Superposition, 56
Supersonic aircraft model, 383
System characteristics, 293–295

T

Taylor series expansion, 37, 73–75, 120,
    153, 165, 170, 178
Thaler, G. J., 88, 302
Thomas, B., 187, 302
Thompson, J. M. T., 273, 302
Time delays, 11, 272–273
Time varying systems, 53–56, 292
Torpedo model, 382
Torus, 247
Transfer functions, 12–14
    linear multistep methods, 68
    Runge-Kutta methods, 177–179
    z-domain, 12–13
Trapezoidal rule, 31, 35–37, 38, 69, 75,
    111–114, 146–148
    as Tustin's method, 35–37, 60
    delayed, 128
    implementation, 111–114
    root locus, 89–91
    stability region, 99–101, 146–148, 319
Triangular hold, 18–20
Triple integration, 39, 282–284
Truncation error (see local truncation
    error)
Tuned method, 29, 95, 220
Tustin, A., 35, 302
Tustin's method, 31, 60, 285
Two body problem, 277

U

Uniqueness, 67

V

Van der Pol's equation, 246–247,
    266–268, 274
Van de Vegte, J., 224, 302
Van Valkenburg, M. E., 104, 302
Variable timestep, 124–125, 184–188,
    274
Variable order, 124–126
Verner's method, 302, 312, 366–367
Vibrations of a string, 381

W

Web plots, 252–257
Weak stability, 94–97
    predictor-corrector methods, 116–120
    Runge-Kutta methods, 178–182

X

X-15 longitudinal dynamics, 289

Z

Zero-order hold (ZOH), 17–21, 24,
    29–30, 45–46, 57, 156, 161, 292
Zero-order mean hold, 20
Zero-order midinterval hold, 20
Zero stability, 77, 81
z-forms (see Boxer-Thaler method)
z-plane, 11–12
z-transform substitution method, 31, 39,
    61, 285
    explicit, 31, 285
    implicit, 31, 61, 285
z-transforms, 8–14, 24, 292
    correspondence between z and s,
        11–12, 23, 40
    definition, 9

final value theorem, 9–10
initial value theorem, 10–11
inverse, 9
pole location and time response, 11, 13
shift theorem, 9

stability, 11
table of properties, 10
table of pairs, 10
transfer functions, 12–14